Egg Science and Technology
Third Edition

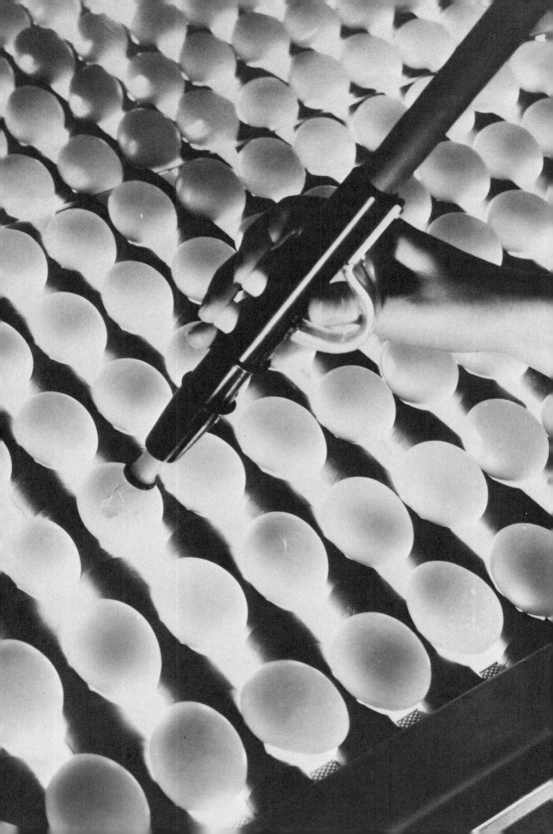

Egg Science and Technology
Third Edition

Edited by

William J. Stadelman
Department of Food Science
Purdue University
West Lafayette, Indiana

Owen J. Cotterill
Department of Food Science
and Nutrition
University of Missouri-Columbia
Columbia, Missouri

AVI PUBLISHING COMPANY, INC.
Westport, Connecticut

Frontspiece:

Undergrade Programmer Stains, checks, blood spot eggs, etc., are identified by touching the egg with this instrument. The inspector keys in the type of egg and this information is stored in a computer. Later, these eggs are removed automatically. Manual removal of undergrade eggs is eliminated. Two operators can handle 200 cases per hour.

Illustration courtesy of
Diamond Automation, Farmington Hills, Michigan 48024

Copyright © 1986 by
THE AVI PUBLISHING COMPANY, INC.
P.O. Box 831
250 Post Road East
Westport, Connecticut 06881

Library of Congress Cataloging-in-Publication Data

Egg science and technology.

 Includes bibliographies and index.
 1. Eggs. 2. Egg trade. 3. Egg processing.
4. Egg products industry. 5. Egg processing—Patents.
I. Stadelman, William J. II. Cotterill, Owen J.
SF490.E37   1986   637.5   86–8035
ISBN 0-87055-516-2

Printed in the United States of America
A B C D 5432109876

Contents

Contributors

RUTH E. BALDWIN Department of Food Science and Nutrition, University of Missouri-Columbia, Columbia, MO 65211

HERSHELL R. BALL, JR. Department of Food Science, North Carolina State University, Raleigh, NC 27606

DWIGHT H. BERGQUIST Henningsen Research and Development Center, Henningsen Foods, Inc., Omaha, NE 68137

RON G. BOARD School of Biological Sciences, Bath University of Technology, Claverton Down, Bath, England BA2 7AY

GEORGE M. BRIGGS Department of Nutritional Sciences, University of California, Berkeley, CA 94720

FRANCES COOK 5372 Punta Alta 1-A, Laguna Hills, CA 92652

OWEN J. COTTERILL Department of Food Science and Nutrition, University of Missouri-Columbia, Columbia, MO 65211

FRANKLIN E. CUNNINGHAM Department of Animal Science, Kansas State University, Manhattan, KS 66502

RONALD D. GALYEAN Griffith Laboratories, 1 Griffith Center, Alsap, IL 60658-3495

JAMES M. GORMAN* Technical Services, Seymour Foods, Inc., Topeka, KS

W. MIKE HILL 6709 South Marks, Fresno, CA 93706

SHURYO NAKAI Department of Food Science, University of British Columbia, Vancouver, Canada V6T 1W5

ERIC C. OESTERLE Department of Agriculture Economics, Purdue University, Lafayette, IN 47907

WILLIAM D. POWRIE Department of Food Science, University of British Columbia, Vancouver, Canada V6T 1W5

MIKE SEBRING National Egg Products Corporation, P.O. Box 608, Social Circle, GA 30279

WILLIAM J. STADELMAN Department of Food Science, Purdue University, Lafayette, IN 47907

* Deceased.

MILO H. SWANSON   2804 Iron Hill Way, Riverside, CA 92506

H. S. TRANTER   School of Biological Sciences, Bath University of Technology, Claverton Down, Bath, England BA2 7AY

Preface

The first and second editions of *Egg Science and Technology* served a wide segment of people interested in shell eggs and egg products. The publishing of three editions in thirteen years indicates the need for information concerning this subject and the progress that has been made. Interest in this subject matter is both national and international in scope, as is evidenced by the many copies purchased overseas. Also, the second edition was translated into Japanese. Two very common comments about this book were: "I didn't realize there is so much known about an egg" and "My copy keeps disappearing."

This volume is directed toward those people involved in the following areas:

1. *Industry*—management, technical supervisors and quality assurance personnel, trade associations, libraries, and foodservice units.
2. *Universities*—libraries, students, and professors.
3. *Government*—regulatory agencies and research laboratories.

In this edition, three new authors have been added to the list of contributors: Dr. R. D. Galyean, Dr. S. Nakai, and Dr. H. S. Tranter. Unfortunately, we have lost the help of James Gorman and Dave Jackson (both deceased) whose dedication to egg science and technology will forever be appreciated.

Many recent photographs of egg production and processing equipment have been added. Tables have been updated. A new nutrient composition table is included, which features the best nutrient value where only one value can be cited. Eighty-eight components are listed for shell, liquid/frozen, and egg solids.

One new chapter on hard-cooked peeled eggs has been added. Pickled eggs, preserved in a brine of vinegar, salt, etc., have been available for many years. However, a different merchandising concept of hard-cooked peeled eggs has now evolved. These eggs are packed in a citric acid-sodium benzoate solution (refrigerated) and marketed through institutional foodservice units.

Additional information has been included on grades of shell eggs and standards of identity for egg products, including mayonnaise and salad dressings. Over 90 new patents and 30 Ph.D. theses have been added since the second edition. We feel that this edition of *Egg Science and Technology* will be as useful to you as were the first two editions.

W. J. STADELMAN
O. J. COTTERILL

Related AVI Books

The Egg Industry

W. J. Stadelman

The egg industry in the Western world is almost exclusively based on chicken eggs. Although the ability of some strains of ducks to produce large numbers of eggs has led to suggestions that such eggs be used for human consumption, this has not been accepted in most areas thus far. Coturnix quail eggs are being marketed in Japan and some Western countries and are generally sold as hard-cooked eggs packaged in jars in a preservative brine solution. However, throughout this text all references to eggs will mean chicken eggs, unless specifically identified otherwise.

Eggs are one of the few foods that are used throughout the world; thus the egg industry is an important segment of the world food industry. Eggs have been an important part of the human diet since the dawn of recorded history. In modern times eggs have been an important commodity in international trade. The principal egg-producing countries of the world are listed in Table 1.1. Although the countries in the North Temperate Zone are the primary producers, there is also substantial egg production in Australia and South America. The egg industry of many developing countries is expanding rapidly to help meet their protein needs, but limited feed supplies and the low production of native chickens hamper progress. When high-production strains are introduced, the stress conditions of disease and poor management frequently lead to failure.

Within the United States the egg industry was historically a Corn Belt industry. More recently the West Coast and Southeastern states have become major production areas. The changes from 1945 through 1980 are indicated in Tables 1.2 and 1.3 for states and regions. The

EGG SCIENCE & TECHNOLOGY,
3rd Edition

TABLE 1.1. Egg-Producing Countries of the World Ranked on the Basis of 1966 Production Figures

Country	Millions of eggs						
	1961	1964	1969	1972	1975	1979	1982[a]
United States	62,423	62,215	68,700	69,540	64,341	69,228	68,940
USSR	29,300	26,700	37,000	47,900	57,700	65,585	73,200
Japan	12,863	17,898	27,565	30,524	29,789	33,150	33,150
United Kingdom	13,584	15,204	15,036	14,628	13,000	14,962	13,475
West Germany	7,895	9,997	14,400	16,143	15,200	13,336	13,500
Italy	6,690	8,287	10,281	10,612	11,279	11,037	11,500
France	8,955	9,740	11,200	11,490	13,120	13,840	15,600
Brazil	6,527	7,772	9,670	5,820	6,000	7,200	10,000
Spain	4,588	6,379	7,140	9,007	9,780	11,035	11,700
Poland	6,141	6,000	6,400	7,475	8,001	8,670	8,000
Mexico	4,200	4,750	5,657	—[b]	7,446	10,390	12,750
Canada	5,159	5,255	5,655	5,620	5,339	5,551	5,900
East Germany	3,602	3,696	4,050	4,425	5,030	5,219	5,750
Netherlands	5,999	5,095	4,370	4,421	5,320	8,187	9,600
Belgium & Luxembourg	3,103	3,167	3,900	3,946	3,775	3,453	3,150
Czechoslovakia	2,351	2,695	3,410	4,120	4,500	4,265	4,650
Rumania	2,600	2,456	3,200	4,300	5,400	7,085	6,400
Argentina	3,480	2,400	2,940	3,390	3,300	3,350	3,300
Hungary	1,885	2,200	2,600	3,217	3,800	4,450	4,400
Australia	2,544	2,520	2,380	3,552	3,370	3,395	3,470
Republic of South Africa	—	1,161	2,028	—	2,519	3,083	3,418
Venezuela	—	—	—	—	—	2,346	2,965
Egypt	—	—	—	—	—	2,000	2,140
Turkey	—	—	—	—	—	4,323	5,000

SOURCE: U.S. Department of Agriculture (1967, 1970).
[a] Estimated data.
[b] Data not reported.

TABLE 1.2. Egg Production by States and Regions

States and regions	Millions of eggs					
	1945	1955	1965	1970	1975	1980
Maine	404	695	1,031	1,462	1,650	1,793
New Hampshire	350	440	359	300	283	182
Vermont	175	201	139	129	105	100
Massachusetts	932	704	553	521	541	326
Rhode Island	77	80	80	77	69.5	84
Connecticut	483	672	797	927	798	1,004
New England	2,421	2,792	2,959	3,416	3,446.5	3,489
New York	1,935	2,121	2,257	2,297	1,984	1,776
New Jersey	888	2,433	1,378	779	620	279
Pennsylvania	2,491	3,654	3,392	3,328	3,299	4,251
Delaware	119	126	132	126	115	138
Maryland	427	406	332	330	331	381
Virginia	1,039	843	1,091	1,092	765	913
West Virginia	453	398	325	319	256	149
Mid-Atlantic	7,352	9,981	8,907	8,271	7,370	7,887
North Carolina	1,190	1,469	2,632	3,671	2,802	3,174
South Carolina	384	510	1,001	1,296	1,385	1,679
Georgia	655	1,213	4,042	5,396	5,284	5,637
Florida	196	505	1,778	2,547	2,779	3,044
Alabama	651	781	2,291	2,720	2,951	3,354
Mississippi	616	584	2,222	2,470	1,707	1,584
Tennessee	1,084	955	972	1,264	949	962
Kentucky	1,187	1,029	687	659	518	536
Arkansas	740	544	2,383	3,482	3,594	4,153
Louisiana	383	363	712	769	658	553
Southeast	7,086	7,952	18,711	24,220	22,627	24,676

(Continued)

3

TABLE 1.2. (Continued)

States and regions	Millions of eggs					
	1945	1955	1965	1970	1975	1980
Ohio	2,781	2,366	2,370	2,071	1,999	2,333
Indiana	2,012	2,289	2,469	2,880	2,609	3,697
Illinois	2,757	3,064	1,821	1,905	1,483	1,267
Michigan	1,616	1,690	1,543	1,417	1,303	1,459
Minnesota	3,757	4,287	2,622	2,247	2,209	2,223
Wisconsin	2,315	2,328	1,470	1,216	1,194	946
Iowa	4,327	4,859	3,854	3,034	2,006	1,784
Missouri	2,890	2,132	1,375	1,469	1,241	1,460
North Central	22,455	23,015	27,524	16,239	14,044	15,169
Kansas	2,136	1,723	967	842	599	427
Nebraska	2,014	1,790	1,346	1,071	782	847
South Dakota	1,071	1,234	1,374	1,053	702	464
North Dakota	665	559	322	218	132	82
Colorado	438	336	238	335	473	464
Western Plains	6,324	5,642	4,247	3,519	2,688	2,284
Texas	3,309	2,249	2,606	2,679	2,360	3,092
Oklahoma	1,546	829	518	569	430	839
Arizona	61	87	208	224	159	113
New Mexico	119	114	145	224	234	378
Southwest	5,035	3,279	3,477	3,696	3,183	4,422

Montana	246	228	178	240	197	170
Idaho	280	281	225	189	184	202
Wyoming	88	78	49	40	29.7	10.5
Nevada	46	20	9	4	4.3	1.8
Utah	376	380	241	263	321	416
Mountain	1,036	987	702	736	736	800.3
California	2,087	4,404	7,406	8,380	8,467	8,796
Oregon	480	616	498	533	519	638
Washington	921	828	1,062	1,046	1,068	1,295
Alaska	—	—	9	5	5	4.4
Hawaii	—	—	190	197	209	222
Western or Pacific	3,488	5,848	9,165	10,161	10,268	10,955.4
Total United States	55,197	59,496	65,592	70,312	64,362	69,683
Av. no. of hens and pullets (in thousands)	365,839	309,104	301,687	322,307	276,905	287,834
Eggs per hen	151	192	218	218	232	242

SOURCES: *Poultry Business Facts, Watt Publishing Co., Mount Morris, IL; U.S. Department of Agriculture (1971, 1976, 1981)*

5

TABLE 1.3. Size of Laying Flocks Producing Market Eggs[a]

Area[b]	1959 Census				1964 Census				
	Less than 400	400 to 1,599	1,600 to 3,199	Over[c] 3,200	Less than 400	400 to 1,599	1,600 to 3,199	Over 3,200	Over 20,000
New England	4.85	12.10	18.20	64.85	1.60	5.35	9.05	84.00	27.10
Middle Atlantic	12.01	23.17	19.67	45.04	5.16	12.96	11.42	70.46	25.70
Southeast	12.87	15.21	17.94	53.98	2.00	3.09	5.94	88.97	37.88
North Central	48.16	33.57	7.54	10.72	24.77	26.45	7.43	41.35	12.83
Western Plains	61.73	27.61	5.18	5.47	41.02	33.57	7.40	18.01	5.69
Southwest	24.61	19.25	13.74	42.40	8.47	8.43	6.96	76.14	44.89
Mountain	23.30	18.78	11.66	46.25	15.27	14.93	15.55	54.25	15.61
Pacific	3.19	8.20	12.73	75.87	0.65	1.70	3.18	94.48	61.81
United States	26.78	22.58	12.83	37.81	10.87	12.64	7.39	69.10	30.10

SOURCE: *Poultry Business Facts, Watt Publishing Co., Mount Morris, IL.*
[a] Percentage of all eggs sold in the area.
[b] States in each area given in Table 1.2.
[c] Data not available for 1959. These flocks included in percentages of flocks of over 3,200 in 1964 census.

declining hen population and the increased average hen production are also indicated.

The production units are also changing rapidly. Before World War II egg production in the Midwest was largely a farm-flock business. Most of the eggs produced were from hens in flocks of less than 400. In 1949 2,422,000 farms reported selling eggs. By 1964, this number had decreased to 527,000, and in 1974, only 198,577 farms reported selling eggs. Of this reduced number in 1974 almost 73% reported flocks of less than 100 hens (Table 1.4). These 144,911 farms accounted for only 0.16% of all layers kept in the United States. On the other end of farm size, 354 farms or 0.18% of all poultry farms accounted for over 30% of all laying hens in the 1974 census. Another indication of commercialization in the egg industry is the report that in 1978, 34 firms had 69,650,000 layers, or about 25% of the hen population. In 1981, 47 egg-production companies owned 96,389,000 hens or 37.1% of the laying hens in the U.S. Many of these companies had several farms; the largest had farms in 13 states. The greatest number of hens on a single farm in 1981 was 3,100,000.

As indicated in Table 1.1, the major egg-producing countries of the world have shifted rankings during the last 18 years. The greatest expansion has occurred in the USSR, with 1982 figures indicating that the Soviet Union is now the major egg producer of the world. Production figures for mainland China are not available. Other countries with big increases include France, Italy, Spain, Mexico, East Germany, The Netherlands, Czechoslovakia, Roumania, Hungary, The Republic of South Africa, Turkey, Egypt, and Venezuela. While there is no clear pattern, it is apparent that most of the eastern European countries have greatly expanded their egg-production facilities.

The trend to commercial size egg-production flocks is not limited to the United States; similar changes are occurring in all parts of the world. Even the flocks of 100,000 hens or more are found in many countries. One of the reasons for the increases in flock size is the development of mechanical equipment to aid the caretaker in watering, feeding, ventilation, egg handling, and litter management. On fully mechanized farms a single caretaker can manage over 100,000 laying hens. Assistance is needed only in egg packaging and when hens are sold or pullets are put into the facility. Such operations should be an all-in, all-out situation so that only dead or ill birds are removed, and no replacements are added.

TABLE 1.4. Size of Laying Flocks Producing Eggs (from 1974 Census)[a]

Area	Less than 400	400 to 1599	1600 to 3199	3200 to 9,999	10,000 to 19,999	20,000 to 49,999	50,000 to 99,999	Over 100,000
New England	0.42	0.77	1.34	6.27	13.82	24.91	7.15	45.32
Middle Atlantic	2.41	3.75	3.35	10.17	18.03	26.74	13.37	22.17
East North Central	6.02	2.82	1.67	14.96	22.49	25.40	11.92	16.31
West North Central	14.99	6.69	1.79	13.17	18.59	19.90	9.17	15.71
South Atlantic	1.08	0.65	0.81	10.75	19.40	32.48	15.94	18.86
East South Central	2.16	0.38	0.61	11.13	17.65	23.90	12.89	31.27
West South Central	2.03	0.59	0.55	9.58	17.81	25.34	14.66	29.44
Mountain	5.71	1.48	1.17	3.89	7.86	13.26	7.78	58.85
Pacific	0.71	0.16	0.28	1.92	5.28	14.53	15.17	62.37
United States	3.64	1.80	1.14	9.59	16.10	24.02	13.14	30.58

SOURCE: *Poultry Business Facts, Watt Publishing Co., Mount Morris, IL.*
[a] Percentage of laying hens on farms.

DIVISIONS OF THE INDUSTRY

The egg industry is divided into areas of specialization, as is true of most production industries. A relatively few poultry breeders supply the hatching eggs for the commercial laying flocks. The number of hatcheries operating today is a small fraction of those in business only 20 years ago. Similarly, there are specialists in feed formulation and manufacture, replacement pullet rearing, equipment and house construction, laying flock management, and egg marketing. Eggs are usually marketed as shell eggs, although an ever-increasing percentage of the total production is sold to the consumer in the form of egg products or formulated foods.

The egg-products segment of the egg industry in the United States dates from about 1900. The first product was frozen whole eggs, followed closely by separated whites and yolks (frozen). With improvements in technology, equipment, and user acceptance, the industry progressed slowly until about 1940.

Utilization of egg products, especially dried whole eggs, expanded very rapidly during World War II, but after the war production cut back to civilian needs. There was a steady growth in volume from 1952 to 1967 as shown in Table 1.5. A decline followed but 1980 shows a resurgence of the industry. In addition, a number of small operations have come into existence to produce hard-cooked eggs and other egg products primarily for the institutional trade.

Historically, the egg-breaking industry has been concentrated in the North Central and Western Plains states. However, these plants have gradually shifted operations to the West Coast and the Southeast, following the relocation of egg production to these areas (Table 1.2). For example, in 1961 there were 105 (42%) plants in the North Central and 36

TABLE 1.5. Annual Production of Egg Products in the United States

Year	Production in millions of pounds			
	Yolks	Whites	Whole eggs	Total
1938	40.5	46.7	58.9	146.1
1944	82.0	84.3	1380.7	1547.0
1951	89.6	116.4	190.4	396.4
1956	118.7	172.5	161.2	452.4
1961	121.2	184.6	299.6	605.4
1964	137.1	210.6	311.3	659.0
1967	164.5	236.0	401.2	801.7
1970	153.3	208.1	386.5	747.9
1975	141.0	204.0	358.0	703.0
1980	170.3	266.9	445.8	883.0

SOURCE: *Koudele and Heinsohn (1964); U.S. Department of Agriculture (1971, 1976, 1981).*

(15%) plants in the Western Plains states. By 1982, 24 (27%) of the plants were in California, Oregon, and Washington, as compared to 29 (12%) in 1961. At the same time (1982), there were still 34 (39%) in the North Central states. On July 1, 1971, the establishment of mandatory USDA inspection of all egg-breaking operations also caused changes in the industry. Thus, in 1961 there were 248 egg-breaking operations processing about 511 million dozen eggs, but only 108 of these plants were operating under USDA inspection. The large number of plants not under federal supervision were mostly smaller plants situated near population centers. Many of these smaller plants, which up until 1971 operated under state inspection programs, closed after federal inspection became mandatory.

REFERENCES

Koudele, J. W., and Heinsohn, E. C. 1960. The egg products industry of the United States. I. Historical highlights, 1900–1959. Kans. Agric. Exp. Stn., Bull. *423*.

Koudele, J. W., and Heinsohn, E. C. 1964. The egg products industry of the United States. II. Economic and technological trends, 1936–1961. Kans. Agric. Exp. Stn., Bull. *466*.

U.S. Department of Agriculture 1967. Foreign Agric. Circ. *FPE 2-67*, December.

U.S. Department of Agriculture 1970. Foreign Agric. Circ. *FPE 2-70*, October.

U.S. Department of Agriculture 1971. Chickens and eggs. Crop Reporting Board, SRS, USDA, Washington, DC.

U.S. Department of Agriculture 1976. Agricultural Statistics. USDA, Washington, DC.

U.S. Department of Agriculture 1981. Agricultural Statistics. USDA, Washington, DC.

Egg Production Practices

W. J. Stadelman
M. H. Swanson

Agriculture is in a major technological and sociological revolution. There is no better example of this than the dramatic changes occurring in commercial egg production. Less than a generation ago the main supply of eggs came from small units of a few hundred birds maintained as general farm and backyard flocks. Today, production is becoming highly specialized with units of 30,000 or more layers not uncommon in most parts of the United States (Fig. 2.1). Several operations have in excess of one million hens each.

Along with this rapid increase in flock size have come extraordinary changes in production practices. Technology generated by such basic disciplines as genetics, nutrition, and physiology has replaced the management art of former years. The results are manifested in higher rates of lay, improved efficiency in converting feed to eggs, lowered costs of production, better product quality, and reduced retail prices to consumers.

The changes in production practices from the small flock allowed an outside runway or complete freedom to roam the barnyard to multiple-bird caged layers has become a major concern of some people as to the welfare and rights of the hens. The pressure from animal welfare groups in Switzerland has led to passage of laws that will phase out caged-layer production facilities in that country during this decade. Similar regulations have been proposed in several other countries. Such legislation has been discussed at state and federal levels in the United States. Some further changes in production practices will likely result from this social pressure. Duncan (1981) has reviewed the animal rights-animal welfare issue.

11

EGG SCIENCE & TECHNOLOGY,
3rd Edition

FIG. 2.1.   Large Southern California egg ranch.
Courtesy University of California.

Processors, handlers, and distributors of eggs and egg products need an understanding of egg-production practices to cope more intelligently with problems of product quality and character. Likewise, food scientists involved with quality control, new product development, or composition studies can benefit from the knowledge of how eggs are produced. Therefore, the purpose of this chapter is to sketch briefly the background of egg-production practices, and particularly to relate such practices to initial product quality. Those needing detailed information on how to manage and operate a commercial egg farm should seek references specifically devoted to that subject. The selected references at the end of the chapter will be helpful in this respect. With such a rapidly changing industry, the beginner would want to be sure of getting the most current information

LAYING STOCK

Records indicate that wild jungle fowl were domesticated in southeast Asia over 3,000 years ago. Since then, nearly 200 breeds and varieties of chickens have been established around the world, although today only a few are of economic importance as egg producers.

Today, the commercial egg industry in the United States is dominated by the Single Comb White Leghorn (Fig. 2.2). This breed has found favor because of its high rate of lay, early maturity, good feed efficiency, relatively small body size, and adaptability to diverse climates. Furthermore, Leghorns lay eggs with white shells, the most widely demanded shell color among consumers. In those limited areas of the United States (principally New England) where brown eggs often bring a premium price, such brown-shell breeds as Rhode Island Red, New Hampshire, and Plymouth Rock may predominate. Much of the European market prefers a light brown egg. Strains of layers have been developed to supply this market.

FIG. 2.2.   Single Comb White Leghorn, a typical layer.
Courtesy Heisdorf & Nelson Farms, Inc., Redmond, Wash.

For many years the conventional system of poultry breeding involved mass selection combined with progeny testing and family selection. Further genetic improvement in performance of layers over the past several decades has been achieved by primary breeders using several different breeding concepts. One of the earlier approaches was cross-breeding, which brought together dominant genes from widely separated sources and resulted in the progeny being superior in many characteristics to either parent breed. This superiority was called heterosis or hybrid vigor. "Austra-Whites," a cross between Australorps and White Leghorns, and crosses between Rhode Island Reds and White Leghorns were very popular at one time because of their viability under adverse disease conditions. However, in the United States the tinted egg shells produced by these crosses did not find wide consumer acceptance, and their popularity waned.

Another breeding system utilizing heterosis is one in which closed flock selection is practiced for several generations to develop distinct strains within a single breed. When two such strains are crossed, the progeny exhibit hybrid vigor. Several Leghorn breeders today use strain crosses to produce chicks for market egg production.

Following World War II, some breeders turned to methods similar to those used so successfully in producing hybrid seed corn. Selected inbred lines of chickens, developed by several generations of mating closely related birds, were crossed. The very uniform progeny were called "inbred hybrids." This breeding system is still in use today.

Important in the development of a commercial line of chickens is field testing of the progeny not only from test crosses in reciprocal recurrent (parent selection on the basis of progeny averages), or closed flock selection programs, but also from the final cross, as in the production of inbred hybrids. The most valuable information gained in these tests is in the area of disease resistance and general viability of the progenies.

The average annual egg production per bird in 1940 was about 11 dozen. Today, 20 dozen eggs per layer is common, and some well-managed units consistently exceed this level. Only a portion of this improvement, of course, can be claimed by the breeder, since better nutrition, housing, and general management have also played important roles. Poultrymen cite the improved production as evidence that the birds are being well cared for in their attempt to keep animal welfare activists from dictating production practices.

The application of basic genetic principles to breeding programs has paid dividends. Broodiness and so-called winter pause have been virtually eliminated from commercial stocks. Age at sexual maturity has been lowered. Viability has been increased by breeding for genetic disease resistance.

Negative associations between desirable hereditary characteristics have made progress difficult (Fig. 2.3). Egg size, for instance, tends to decline as production rate increases. Reducing body weight to save on maintenance costs also may reduce egg size. Additional problems are raised by the interaction of genetic factors with environmental conditions. Good management is always essential for the full expression of a bird's genetic potential.

An area to which breeders have been giving increased attention is the quality of the freshly laid egg. Research and data from Random Sample Tests have established that both interior and shell quality are heritable. The physiological mechanisms through which these factors are expressed are less well understood. Albumen quality in the broken-out egg is associated mainly with the quantity and viscosity of the thick white. Apparently, strains differ in ability to secrete in the magnum large quantities of albumen higher than average in mucin content. Unfortunately, there appears to be some negative association between albumen quality and egg number. As in the case of egg size and egg number, this negative genetic correlation slows down improvement, but by simultaneous selection for the two desirable traits, progress can gradually be made in both.

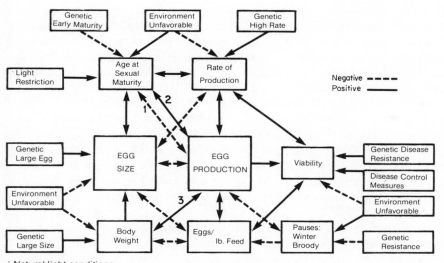

¹ Natural light conditions.
² Light controlled.
³ At low body weight ranges, with exceptions.

Fig. 2.3. Relationships determining egg size and production.
Courtesy L. W. Taylor, University of California.

Interior egg quality is dependent also on the incidence of blood and meat spots. There is still much to be learned about the causes of these defects, but genetic factors are definitely involved. This has been demonstrated by developing, through selection from a common stock, one line with a high incidence of spots and a second line with a low incidence. Breeders have applied considerable selection pressure for a number of years to reduce blood and meat spots. Although some improvement has been made, nutritional and environmental interactions with the genetic factors hinder more rapid advances.

Shell quality is a serious economic problem of the industry because of losses from cracked and broken eggs. Every approach to its solution must be used, including breeding for increased shell strength. Again, a negative relationship exists, in that shell thickness tends to decline as egg numbers increase. Genetic progress in shell-quality improvement, therefore, is somewhat difficult.

Raising Replacements

Egg producers have a choice of brooding and rearing their own pullets for layer replacement or buying 16- to 20-week-old birds from growers who specialize in raising "started pullets." This choice is dependent, in part, on the availability of labor and capital as well as on the ability of the operator to produce healthy, well-developed pullets at a reasonable cost. Where mortality in the young stock has been a particular problem, the disease cycle often can be broken only by a complete cessation of brooding and rearing operations. In such cases, layer replacement must be made by either purchasing started pullets or raising them at a location geographically isolated from any laying stock.

The brooding period covers the first 6 to 8 weeks of the chick's life. During this time, the bird should be subjected to a minimum of environmental and social stresses so that it can develop to its fullest genetic potential. The first requirement is for housing and equipment that can be thoroughly cleaned and disinfected between broods. Cement floors, smooth wall and ceiling surfaces, and equipment that can be cleaned in place or dismantled for cleaning will facilitate good sanitation practices.

Temperature is a most important environmental factor in the early life of the bird. Heat is supplied either by the hover system or by warm-room brooding. In the hover system each unit is controlled by a thermostat set to maintain a temperature for day-old chicks of 32° to 35°C at the edge of the brooder. Gas, oil, and electricity are commonly used. At the end of each week, the temperature can be reduced by 3°C until the 21°C mark is reached. In warm-room brooding, circulating hot air or hot water is used, but the starting temperature is usually lower, since the chicks cannot escape to a cooler area. An even distribution of chicks under the hover or over the area of confinement is an indication of bird comfort and proper temperature. Such treatment of the chicks is further evidence that poultrymen are concerned about the welfare of their animals.

Although floor brooding and rearing are more common (Fig. 2.4), an increasing number of replacements are being maintained on wire from the day-old stage through the growing period. Much of the new construction for replacement rearing provides for cage brooding in insulated, windowless buildings in which both light and temperature are controlled or modified (Fig. 2.5). Birds may be maintained in the same cage until transferred to the lay house, or they may be placed in grow cages when artificial heat is no longer required (Fig. 2.6).

Feed and water should be placed in the brooding areas before the chicks arrive. It is important that birds have ready access to both to prevent starvation and dehydration. As the birds grow, the amount of

FIG. 2.4.   Replacement pullets brooded and reared on litter. Hovers have been raised until needed.
Courtesy University of California.

FIG. 2.5. Interior of environmentally controlled cage brooder house. Warm air enters near the ceiling causing uniform temperature.
Courtesy Watt Publishing Co.

FIG. 2.6. Replacement pullets reared in cages.
Courtesy University of California.

feeder and water space must be increased. The use of mechanical feeders and automatic water cups or troughs reduces labor requirements. Routine cleaning of waterers is required for good sanitation.

A good feeding and management program will result in birds of uniform size and body growth typical for the strain being raised. Samples of birds are commonly weighed during the rearing period and the results compared with a standard growth curve supplied by the breeder. Underweight birds can result from disease, overcrowding, improper feed formulation, feed shortage, and other mismanagement practices. Overweight pullets are also possible as a result of excessive feed consumption.

The intensity and daily duration of light during the growing period significantly affect age of sexual maturity and response to lighting after egg-laying begins. Light has a stimulatory effect on the pituitary gland, which in turn controls production and release of other hormones affecting egg formation. Excessive light stimulation results in too early sexual maturity and the laying of many small eggs. With controlled lighting, sexual maturity can be delayed until body size is sufficient to permit the laying of larger eggs. Several types of lighting programs can accomplish this objective. Either constant short days or decreasing photoperiods may be used; increasing photoperiods are stimulatory. When replacements are raised in windowless houses, a constant day length of 10 or less hours is common. For houses using natural daylight, a step-down lighting program is required, particularly for chicks hatched in the fall and early winter, which mature during the spring months of increasing day length.

A light intensity of 1 foot-candle from either incandescent bulbs or fluorescent tubes is adequate. Levels below 0.04 foot-candle are non-stimulating. By keeping light intensity to a minimum, bird activity is reduced, cannibalism is discouraged, and social stresses are lowered. The use of red light during the grow period is quite popular because of its quieting effects.

Housing and Equipment for Layers

The primary purpose of housing is to provide a comfortable environment for the layers and thereby maximize egg production. In such an environment temperature and light can be controlled, humidity maintained at a moderate level, and circulated air kept free of excessive dust and ammonia gas.

Adult birds can withstand a rather wide temperature range, if the changes in temperature are not too sudden. The zone of thermal neutrality has been reported to be in the range of 14° to 26°C. Below 14°C additional feed must be consumed to maintain body temperature, resulting in increased production costs. Above 26°C water consumption rises dramatically and feed consumption decreases. High temperatures not only depress production but also adversely affect shell quality and egg size.

Humidity is less critical than temperature for bird comfort. However, a relative humidity of 40 to 60% is preferred. If the air is too dry, dust will be a problem, especially for floor operations using litter. Too humid conditions may distress both birds and workers, particularly at elevated temperatures.

So-called "conventional housing" of single-wall construction with no insulation and with windows for natural lighting and ventilation falls short of providing the ideal bird environment. Much of the older layer housing is of this type, but it is rapidly being replaced by well-insulated, windowless, force-ventilated buildings in which light can be controlled and temperature modified (Fig. 2.7). In cold climates insulation is re-

FIG. 2.7.   Environmentally controlled layer houses. The upper two windowless deep-pit houses are cooled by fans pulling air through the two wetted-pad towers for each house. Air is expelled through the light-trapped chimneys. Tanks for feed storage are at the far end of each house.
Courtesy Watt Publishing Co.

quired to keep house temperatures from falling below critical levels; in warmer areas insulation aids in preventing abnormal elevation of temperature.

Forced ventilation systems can be either the positive- or negative-pressure type. The former uses fans to blow air into the house to create a small rise in pressure; the latter uses fans to expel air from the house, generally with a slot inlet around the perimeter of the building for incoming air. Fans operated by thermostats provide maximum ventilation during hot weather and minimum operation during cold periods.

In areas where high summer temperatures prevail, some method for cooling incoming air must be employed. Since mechanical refrigeration is too expensive, most systems rely on the principle of evaporative cooling. Air is pushed or pulled through a wet pad or past a series of spray nozzles. As the water evaporates, energy is removed from the air and the air temperature drops. Efficiency of cooling is dependent on the relative humidity. In such arid areas as the Southwest, laying-house temperatures can be lowered by as much as 11°C below the outside. Less sophisticated cooling systems use sprinklers on the roof or foggers directly over birds. Such systems are most common in warm climates, where "open" houses with walls of wood lath or with adjustable curtains are adequate weather protection.

Within the layer house, birds are maintained either on the floor or in cages. Although most new construction for commercial egg production is based on multiple-bird cages for labor efficiency and high bird density, floor operations are still used for small flocks and most breeding flocks (Fig. 2.8). A litter of wood shavings, ground corncobs, or other suitable material may be used over the entire floor, or raised wood-slat platforms through which the droppings fall may be substituted. Often a combination is used, where feeders and waterers are placed on wood slats over a mechanically cleaned pit in the center of the house with litter on either side. The nests should be located in the cleanest and driest portion of the house to obtain the fewest number of soiled eggs.

A wide variety of cage systems is in use. Cage size, for example, may vary from one 8 in. wide and 16 in. deep housing a single bird to "colony" cages several feet in width and length holding 10 to 25 layers. A very common size is the 12 × 18-inch cage for three birds. Mortality due to cannibalism tends to increase as the number of layers per cage increases.

Cage shape varies from a narrow-front deep cage to a wide-front shallow cage, often called the reverse cage. There are advantages to each type. In some instances the reverse-caged hens produce a greater number of eggs than similar hens in deep cages. However, feed consumption is greater in the reverse cage so that the cost of egg production is similar for both types of cages.

FIG. 2.8.   Floor housing for a breeder flock with individual nests on side wall. Roosts, waterers, and feeders are located over a raised wire-covered pit.
Courtesy Watt Publishing Co.

Cage arrangements also vary. Single-deck cages simplify bird care and permit droppings to accumulate on the floor, where they require less frequent removal. Truss-type houses with no supporting posts facilitate manure removal (Fig. 2.9). Higher bird density per square foot of house can be obtained by double- or triple-decking the cages. In these cases, if the cage rows are mounted directly above one another, dropping boards and frequent manure removal are required (Fig. 2.10). Multiple-decked cages may also be stair-stepped to eliminate the necessity for dropping boards.

Supplying feed and water to the birds and gathering the eggs require considerable labor, if they are done by hand; therefore, the trend in larger commercial operations is toward automating these tasks. Water may be easily supplied to floor birds by troughs provided with float valves. More sanitary systems employing self-cleaning drinking cups or nipple valves have been developed for cage birds. These have now been adapted for floor birds, too.

Automatic feeders in which the feed is conveyed and evenly distributed in a trough are popular for both floor and cage operations (Fig. 2.11). Time clocks are used to operate the feeder at periodic intervals.

FIG. 2.9. Single-deck open-type cage layer house with side walls of wood lath to protect birds from the wind. Suspension of cages keeps area beneath free of posts. *Courtesy University of California.*

FIG. 2.10. Double-decked cages equipped with dropping boards and a battery-operated feed cart. Continuous-flow V-shaped water trough supplies two rows of back-to-back cages. *Courtesy Watt Publishing Co.*

FIG. 2.11.   Automated single-deck caged layer house. A mechanical feeder oper-
ates intermittently to bring feed to the hens. A continuous-flow water trough
supplies cool water, and an intermittent egg belt gathers eggs from the roll-out
cages and transports them to a central egg room.
Courtesy of Indiana State Poultry Association.

Somewhat less automatic, but quite efficient for feeding cage birds, is
the electric cart equipped with a power auger (Fig. 2.10).

Hand-gathering of eggs into wire baskets or directly onto filler-flats is
still the most common way of handling eggs in the lay-house. For floor
operations, the eggs must be collected from individual or community
nests. Cages are constructed with a sloping floor so that the eggs will roll
out into a tray next to the aisle for easy gathering. The use of an
overhead rail conveyor or an electric cart facilitates the hand-gathering
operation. For larger operations, particularly where labor is scarce and
high in cost, automatic egg-gathering belts may be justified. This
equipment can be adapted to floor, single-deck cage, or multiple-deck
cage systems (Fig. 2.12). The eggs are gathered from each row of roll-out
nests or cages to a cross conveyor which takes them to a central assembly point. In some large operations the eggs are washed, candled, sized,
and packaged on a continuing basis as the eggs are assembled by the
automated conveyors. Such eggs are untouched by humans until they
are either in dozen cartons or tray-packed for bulk shipment.

FIG. 2.12.   Automated egg gathering from triple-deck cages on the right. Egg-
gathering belts bring the eggs to de-escalators, which deliver them to a cross-
conveyor belt at the lower left. These belts move the eggs out of the house to a
central collection point.
Courtesy Watt Publishing Co.

POULTRY NUTRITION AND FEEDING PRACTICES

More is known about the nutritional requirements of the chicken than
of any other domestic animal. This knowledge has made possible dramatic gains in feed efficiency for both poultry meat and egg production.
Only a few years ago more than 3 lb of feed were required to produce 1 lb
of live broiler and more than 5 lb were consumed for each dozen eggs
laid. Today, better poultry managers can produce 1 lb of broiler on less
than 2 lb of feed and a dozen large-sized eggs on less than 4 lb of feed
(Patrick and Schaible 1980).

The efficiency of any feeding program is dependent on its meeting the
nutrient requirements of the bird. Therefore, before feed formulation is
undertaken, specific nutrient requirements must be determined. Tables
2.1 and 2.2 list the requirements for starting, growing, and laying birds
as published by the National Research Council. Feed manufacturers
must include added safety allowances for variations in ingredient composition and nutrient losses during processing and storage.

TABLE 2.1. Energy, Protein, and Amino Acid Requirements of Egg-Type Chickens

	Replacement pullets			Laying and breeding hens
	0–6 weeks	6–14 weeks	14–20 weeks	
Metabolizable energy (kcal/kg)	2900	2900	2900	2850
Protein, %	18	15	12	15
Arginine, %	1.00	0.83	0.67	0.8
Glycine and/or serine, %	0.70	0.58	0.47	0.5[a]
Histidine, %	0.26	0.22	0.17	0.22[a]
Isoleucine, %	0.60	0.50	0.40	0.5
Leucine, %	1.00	0.83	0.67	1.2
Lysine, %	0.85	0.60	0.45	0.60
Methionine + cystine, %	0.60	0.50	0.40	0.50
or				
Methionine, %	0.32	0.27	0.21	0.27
Phenylalanine + tyrosine, %	1.00	0.83	0.67	0.8[a]
or				
Phenylalanine, %	0.54	0.45	0.36	0.4[a]
Threonine, %	0.56	0.47	0.37	0.4
Tryptophan, %	0.17	0.14	0.11	0.11
Valine, %	0.62	0.52	0.41	0.5[a]

[a] Estimated values.

TABLE 2.2. Nutrient Requirements of Chickens[a]

Nutrient	Starting chickens 0–8 weeks	Growing chickens 8–18 weeks	Laying hens	Breeding hens
Vitamin A activity				
(USP Units)	1,500	1,500	4,000	4,000
Vitamin D (ICU)	200	200	500	500
Vitamin K_1, mg	0.50	0.50[b]	0.50[b]	0.50[b]
Thiamine, mg	1.8	1.3[b]	0.8[b]	0.8
Riboflavin, mg	3.6	1.8	2.2	3.8
Pantothenic acid, mg	10	10	2.2	10
Niacin, mg	27	11	10[b]	10[b]
Pyridoxine, mg	3	3.0[b]	3	4.5
Biotin, mg	0.09	0.10[b]	0.10[b]	0.15
Choline, mg	1,300	500[b]	500[b]	500[b]
Folacin, mg	0.55	0.25[b]	0.25	0.35
Vitamin B_{12}, mg	0.009	0.003[b]	0.003[b]	0.003
Linoleic acid, %	1.0	0.8[b]	1.0	1.0
Minerals				
Calcium, %	0.9	0.6	3.25	2.75
Phosphorus, %	0.7	0.4	0.5	0.5
Sodium, %	0.15	0.15	0.15	0.15
Potassium, %	0.2	0.16	0.1	0.1
Manganese, mg	55	0.25[b]	0.25[b]	33
Iodine, mg	0.35	0.35	0.30	0.30
Magnesium, mg	600	400[b]	500	500[b]
Iron, mg	80	40[b]	50[b]	20[b]
Copper, mg	4	3[b]	3[b]	4[b]
Zinc, mg	40	35[b]	50[b]	65[b]
Chlorine, mg	800[b]	800[b]	800[b]	800[b]
Selenium, mg	0.1[b]	0.1[b]	0.1[b]	0.1[b]

SOURCE: *Adapted from National Academy of Sciences, National Research Council (1977).*
[a] Values given are (1) estimates of requirements and include no margins of safety and (2) in percentage or amount per kilogram of feed.
[b] Tentative values.

In addition to the birds' nutrient requirements, specific feed formulas are determined by available feed ingredients, their composition, and their relative costs. "Least-cost feed formulation" is most efficiently done with the aid of a computer. Table 2.3 presents rations typical for those used in the northern states. In the Southwest, the sorghum grains might replace most or all of the corn in these formulas. In the South and other areas of the United States where cottonseed meal is readily available, this protein supplement would probably replace a part of the soybean oil meal. Practically all poultry rations are of the all-mash type, because feeding is simplified and a better balanced diet is assured. Modern mills (Fig. 2.13) are designed to mix rations efficiently and accurately. Bulk delivery using tanker trucks of 12- to 20-ton capacity is common.

The amount of feed consumed is dependent on the energy level of the feed, egg production rate, body size, and environmental temperature. In

TABLE 2.3. All-Mash Chicken Rations[a]

Mash ingredients	Units	Starter	Grower	Layer	Breeder
Ground yellow corn	lb	1,129	1,206	1,250	1,326
Standard wheat middlings	lb	100	300	100	100
Soybean oil meal (44%)	lb	550	350	375	250
Fish meal (60%)	lb	50	—	—	75
Stabilized fat	lb	20	35	35	20
Alfalfa meal (100,000 A/lb)	lb	50	50	50	50
Distillers dried solubles	lb	50	—	50	50
Dicalcium phosphate	lb	22	32	30	24
Ground limestone	lb	24	22	105	100
Iodized salt	lb	5	5	5	5
Antibiotic supplement		b	—	—	—
Antioxidant		b	b	b	b
Coccidiostat		b	b		
Manganese	gm	52	52	52	52
Vitamin supplements					
Vitamin A	I.U.	1,000,000	1,000,000	3,000,000	4,000,000
Vitamin D$_3$	I.C.U.	681,000	681,000	1,362,000	1,362,000
Vitamin B$_{12}$	mg	6	6	6	6
Choline chloride	mg	227,000	—	—	—
Riboflavin	mg	1,500	1,000	1,000	2,000
Calcium pantothenate	mg	5,000	5,000		5,000
Niacin	mg	20,000	10,000	20,000	20,000
Totals	lb	2,000	2,000	2,000	2,000

[a] Practical formulas recommended by the New England College Conference Board, based on experimental work in New England.
[b] Use according to manufacturer's directions.

general, poultry eat to satisfy their energy needs. The energy level of a ration can be varied by reducing the fiber content or by adding fat. If the energy level of the feed is increased, a corresponding increase in protein, vitamins, and minerals is required for a balanced ration, since the birds will consume proportionately less of the high-energy feed. Four-pound hens laying at a rate of 70% and fed a ration moderate in energy level will consume daily about 22 lb of feed per 100 layers under mild temperature conditions. Such a ration should contain approximately 17.5% good-quality protein in order to meet daily amino acid requirements. The difference between the 15% protein listed for laying and breeding hens in Table 2.1 and the recommended protein level is to allow a margin for the amino acid deficiencies in some protein sources. A least-cost ration formulated to the amino acid levels listed in Table 2.1 would be suitable but the protein level would vary from the 15% suggested in Table 2.1 based on feed ingredients used in the formulation.

The layer ration can have a pronounced effect on the quality of the freshly laid egg. Shell strength, for example, is highly dependent on mineral and vitamin levels, including those of calcium, phosphorus, manganese, and vitamin D. Yolk color can be varied by the amount of

FIG. 2.13.   Automated feed mill and bulk delivery truck.
Courtesy Watt Publishing Co.

alfalfa meal, yellow corn, and other xanthophyll-containing ingredients. Low levels of vitamin A may increase the blood spots. For maximum egg size, the ration must contain an adequate level of essential fatty acids.

Olive-brown mottled yolks are produced by rations containing excessive amounts of gossypol, a fat-soluble compound found in cottonseed meal and oil. Gossypol reacts with iron in the yolk to give the mottled effect. Cottonseed products also contain cyclopropenoid compounds which increase vitelline membrane permeability. As a result, iron from the yolk passes through and reacts with the conalbumin of the white to produce a pink pigment. In addition, the cyclopropenoid compounds cause the hen to deposit a higher proportion of saturated fats than normal in the yolk. The result is a pasty, custardlike yolk, particularly in the cooled egg.

Eggs from a confined bird on standard commercial rations have a remarkably uniform and mild flavor. However, off-flavors in the egg can result from poor-quality fish meal or from the accidental inclusion of garlic, certain weed seeds, and other foreign materials in the ration.

GENERAL MANAGEMENT

Lighting for Layers

Day length, as well as changes in length of day, influence the rate of egg production. Just as a diminishing day length retards sexual maturity in young pullets, an increasing day length tends to stimulate egg production in layers. With light-tight lay houses, advantage can be taken of this phenomenon by the use of time clocks to gradually lengthen the daily exposure to artificial light. When entrance of daylight into the house cannot be controlled, a day length must be selected equal to the longest natural day of the laying year, and supplementary artificial lights used to maintain this schedule. The minimum daily light requirement is 12–13 hr at an intensity of 0.5 to 1 foot-candle. A day length of 14–16 hr is commonly used.

Cannibalism Control

Birds maintained in cages or at high-density rates on the floor may develop cannibalistic habits, which can lead to increased mortality and loss in performance. The light breeds, such as the Leghorn, are more subject to this problem than the heavy breeds. Strain differences within breeds also exist. Some poultrymen are able to control cannibalism by using a minimum light intensity during both the grow and lay periods. Others routinely "debeak" their birds by cutting off a portion of either the upper beak or both upper and lower beaks. If this is done at an early age, it may be necessary to repeat the operation before placing the pullet in the lay house. Special debeaking machines with a hot cutting blade to cauterize the beak have been developed.

Force Molting

A majority of flocks are sold for slaughter at the end of their first "laying year." Since most strains become sexually mature at 22–24 weeks and lay well for 12–14 months, the age at this point is generally 18–20 months. However, in an effort to reduce replacement costs, an increasingly popular practice is to "force molt" the birds and keep them for a second period of lay. Periodic molting of feathers, during which egg production generally ceases, is natural to all species of birds. Poultrymen can bring the entire flock into a controlled molt at one time by withholding feed for several days. This throws the birds out of production, and a molt follows. A low-protein feed may be used for 1–4 weeks to permit a rest period before returning to production. Detailed procedures for force molting of layers are given by Swanson and Bell (1974–1976).

Rate of lay during the second year does not generally reach first-year levels, but egg quality following the molt is substantially improved over that prior to molting. Shell thickness decreases, shell roughness increases, and albumen quality falls rather markedly during the last few months of the pullet year. Considerable quality is regained by molting and then declines again after several months of lay. By molting pullets at 14 months of age and again at 22 months of age so that each lay period is limited to 8 months, egg quality can be maintained at a higher average level than by a single molt at 18–20 months of age.

Waste Management

A problem of increasing seriousness for poultrymen is the handling and disposal of waste materials, particularly where long-established production units have become encircled by expanding urban housing. Poultry droppings have some fertilizer value, but costs of removal, hauling, and spreading on crop land often exceeded the costs of the same nutrients supplied by inorganic fertilizers. This situation suddenly changed in the mid-1970s, when energy costs spiraled upward and commercial fertilizer prices increased substantially. Poultry waste is now in demand as an organic soil amendment and brings the poultryman a return rather than being a cost factor in egg production. A limited amount of processing of poultry manure is being done in which fresh droppings are rapidly dehydrated and the resulting dry, nonodorous product finds an outlet in nursery and home garden use. Some states also permit this product to be sold as a feed ingredient for ruminants.

Where droppings are accumulated within the poultry house or on the floor for periodic removal, care must be taken to prevent fly breeding. Adequate ventilation is important to rapidly lower the moisture content of the droppings below the level supporting maggot development. Screening of open-type houses and the inlets of environmental houses is effective. The use of chemical pesticides for fly control should be limited to emergency situations. Matthew (1976) gives details for control of several species of flies around caged layer houses. Improper management of poultry wastes can also result in production of obnoxious odors. Ammonia fumes not only irritate workers but cause birds to be more susceptible to respiratory infections; also, eggs exposed to ammonia lose albumen quality at a more rapid rate.

Business Management

Skills in both production and business management are required for a successful egg operation. Capital investments in land, buildings, and equipment may run as high as $3 to $5 per layer, and substantial operating capital is needed in addition. Thus, several hundred thousand dollars may be involved in even a modest-size unit. It is essential that accurate records be kept in such detail that all phases of the operation can be evaluated for efficiency and productivity against past performance and attainable standards. For each flock it should be possible at any given time to determine, both currently and cumulatively, such criteria as egg production rate, feed intake per layer, pounds of feed per dozen eggs, mortality, feed cost per dozen, total cost per dozen, and distribution of egg size and quality—all of which affect daily management decisions. A good record system enables the alert manager to spot and correct production problems in their early stages.

POULTRY HEALTH

Success in egg production is highly dependent on the poultryman's ability to raise strong, healthy pullets and keep stresses of disease and parasites to a minimum in the laying stock. Mortality rates of 5% or less during the growing period and 1% or less per month of egg production are possible with a carefully planned and executed disease prevention program. Poultry farms should be as isolated as possible from other poultry operations and should be fenced to control traffic. Diseases tend to be spread more by the movement of people and equipment than by the birds themselves.

Day-old chicks can now be purchased free of pullorum and typhoid, two egg-borne diseases which at one time resulted in high death losses during the first few days of brooding. Marek's disease, a virus infection of young birds, was once the principal cause of mortality up to 6 months of age. In 1971, a vaccine was introduced which has proved to be a very effective control. Selection for genetic resistance to the disease in foundation breeding stock has resulted in considerable progress. Also, brooding and growing of pullets in quarters isolated from older birds is recommended, since Marek's, as well as many other diseases, spreads by contact exposure. For the same reason, only layers of the same age should be housed together. Lymphoid leukosis, a disease closely related to Marek's, produces characteristic tumors in the abdomen and is found more commonly in birds of laying age. Breeding for resistance is the principal approach to control.

Other diseases of poultry commonly controlled by vaccination include fowl pox, laryngotracheitis, infectious bronchitis, and Newcastle disease. The last two, which affect the respiratory tract, have detrimental effects on egg quality. Birds recovering from infectious bronchitis or Newcastle and returning to production will often lay eggs which are odd-shaped, and the shells may be thin and rough. The interior quality is also affected, as indicated by thin, watery albumen and air cells that are loose or bubbly.

Antibiotics have been used by poultrymen for a number of years to control a variety of bacterial diseases. Although antibiotics can be injected directly into the body of the bird, they are usually administered through the feed or drinking water. Prophylactic levels are used to prevent certain diseases from becoming established in the flock; therapeutic levels are used to restore diseased birds to normalcy. Antibiotics are of little value in virus infections.

Coccidiosis, a common intestinal disease of birds of all ages maintained on litter, is caused by a microscopic transmissible oocyst passed through the droppings. Control can be achieved by the addition to the feed of one of several available drugs called "coccidiostats." Roundworms can also infest the intestinal tract. Occasionally a worm may migrate through the cloaca into the upper portion of the oviduct and become enclosed within an egg. Birds maintained in cages have little or no problem with either coccidiosis or worms, since the separation of birds and droppings breaks the life cycle of the parasite. Mites, lice, fleas, and ticks are external parasites of poultry which must be kept under control by the use of appropriate pesticides (Williams 1979).

Poultrymen must be very careful in using both drugs and pesticides to control disease and parasites. Irresponsible use can result in illegal residues in the egg and body tissues. Only registered and authorized medications and chemicals should be employed; label instructions must be explicitly followed.

EGG HANDLING AND PROCESSING BY PRODUCERS

Contrary to popular belief, not every egg gathered from the nest or cage is necessarily of first-rate quality, but a high percentage of them will be if good flock-management practices have been followed. As noted above, these include selecting a strain of birds with genetic capabilities of laying high-quality eggs, feeding well-balanced rations, keeping the flock disease-free by good sanitation and vaccination, and replacing or molting the flock before an excessive decline in egg quality occurs.

Once a good-quality egg has been laid, it must then be properly handled to minimize any loss of that quality. Eggs are a perishable product. The care they receive between the time of laying and delivery to the first buyer is most crucial. Since the physical and chemical changes responsible for quality decline in the egg are accelerated by high temperatures, it is important to cool eggs promptly. This means frequent gathering, especially in the summer months, to minimize exposure of the eggs to laying house temperatures. Eggs are often placed directly in one-piece filler-flats, which may be stacked on open racks for transporting from the lay houses to the refrigerated holding rooms (Fig. 2.14)

Although temperatures just above the freezing point are the most effective in maintaining quality, eggs are generally held between 10° and 16°C where storage times are limited to a few days. Quality decline at this higher level is slightly greater, but the costs of refrigeration are considerably less. Also, the problem of moisture condensation on the cold egg on removal from the cooler, commonly called "sweating," is reduced at the higher holding temperature. A relatively high moisture level in the air of the holding room is desirable to minimize loss of water

FIG. 2.14. Transporting eggs on filler-flats from the laying house to the egg-holding room.
Courtesy of Indiana State Poultry Association.

from the egg. Above 80% relative humidity, mustiness and off-odors may develop.

In addition to holding eggs under refrigeration to preserve quality, they may be oil-processed to seal the pores of the shell. This treatment not only retards loss of moisture but also reduces loss of carbon dioxide from the egg. As a result, the pH of the albumen rises less rapidly. Since thinning of the thick white and deterioration of the yolk membrane are accelerated by a rise in pH, oiling slows down these undesirable changes and improves broken-out quality. If the oiling process is delayed several days, it is less effective in preserving quality, since a high proportion of carbon dioxide has already escaped. Therefore, eggs are often oiled as they are gathered in the laying house or shortly thereafter. A low-viscosity, highly refined mineral oil that is odorless and tasteless is used for this treatment. The process may be repeated later at the assembly plant at the time of grading and cartoning, particularly if the eggs are first washed.

FIG. 2.15.   Washing, sizing, and cartoning equipment in a shell-egg processing plant.
Courtesy University of California.

Whether or not eggs are cleaned, graded for quality, sized, bulk-packed, cartoned, or converted into egg products at the point of production depends on the type of operation and market outlets. Minimum processing would involve only bulk packing for transportation to a central assembly plant. On the other hand, a number of large egg farms or ranches now carry out the complete processing operation in facilities equal in size and efficiency to plants operating independently of production units (Fig. 2.15).

Unless the volume of eggs is sufficiently large to justify the use of modern high-speed processing equipment, it is generally more economical for the producer to concentrate on production problems and sell his bulk eggs to a firm specializing in egg processing. However, some producers do find it profitable to clean, grade, size, and carton eggs for local retail outlets or for direct sale to consumers.

SELECTED REFERENCES

Anon. 1971. Certain factors affecting shell egg quality: Conditions, causes and corrective measures. Maine Coop. Ext. Circ. *402.*

Casey, J. M. and Shutze, J. V. 1975. Light management. Ga. Coop. Ext. Circ. *684.*

Charles, O. W. 1974. Production of commercial layer replacement pullets. Ga. Coop. Ext. Bull. *719.*

Duncan, I. J. H. 1981. Animal rights-animal welfare: A scientist's assessment. Poult. Sci. *60,* 489–499.

Matthew, D. L. 1976. Fly control in poultry houses. Coop. Ext. Serv. Bull. (Purdue Univ.) *E-3.*

National Academy of Sciences, National Research Council. 1977. Nutrient Requirements of Domestic Animals. Nutrient Requirements of Poultry, 7th Revised Edition, Publ. No. 1. NAS/NRC, Washington, D.C.

North, M. O. 1984. Commercial Chicken Production Manual. 3rd Edition. AVI Publishing Co., Westport, CT.

Patrick, H., and Schaible, P. J. 1980. Poultry: Feeds and Nutrition, 2nd Edition. AVI Publishing Co., Westport, CT.

Rosenwald, A. S., *et al.* 1973–1976. Planned disease prevention for poultry. A series of ten Calif. Coop. Ext. Leaflets: 2626, 2627, 2628, 2629, 2630, 2631, 2632, 2633, 2786, 2871.

Scott, M. L., Neisheim, M. C., and Young, R. J. 1976. Nutrition of the Chicken, 2nd Edition. M. L. Scott and Associates, Ithaca, NY.

Swanson, M. H., and Bell, D. D. 1974–1976. Force molting of chickens. A series of six Calif. Coop. Ext. Leaflets: 2649, 2640, 2651, 2810, 2811, 2874.

Whiteman, C. E., and Bickford, A. A. 1979. Avian Disease Manual. American Association of Avian Pathologists.

Williams, R. E. 1979. External parasites of poultry. Coop.Ext. Serv. Bull. (Purdue Univ.) *E-6.*

3

Quality Identification of Shell Eggs

W. J. Stadelman

Egg quality is based on those characteristics of an egg that affect its acceptability to the consumer. Studies have been conducted for many years to develop methods of determining the various quality factors. In order that we may discuss the several quality factors of shell eggs, we must first become acquainted with the formation and structure of an egg. The egg is formed not for human food but to produce a chick. With this in mind, the structure of the egg can be more readily understood.

STRUCTURE OF THE EGG

The parts of the egg are shown schematically in Fig. 3.1. It is generally accepted that a hen forms an egg in about 2 weeks. This is true, except for the very small core of the yolk. At the time of hatching, the ovary of the female chick has many small ova present—the number has been estimated at over 3000. The yolk of the egg is formed in three stages: (1) the part formed during embryonic development of the female chick, (2) the normal slow development of the ovum from time of hatching of the chick to a point in sexual maturity some 10 days before ovulation, and (3) the accelerated growth period during the last 10 days before ovulation (release of the ovum or yolk) into the oviduct, a part of the female reproductive system. In the oviduct, the egg white is secreted to surround the yolk, and finally the shell membranes and shell are deposited to complete the formation process.

EGG SCIENCE & TECHNOLOGY,
3rd Edition

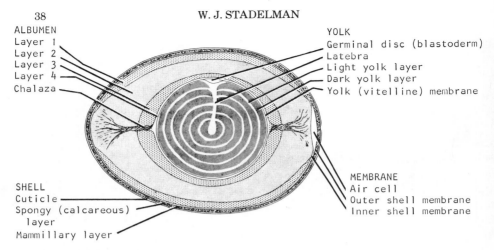

ALBUMEN
Layer 1
Layer 2
Layer 3
Layer 4
Chalaza

YOLK
Germinal disc (blastoderm)
Latebra
Light yolk layer
Dark yolk layer
Yolk (vitelline) membrane

SHELL
Cuticle
Spongy (calcareous)
  layer
Mammillary layer

MEMBRANE
Air cell
Outer shell membrane
Inner shell membrane

FIG. 3.1.　Parts of an egg.
Courtesy USDA.

The yolk is formed during the final 10 to 12 days prior to laying of the egg. Its structure consists of the latebra, germinal disk, and concentric layers of light and dark surrounded by the vitelline membrane. The yolk comprises 30 to 33% of the total egg weight. At the time of ovulation the yolk sac, or follicular membrane, releases the fully developed yolk into the open upper end of the oviduct.

The white, or albumen, of the egg is formed in a matter of a few hours and represents approximately 60% of the total egg weight. The white occurs in four layers (see Fig. 3.1) in most chicken eggs. These are the chalaziferous or inner thick white (layer 4), which is adjacent to the vitelline membrane and continuous with the chalaza (Fig. 3.1). The next outer layer is the inner thin white (layer 3) surrounded by the outer thick white (layer 2). The outer layer of white is the outer thin layer (layer 1). The percentage of the total white found in each of the four layers varies widely, depending on strain of the laying hen, age of the hen, and age of the egg.

The next layers of the egg are the inner and outer shell membranes. These relatively thin keratinlike membranes are one of the egg's chief defenses against bacterial invasion. The inner membrane is thinner than the outer membrane, but together they are only 0.01 – 0.02 mm thick.

The outer covering of the egg, the shell, comprises 9 to 12% of the total egg weight. It consists largely of calcium carbonate (94%) with some magnesium carbonate (1%), calcium phosphate (1%), and organic matter, chiefly protein (4%). In colored eggs the color is due to pigments (ooporphins) deposited on the shell surface. The shell is formed in a distinct pattern with pores for gas exchange. Even though the pores are partially sealed by protein (keratin), they still permit the escape of carbon dioxide and moisture from the egg. Under some conditions, the pores also permit bacterial penetration as far as the shell membranes.

All the structural parts of the egg have been mentioned except the air cell. This develops by a separation of the two shell membranes, usually at the large end of the egg, as the egg contents shrink during cooling. The air cell continues to increase in size as moisture and carbon dioxide are lost throughout the existence of the intact egg (Romanoff and Romanoff 1949).

The process of egg formation is a complex series of hormone-controlled reactions, a thorough study of which requires a knowledge of reproductive physiology. The brief description given here is sufficient for a person interested in eggs primarily as an article for human food or commerce.

GRADING OF SHELL EGGS

Grading involves the sorting of eggs into categories based on size or weight and quality factors of the shell and such internal portions of the egg as the white, yolk, air cell, and possible abnormalities. Historically, and even today in small grading stations, each egg is handled during candling (Fig. 3.2). An easily portable candling device is shown in Fig. 3.3. A similar device is often used by regulatory inspectors verifying the interior quality of eggs in market channels. However, in most large plants eggs are flash-candled, as shown in Fig. 3.4. With such equipment only under-grade eggs are handled by the grader.

Originally candling procedures were developed to separate fresh eggs from storage and partially incubated eggs. With advances in production practices the role of candling changed, so that at present the primary function is to detect and remove cracked or abnormal eggs, such as an egg with a blood spot.

Detailed instructions for egg grading are given in the *USDA Egg Grading Manual* [U.S. Department of Agriculture (USDA) 1978]. The specific standards for each quality grade of eggs were revised in 1981. Current specifications are published in the *Federal Register* (USDA 1981). These changes modified tolerances for errors in grading and eliminated the formerly used grade C and fresh fancy quality control programs. A significant change was the allowance for an increased incidence of checked eggs in destination grading as compared to origin grading. Origin grading was defined as grading made at a plant where eggs are graded and packed.

FIG. 3.2. A typical hand-candling device used by USDA inspectors to verify quality of eggs sold under the USDA shield.
Courtesy of A. R. Rhorer, Indiana State Egg Board.

The following material has been adapted from the federal regulations governing grading of shell eggs [*Code of Federal Regulations* 7, part 56 (USDA 1981)].

U.S. STANDARDS FOR QUALITY OF INDIVIDUAL SHELL EGGS

Application

The U.S. standards for quality of individual shell eggs are applicable only to eggs that are the product of the domesticated chicken hen and are in the shell.

Interior egg quality specifications for these standards are based on the apparent condition of the interior contents of the egg as it is twirled before the candling light. Any type or make of candling light may be used that will enable the particular grader to make consistently accurate determination of the interior quality of shell eggs. It is desirable to break out an occasional egg and by determining the Haugh unit value

FIG. 3.3.   Portable candling device made from a flash-light with a dark sleeve.
Courtesy of A. R. Rhorer, Indiana State Egg Board.

FIG. 3.4.   Eggs passing over flash-candling area where they are mechanically rotated, inspected, and removed, if necessary.

(defined later in this chapter) of the broken-out egg, compare the broken-out and candled appearance, thereby aiding in correlating candled and broken-out appearance.

AA Quality

The shell must be clean, unbroken, and practically normal. The air cell must not exceed 1/8 in. in depth, may show unlimited movement, and may be free or bubbly. The white must be clear and firm so that the yolk is only slightly defined when the egg is twirled before the candling light. The yolk must be practically free from apparent defects.

A Quality

The shell must be clean, unbroken, and practically normal. The air cell must not exceed 3/16 in. in depth, may show unlimited movement, and may be free or bubbly. The white must be clear and at least reasonably firm so that the yolk outline is only fairly well defined when the egg is twirled before the candling light. The yolk must be practically free from apparent defects.

B Quality

The shell must be unbroken, may be abnormal, and may have slightly stained areas. Moderately stained areas are permitted if they do not cover more than 1/32 of the shell surface if localized, or 1/16 of the shell surface if scattered. Eggs having shells with prominent stains or adhering dirt are not permitted. The air cell may be over 3/16 in. in depth, may show unlimited movement, and may be free or bubbly. The white may be weak and watery, so that the yolk outline is plainly visible when the egg is twirled before the candling light. The yolk may appear dark, enlarged, and flattened, and may show clearly visible germ development but no blood due to such development. It may show other serious defects that do not render the egg inedible. Small blood spots or meat spots (aggregating not more than 1/8 in. in diameter) may be present.

Check

The egg that has a broken shell or crack in the shell but with its shell membranes intact and its contents do not leak. A "check" is considered to be lower in quality than a "dirty."

Terms Descriptive of the Shell

1. *Clean*—a shell that is free from foreign material and from stains or discolorations that are readily visible. An egg may be considered clean if it has only very small specks, stains, or cage marks, if such specks, stains, or cage marks are not of sufficient number or intensity to detract from the generally clean appearance of the egg. Eggs that show traces of processing oil on the shell are considered clean unless otherwise soiled.

2. *Dirty*—a shell that is unbroken and that has dirt or foreign material adhering to its surface, that has prominent stains, or moderate stains covering more than 1/32 of the shell surface if localized, or 1/16 of the shell surface if scattered.

3. *Practically Normal (AA or A Quality)*—a shell that approximates the usual shape and that is sound and free from thin spots. Ridges and rough areas that do not materially affect the shape and strength of the shell are permitted.

4. *Abnormal (B Quality)*—a shell that may be somewhat unusual or decidedly misshapen or faulty in soundness or strength or that may show pronounced ridges or thin spots.

Terms Descriptive of the Air Cell

1. *Depth of Air Cell*—the depth of the air cell is the distance from its top to its bottom when the egg is held air-cell upward.

2. *Free Air Cell*—an air cell that moves freely toward the uppermost point in the egg as the egg is rotated slowly.

3. *Bubbly Air Cell*—a ruptured air cell resulting in one or more small separate air bubbles usually floating beneath the main air cell.

Terms Descriptive of the White

1. *Clear*—a white that is free from discolorations or from any foreign bodies floating in it. (Prominent chalazas should not be confused with such foreign bodies as spots or blood clots.)

2. *Firm (AA Quality)*—a white that is sufficiently thick or viscous to prevent the yolk outline from being more than slightly defined or indistinctly indicated when the egg is twirled. With respect to a broken-out egg, a firm white has a Haugh unit value of 72 or higher when measured at a temperature between 45° and 60°F.

3. *Reasonably Firm (A Quality)*—a white that is somewhat less thick or viscous than a firm white. A reasonably firm white permits the yolk to approach the shell more closely which results in a fairly well-defined

yolk outline when the egg is twirled. With respect to a broken-out egg, a reasonably firm white has a Haugh unit value of 60 to 72 when measured at a temperature between 45° and 60°F.

4. *Weak and Watery (B Quality)*—a white that is weak, thin, and generally lacking in viscosity. A weak and watery white permits the yolk to approach the shell closely, thus causing the yolk outline to appear plainly visible and dark when the egg is twirled. With respect to a broken-out egg, a weak and watery white has a Haugh unit value lower than 60 when measured at a temperature between 45° and 60°F.

5. *Blood Spots or Meat Spots*—small blood spots or meat spots (aggregating not more than 1/8 in. in diameter) may be classified as B quality. If larger, or showing diffusion of blood into the white surrounding a blood spot, the egg shall be classified as "loss." Blood spots shall not be due to germ development. They may be on the yolk or in the white. Meat spots may be blood spots that have lost their characteristic red color or tissue from the reproductive organs.

6. *Bloody White*—an egg that has blood diffused through the white. Eggs with bloody whites are classed as "loss." Eggs with blood spots that show a slight diffusion into the white around the localized spot are not to be classed as bloody whites.

Terms Descriptive of the Yolk

1. *Outline Slightly Defined (AA Quality)*—a yolk outline that is indistinctly indicated and appears to blend into the surrounding white as the egg is twirled.

2. *Outline Fairly Well Defined (A Quality)*—a yolk outline that is discernible but not clearly outlined as the egg is twirled.

3. *Outline Plainly Visible (B Quality)*—a yolk outline that is clearly visible as a dark shadow when the egg is twirled.

4. *Enlarged and Flattened (B Quality)*—a yolk in which the yolk membranes and tissues have weakened and/or moisture has been absorbed from the white to such an extent that the yolk appears definitely enlarged and flat.

5. *Practically Free from Defects (AA or A Quality)*—a yolk that shows no germ development but may show other very slight defects on its surface.

6. *Serious Defects (B Quality)*—a yolk that shows well-developed spots or areas and other serious defects, such as olive yolks, that do not render the egg inedible.

7. *Clearly Visible Germ Development (B Quality)*—a development of the germ spot on the yolk of a fertile egg that has progressed to a point where it is plainly visible as a definite circular area or spot with no blood in evidence.

8. *Blood Due to Germ Development*—blood caused by development of the germ in a fertile egg to the point where it is visible as definite lines or as a blood ring. Such an egg is classified as inedible.

General Terms

1. *"Loss"*—an egg that is inedible, smashed, or broken so that its contents are leaking, cooked, frozen, contaminated, or containing bloody whites, large blood spots, large unsightly meat spots, or other foreign material.

2. *Inedible Eggs*—Eggs of the following descriptions are classed as inedible: black rots, yellow rots, white rots, mixed rots (addled eggs), sour eggs, eggs with green whites, eggs with stuck yolks, moldy eggs, musty eggs, eggs showing blood rings, eggs containing embryo chicks (at or beyond the blood-ring state), and any eggs that are adulterated as such term is defined by the Federal Food, Drug, and Cosmetic Act.

3. *Leaker*—an individual egg that has a crack or break in the shell and shell membranes to the extent that the egg contents are exuding or free to exude through the shell.

U.S. GRADES AND WEIGHT CLASSES FOR SHELL EGGS

General

1. These grades are applicable to edible shell eggs in "lot" quantities rather than on an "individual" egg basis. A lot may contain any quantity of two or more eggs. The term "case" in these standards means 30-dozen egg cases as used in commercial practices in the United States. The size of the sample used to determine grade shall be on the basis of the requirements listed as standards of quality in this chapter or as determined by the national Supervisor.

2. The terms used in this part that are defined in the U.S. Standards for Quality of Individual Shell Eggs have the same meaning in this part as in those standards.

3. Aggregate tolerances are permitted within each grade only as an allowance for variable efficiency and interpretation of graders, normal changes under favorable conditions during reasonable periods between grading, and reasonable variation of graders' interpretation.

4. Substitution of higher qualities for the lower qualities specified is permitted.

5. "No grade" means eggs of possible edible quality that fail to meet the requirements of an official U.S. Grade or that have been contami-

nated by smoke, chemicals, or other foreign materials, which have seriously affected the character, appearance, or flavor of the eggs.

U.S. CONSUMER GRADES AND WEIGHT CLASSES FOR SHELL EGGS

Grades

1. *U.S. Grade AA.* (a) The *U.S. Consumer Grade AA (at origin)* shall consist of eggs that are at least 87% AA quality. The maximum tolerance of 13% that are permitted to be below AA quality may consist of A or B quality in any combination, except that within the tolerance for B quality not more than 1% may be B quality due to air cells over 3/8 in., blood spots (aggregating not more than 1/8 in. in diameter), or serious yolk defects. Not more than 5% (7% for Jumbo size) "checks" are permitted and not more than 0.50% "leakers," "dirties," or "loss" (due to meat or blood spots) in any combination, except that such "loss" may not exceed 0.30%. Other types of "loss" are not permitted. The *U.S. Consumer Grade AA (destination)* shall consist of eggs that are at least 72% AA quality. The remaining tolerance of 28% shall consist of at least 10% A quality and the remainder shall be B quality, except that within the tolerance for B quality not more than 1% may be B quality due to air cells over 3/8 in., blood spots (aggregating not more than 1/8 in. in diameter), or serious yolk defects. Not more than 7% (9% for Jumbo size) checks are permitted and not more than 1% leakers, dirties, or loss (due to meat or blood spots) in any combination, except that such loss may not exceed 0.30%. Other types of loss are not permitted.

2. *U.S. Grade A.* (a) The *U.S. Consumer Grade A (at origin)* shall consist of eggs that are at least 87% A quality or better. Within the maximum tolerance of 13% that may be below A quality, not more than 1% may be B quality due to air cells over 3/8 in., blood spots (aggregating not more than 1/8 in. in diameter), or serious yolk defects. Not more than 5% (7% for Jumbo size) "checks" are permitted and not more than 0.50% "leakers," "dirties," or "loss" (due to meat or blood spots) in any combination, except that such "loss" may not exceed 0.30%. Other types of "loss" are not permitted. (b) The *U.S. Consumer Grade A (at destination)* shall consist of eggs that are at least 82% A quality or better. Within the maximum tolerance of 18% permitted to be below A quality, not more than 1% may be B quality due to air cells over 3/8 in., blood spots (aggregating not more than 1/8 in. in diameter), or serious yolk defects. Not more than 7% (9% for Jumbo size) "checks" are permitted and not more than 1% "leakers," "dirties," or "loss" (due to meat or blood spots)

in any combination, except that such Loss may not exceed 0.30%. Other types of "loss" are not permitted.

3. *U.S. Grade B.* (a) The *U.S. Consumer Grade B (at origin)* shall consist of eggs that are at least 90% B quality or better, not more than 10% may be "checks" and not more than 0.50% "leakers." "Dirties," or "loss" (due to meat or blood spots) in any combination, except that such Loss may not exceed 0.30%. Other types of "loss" are not permitted. (b) The *U.S. Consumer Grade B (at destination)* shall consist of eggs that are at least 90% B quality or better; not more than 10% may be "checks" and not more than 1% "leakers," "dirties," or "loss" (due to meat or blood spots) in any combination, except that such Loss may not exceed 0.30%. Other types of "loss" are not permitted.

4. *Additional Tolerances.* (a) *In lots of two or more cases.* (1) *For Grade AA*—No individual case may exceed 10% less AA quality eggs than the minimum permitted for the lot average. (2) *For Grade A*—No individual case may exceed 10% less A quality eggs than the minimum permitted for the lot average. (3) *For Grade B*—No individual case may exceed 10% less B quality eggs than the minimum permitted for the lot average. (b) *For Grades AA, A and B,* no lot shall be rejected or downgraded due to the quality of a single egg, except for "loss" other than blood or meat spots.

Summary of Grades

A summary of U.S. Consumer Grades for Shell Eggs is given in Tables 3.1 and 3.2 of this section.

Weight Classes

1. The weight classes for U.S. Consumer Grades for Shell Eggs shall be as indicated in Table 3.3 of this section and shall apply to all consumer grades.

2. A lot average tolerance of 3.3% for individual eggs in the next lower weight class is permitted as long as no individual case within the lot exceeds 5%.

U.S. WHOLESALE GRADES AND WEIGHT CLASSES FOR SHELL EGGS

Grades

1. *U.S. Specials—% AA Quality* shall consist of eggs of which at least 20% are AA quality, and the actual percentage of AA quality eggs shall

TABLE 3.1. Summary of U.S. Consumer Grades for Shell Eggs

U.S. consumer grade	Quality required[a]	Tolerance permitted	
		%	Quality
At origin[b]			
AA	87% AA	Up to 13	A or B[f]
A		Not over 5	Checks[f]
B	87% A or better	Up to 13	B[f]
		Not over 5	Checks[f]
	90% B or better	Not over 10	Checks
At destination[c]			
AA	72% AA	Up to 28[d]	A or B[e]
		Not over 7	Checks[f]
A	82% A or better	Up to 18	B[f]
		Not over 7	Checks[f]
B	90% B or better	Not over 10	Checks

[a] In lots of two or more cases, see Table 3.2 of this section for tolerances for an individual case within a lot.

[b] For the U.S. Consumer grades at origin, a tolerance of 0.50% Leakers, Dirties, or Loss (due to meat or blood spots) in any combination is permitted, except that such Loss may not exceed 0.30%. Other types of Loss are not permitted.

[c] For the U.S. Consumer grades at destination, a tolerance of 1% Leakers, Dirties, or Loss (due to meat or blood spots) in any combination is permitted, except that such Loss may not exceed 0.30%. Other types of Loss are not permitted.

[d] For U.S. Grade AA at destination, at least 10% must be A quality or better.

[e] For U.S. Grade AA and A at origin and destination within the tolerances permitted for B quality, not more than 1% may be B quality due to air cells over 3/8 in., blood spots (aggregating not more than 1/8 in. in diameter), or serious yolk defects.

[f] For U.S. Grades AA and A Jumbo size eggs, the tolerance for Checks at origin and destination is 7% and 9%, respectively.

be stated in the grade name. Within the maximum of 80% that may be below AA quality, not more than 7.5% may be B quality, "dirties," or "checks" in any combination and not more than 2.0% may be Loss.

2. *U.S. Extras—% A Quality* shall consist of eggs of which at least 20% are A quality, and the actual total percentage of A quality and better shall be stated in the grade name. Within the maximum of 80% that may be below A quality, not more than 11.7% may be dirties or checks in any combination and not more than 3.0% may be Loss.

TABLE 3.2. Tolerance for Individual Case Within a Lot

U.S. consumer grade	Case quality	Origin (%)	Destination (%)
AA	AA (min.)	77	62
	A or B	13	28
	Check (max.)	10	10
A	A (min.)	77	72
	B	13	18
	Check (max.)	10	10
B	B (min.)	80	80
	Check (max.)	20	20

3. *U.S. Standards—% B Quality* shall consist of eggs of which at least 84.3% are B quality; and the actual total percentage of B quality and better shall be stated in the grade name. Within the maximum of 15.7% that may be below B quality, not more than 11.7% may be "dirties" or "checks" in any combination and not more than 4.0% may be Loss.

Weight Classes

The weight classes for U.S. wholesale grades of shell eggs are given in Table 3.4.

U.S. NEST-RUN GRADE AND WEIGHT CLASSES FOR SHELL EGGS

Grade

U.S. Nest-Run—% AA Quality shall consist of eggs of current production of which at least 20% are AA quality, and the actual percentage of AA quality eggs shall be stated in the grade name. Within the maximum of 15% which may be below A quality, not more than 10% may be B quality for shell shape, interior quality (including meat or blood spots), or due to rusty or black-appearing cage marks or blood stains, not more than 5% may have adhering dirt or foreign material on the shell 0.5 in. or larger in diameter, not more than 6% may be "checks," and not more than 3% may be "Loss." Marks that are slightly gray in appearance and adhering dirt or foreign material on the shell less than 0.5 in. in diameter are not considered quality factors. The eggs shall be officially graded for all other quality factors. No case may contain less than 75% A quality and AA quality eggs in any combination.

TABLE 3.3. U.S. Weight Classes for Consumer Grades for Shell Eggs

Size or weight class	Minimum net weight per dozen (oz)	Minimum net weight 30 per dozen (lb)	Minimum net weight for individual eggs at rate per dozen (oz)
Jumbo	30	56	29
Extra large	27	50½	26
Large	24	45	23
Medium	21	39½	20
Small	18	34	17
Peewee	15	28	

TABLE 3.4. Weight Classes for U.S. Wholesale Grades for Shell Eggs

Weight classes	Per 30 dozen eggs		Weights for individual eggs at rate per dozen	
	Average net weight on a lot[a] basis (lb)	Minimum net weight individual case basis[b] (lb)	Minimum weight (oz)	Weight variation tolerance for not more than 10%, by count, of individual eggs
	At least —			
Extra large	50.5	50	26	Under 26, but not under 24 oz
Large	45	44	23	Under 23, but not under 21 oz
Medium	39.5	39	20	Under 20, but not under 18 oz
Small	34	None	None	None

[a] Lot means any quantity of 30 dozen or more eggs.
[b] Case means standard 30-dozen egg case as used in U.S. commercial practice.

Weight Classes

The weight classes for the U.S. Nest-Run Grade for Shell Eggs shall be as indicated in Table 3.5 of this section and shall apply to Nest-Run Grade.

Egg grading is a subjective classification of the eggs into categories (Table 3.6). The actual grade is determined by that factor rating the egg in the lowest grade. The quality factors relating to albumen and yolk condition (internal quality) do not break into sharp divisions but rather follow a linear or curvilinear change. The problem is depicted in Fig. 3.5 relating albumen height to candled grade. From values given an A grade egg with an albumen height of 45 or 30 mm would be equal in grade but quite different in albumen height.

MEASUREMENTS OF SHELL QUALITY

A number of tests may be used to determine shell quality. Commercially, the evaluation is a judgment consideration based on observations relative to cleanliness and freedom from cracks or imperfections that might weaken the shell. The *USDA Egg Grading Manual* (USDA 1978) provides for three classifications of eggshells based on shape and texture, as follows:

Practically Normal—a shell that approximates the usual shape and that is of good, even texture and strength and free from rough areas that do not materially affect its shape, texture and strength . . . of the shell are permitted.

Slightly Abnormal—a shell that may be somewhat unusual in shape or that may be slightly faulty in texture or strength. It may show definite ridges but no pronounced thin spots or rough areas.

Abnormal—a shell that may be decidedly misshapen or faulty in

TABLE 3.5. Weight Classes for U.S. Nest-Run Grade for Shell Eggs

Weight class	Minimum average net weight on lot basis 30-dozen cases[a] (lb)
XI	51
1	48
2	45
3	42
4	39

[a] No individual sample case may vary more than 2 lb (plus or minus) from the lot average.

TABLE 3.6. Summary of U.S. Standards for Quality of Individual Shell Eggs

Quality factor	Specifications for Each Quality Factor		
	AA quality	A quality	B quality
Shell	Clean Unbroken Practically normal	Clean Unbroken Practically normal	Clean to slightly stained.[a] Unbroken Abnormal
Air cell	1/8 in. or less in depth Unlimited movement and free or bubbly	3/16 in. or less in depth Unlimited movement and free or bubbly	Over 3/16 in. in depth. Unlimited movement and free or bubbly
White	Clear Firm	Clear Reasonably firm	Weak and watery Small blood and meat spots present[b]
Yolk	Outline slightly defined Practically free from defects	Outline fairly well defined Practically free from defects	Outline plainly visible Enlarged and flattened Clearly visible germ development but no blood Other serious defects

[a] Moderately stained areas permitted (1/32 of surface, if localized, or 1/16, if scattered).
[b] If they are small (aggregating not more than 1/8 inch in diameter).

Note: For eggs with dirty or broken shells, the standards of quality provide two additional qualities. These are:

Dirty	Check
Unbroken. Adhering dirt or foreign material, prominent strains, moderate stained areas in excess of B quality	Broken or cracked shell, but membranes intact, not leaking[c]

[c] Leaker has broken or cracked shell and membranes, and contents leaking or free to leak.

texture or strength or that may show pronounced ridges, thin spots, or rough areas.

A shell may also be classified as sound, checked or cracked, leaking, or smashed. A checked or cracked egg is one in which the shell has been broken but the shell membrane remains intact. Leakers are eggs with both the shell and shell membranes broken so that egg contents may be lost.

In the study of eggshell quality, research methods have been developed to measure the quality. Most of the measurements are based on shell strength, as cracked and leaking eggshells are a major economic loss to the egg industry.

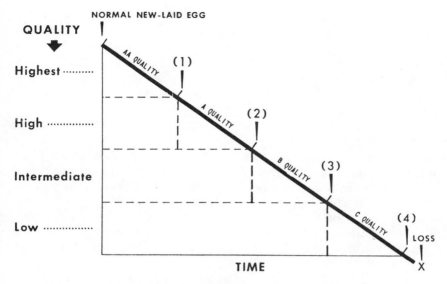

FIG. 3.5.   Egg quality standards and albumen quality are affected by time.
Courtesy of U.S. Department of Agriculture.

One of the most direct measurements is that of shell thickness. It has been estimated that a shell thickness of at least 0.33 mm is needed if the egg is to have a better than 50% chance of moving through normal market handling without breaking. A paper-thickness gauge is most often used to measure shell thickness, but care should be taken to order a convex anvil on one leg of the device, so that errors resulting from shell curvature can be eliminated.

A beta-ray back-scattering device has been developed to estimate shell thickness, but the instrument is not yet commercially available. This procedure has the advantage of being a nondestructive measurement (James and Retzer 1967).

Since the specific gravity of a freshly laid egg is closely correlated with shell thickness, specific gravity measurements have been used to determine shell quality. Its principal application has been in the selection of hatching eggs from breeding flocks. The procedure requires a series of sodium chloride solutions of varying specific gravity and a good hydrometer that can detect differences between adjacent pairs of solutions of not more than 0.005 g per milliliter. After some preliminary testing to establish ranges of values for the flock being studied, a final classification of "keep" or "discard" can be made with a single solution. Eggs are immersed in the test solution—those that float are marketed as table eggs and those that sink are used for reproduction. The range of specific gravities found in most flocks is from 1.065 to 1.100. Care must be taken to make specific gravity measurements on freshly laid eggs only. Holding eggs for a day or two prior to measurement allows the development of an air cell sufficiently large to invalidate results. The specific gravity method for shell thickness is the only objective measurement commercially used for screening flocks used for hatching-egg production (Table 3.7).

A number of methods exist for measuring eggshell strength, all of which correlate well with specific gravity and shell-thickness measurements. Attempts have been made to relate various measurements to the actual observed cracking of eggshells while in market channels, but to date no strength measurement has been found that is more precise than the method for shell thickness. Special equipment has been developed for measuring eggshell strength. One device for example, determines shell strength by crushing, piercing or snapping (Tyler and Coundon 1965); other devices described in the journal literature make these measurements individually.

Crushing strength is determined by using a flat metal plate, a square-ended rod or a round-ended rod rigged so that an increasing load may be applied to the time of shell fracture. Piercing strength is determined by using a steel needle in place of the rod or plate mentioned above. Another method employs a falling steel ball to determine breaking strength. The force of impact can be calculated by knowing the distance of free fall and the weight of the ball. A simple device for such measurements is a calibrated vertical glass column and a portable electromagnet used for holding and releasing the steel ball in the column from a predetermined height above the eggshell.

Some reported correlations among shell measurements are listed in Table 3.7. The high degree of relationship suggests that for any particular shell-quality study any one of the measurements would probably be suitable. Selection of method could be based on available equipment and

TABLE 3.7. Simple Correlation Coefficients among Several Measures of Shell Quality

	B	C	D	E	F	G
Shell thickness (A)	0.78	0.80	0.78	0.26	0.73	0.54
Specific gravity (B)		0.81	0.69	0.14	0.70	0.61
Percentage egg as shell (C)			0.76	0.08	0.78	0.37
Shell weight (D)				0.67	0.62	0.55
Egg weight (E)					0.10	0.45
Force to crush shell (F)[a]						
Impact device force (G)[a]						

SOURCE: *Frank et al. (1964).*
[a] Relationship between F and G not calculated, as both measures are destructive and cannot be made on same egg.

the worker's preference. No attempt has been made here to discuss factors affecting shell-quality, since these were presented by Petersen (1965) in an excellent review.

A shell-quality factor that varies widely in importance among markets is shell color, which is derived from an outer pigmented layer on the shell of colored eggs. No relationship has been found between shell color and interior quality characteristics or other shell-quality factors. An automated machine for sorting eggs by color has been developed (Brant *et al.* 1953). The final shell quality characteristic is cleanliness, which will be discussed in some detail in Chapter 4. The production of clean, shelled eggs was discussed in Chapter 2.

MEASUREMENTS OF ALBUMEN QUALITY

The most widely used measurement of albumen quality is the Haugh unit, proposed by Raymond Haugh in 1937. This method consists of measuring height of the thick albumen as shown in Fig. 3.6. Care must be taken to get a reading with the contact arm not touching the chalaza; otherwise the reading will be too high. The Haugh unit is an expression relating egg weight and height of thick albumen. "The higher the Haugh value, the better the albumen quality of the egg," is a generally accepted fact. Originally, the determination of Haugh units (H.U.) was a time-consuming operation involving collecting measurement data and application of the formula:

$$\text{Haugh units} = 100 \log \left[H - \frac{\sqrt{G}(30\,W^{0.37} - 100)}{100} + 1.9 \right]$$

where H is the albumen height (millimeters), G is 32.2, and W the weight of egg (grams).
The expression

$$\frac{\sqrt{G}(30\,W^{0.37} - 100)}{100} + 1.9$$

equals zero when egg weight is 56.7 g, or 2 oz.

The calculation of Haugh units was speeded up by the development by Brant *et al.* (1951) of an interior egg-quality calculator for the rapid conversion of egg weight and albumen height data to Haugh units.

Other measurements of albumen quality were based on pictures of newly laid eggs with varying heights of thick albumen or of eggs held for varying periods of time to allow for thinning of the white (Van Wagenen and Wilgus 1935). The Van Wagenen chart has nine pictures

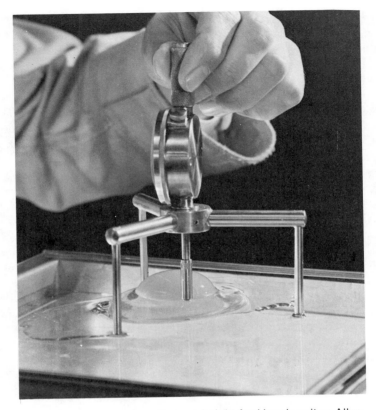

FIG. 3.6.   In measuring albumen height for Haugh unit or Albumen Index calculation, care must be taken to measure at a typical height.

with values in half-units ranging from 1.0 for high-quality to 5.0 for low-quality albumen. The Brant chart, which became known as the USDA Score Chart, consists of 12 pictures with a high, average and low quality of each grade from AA to C being shown. In addition, the Brant chart went further and related the Haugh units with an eight-unit difference between any adjacent pair of pictures (Brant *et al.* 1951).

In 1936 Heiman and Carver proposed an Albumen-Index measurement using measurements similar to those made by Haugh (1937). Since the resulting calculations do not take into account the logarithmic nature of albumen thinning, they yield a curvilinear plot of quality loss on eggs during storage. The formula for calculating albumen index is:

$$\text{Albumen Index} = \frac{H}{\sqrt{G}} - (30W^{0.37} - 100)$$

where H is the height (millimeters), G is 32.2, and W the weight (grams). Wilhelm and Heiman (1936) developed a simple procedure for determining Albumen Index values.

Although other measurements of albumen quality have been developed, they have not been widely used in research. Some of these compared Haugh units with the following parameters: diameter of thick and thin albumen; chalaza size; shape of thick and thin albumen; percentage of outer thin, inner thin, and thick albumen; and total percentage of thin albumen. Generally, Haugh units were more closely correlated to any one of the measurements than they are to each other, as is shown in Table 3.8.

The functional properties of egg albumen can also be used as indications of egg quality, as discussed in Chapter 15. Variations in the chemical constituents of egg white have been studied extensively and it has been observed that the basic composition is not greatly influenced by dietary changes, except for vitamin intake. Thus, the albumen content of riboflavin, in particular, is markedly influenced by the hen's dietary intake of this water-soluble vitamin.

A number of studies have related the interior quality measurements of albumen to candled quality (Parsons and Mink 1937; Baker and Forsythe 1951; Sauter et al. 1953). When a wide range in quality is present, high correlations are found among candled grade and the several albumen quality measurements. Within narrow ranges, the relationships are much less accurate as quality predictors.

TABLE 3.8. Correlation Coefficients Indicating Relationships among Several Measures of Interior Quality of Eggs[a]

	% Thick albumen	Thin albumen diameter	Thick albumen diameter	Yolk index	Haugh unit values
Haugh unit values	—	—	—	—	—
Yolk index	—	—	—	—	—
Thick albumen diameter	—	—	—	0.17	−0.90
Thin albumen diameter	—	—	0.68	0.62	0.82
Chalazae size	−0.10	0.36	0.03	0.09	0.16
Yolk mottling	0.13	0.25	0.12	0.06	0.18
Yolk centering	−0.09	0.25	−0.28	0.53	0.71
Shape of thin albumen	−0.36	0.71	0.59	−0.26	−0.70
Shape of thick albumen	0.71	0.65	0.74	−0.30	−0.61
% inner thin albumen	−0.18	0.23	0.37	0.02	−0.15
% outer thin albumen	−0.31	0.78	0.68	−0.61	−0.91
% thick albumen	—	−0.64	0.93	−0.72	−0.86
Total % thin albumen	−0.71	0.89	−0.84	−0.64	−0.87

SOURCE: Wesley and Stadelman (1959).
[a] One hundred ninety-seven eggs were used for these calculations.

MEASUREMENT OF YOLK QUALITY

Several characteristics of the yolk affect its quality—namely, color, spherical condition, and strength of the membrane. In shell eggs, yolk color is generally determined by using color comparators and numerical values. One of the early developments was the color rotor (Heiman and Carver 1935) which consisted of 24 curved glass disks with the concave surface painted with mixtures varying in color from light yellow to deep orange-red. The disks were mounted on a wheel with the convex side up, so as to give a curved surface of varying color to be compared with the egg yolks of broken-out eggs. Since the appearance of the color rotor, a number of other color-comparison procedures have been suggested. Ashton and Fletcher (1962) prepared a series of 15 metal rings, with openings about the size of an egg yolk, painted to give a range of colors similar to egg-yolk colors. At the present time the Roche Color Fan is the most commonly used comparator. This consists of a series of 15 colored plastic tabs arranged as a fan corresponding to the range of yolk colors found in eggs. Vuillemier (1969) gave a detailed color description of the Roche Color Fan and comparative data with some of the earlier systems.

There have also been a number of chemical methods used for determining yolk color, one of which involves the objective comparison of yolk pigments (extracted with a dichromate solution) against a standard (Bornstein and Bartov 1966).

The spherical nature of egg yolk can be expressed as a Yolk Index, originated by Sharp and Powell in 1930 and based on the separation of yolk and white, keeping the yolk intact. To save time and simplify the calculation, the yolk height and width can be measured without separating it from the albumen (Funk 1948; Sauter *et al.* 1951).

The ability of the yolk to withstand rupture during the egg-breaking operation is a function of vitelline membrane strength. Fromm and Matrone (1962) developed a technique for measuring vitelline membrane strength that is relatively rapid and that correlates well with the pressure required to rupture the yolk, as well as the Yolk Index. The technique involves the placement of a 2-mm capillary tube against the vitelline membrane, the creation of a given vacuum in the tube, and the measurement of the time required to rupture the membrane. For comparative use, the value is then calculated as millimeters of mercury of vacuum to rupture the membrane in 1 sec. This method was later modified, as one of the major variables in vitelline membrane strength is selection of the area in relation to location of the blastodisc—the farthest points from the blastodisc have the greatest membrane strength (Holder *et al.* 1968).

The functional properties of egg yolk are other important quality characteristics. Methods used for their evaluation are given in Chapter

15. The nutritive value of the egg yolk is also of great importance. For example, fatty acid content of the egg yolk can be modified appreciably by varying the diet of the hen. The nutritional value of eggs is discussed in Chapter 7.

DETERMINATION OF INTERIOR QUALITY DEFECTS

Blood spots in eggs are one of the most commonly occurring defects. Nalbandov and Card (1944) and Helbacka and Swanson (1958) reported on the nature of blood and meat spots and the factors affecting their occurrence. Jensen *et al.* (1952) discussed the problems of blood-spot detection by candling. In fresh eggs, the albumen is frequently cloudy, so that it is difficult to detect the smaller blood spots by candling. This is a particularly troublesome problem in the marketing of very fresh eggs. The electronic blood-spot detection equipment, developed by Brant *et al.* in 1953, largely solved this problem, and refinements of Brant's original design have been made, so that the newer equipment will detect blood in eggs of any age and of any shell color.

Meat spots are a second common internal defect. Research has suggested that blood spots that have degenerated in the oviduct appear as meat spots in eggs. Extensive histological and chemical studies of meat spots, on the other hand, have determined that most of them were from sources other than blood spots. The color of meat spots was found to vary in direct relation with eggshell color.

Yolk mottling is a less frequent internal defect. Several scoring systems for expressing the degree of yolk mottling have been proposed. The score developed by Baker *et al.* in 1957, ranging from 1 for no evidence of mottling to 10 for severely mottled and misshapen, is in the widest usage. The severity of mottling increases during the first 3 weeks of egg storage. This characteristic of the blemish makes accurate detection of even slight mottling in the yolks of fresh eggs a necessity, as the normal time in market channels allows a slight blemish to become a reason for consumer complaints.

Mottling in egg yolks could result from a nonuniform distribution of water in the mottled yolk or from a separation of the vitelline membrane and the chalaziferous layer of the albumen. Both conditions are likely to be present at some stage of mottling of egg yolks (Polin 1957; Doran and Muellar 1961).

SELECTED REFERENCES

Ashton, H. E. and Fletcher, D. A. 1962. Development and use of color standards for egg yolks. Poult. Sci. *41*, 1903–1909.

Baker, R. C., Hill, F. W., Van Tienhoven, A., and Bruckner, J. H. 1957. The effect of nicarbazin on egg production and egg quality. Poult. Sci. *36*, 718–726.

Baker, R. L., and Forsythe, R. H. 1951. U.S. standards for quality of individual shell eggs and the relationships between candled appearance and more objective quality measures. Poult. Sci. *30*, 269–279.

Bornstein, A., and Bartov, I. 1966. Studies on egg yolk pigmentation. I. A comparison between visual scoring of yolk color and colorimetric assay of yolk carotenoids. Poult. Sci. *45*, 287–296.

Brant, A. 1957. Machine sorts eggs for shell color. Poult. Process. Market. *59*(1), 12–13, 25.

Brant, A. W., Otte, A. W., and Norris, K. H. 1951. Recommended standards for scoring and measuring opened egg quality. Food Technol. *5*, 356–361.

Brant, A. W., Norris, K. H., and Chin, G. 1953. A spectrophotometric method for detecting blood in white-shell eggs. Poult. Sci. *32*, 357–363.

Doran, B. M., and Muellar, W. J. 1961. The development and structure of the vitelline membrane and their relationship to yolk mottling. Poult. Sci. *40*, 474–478.

Frank, F. R., Swanson, M. H., and Burger, R. E. 1964. The relationships between selected physical characteristics and the resistance to shell failure of *Gallus domesticus* eggs. Poult. Sci. *43*, 1228–1235.

Fromm, D., and Matrone, G. 1962. A rapid method for evaluating the strength of the vitelline membrane of the hen's egg yolk. Poult. Sci. *41*, 1516–1521.

Funk, E. M. 1948. The relation of yolk index determined in natural position to the yolk index as determined after separating the yolk from the albumen. Poult. Sci. *27*, 367.

Haugh, R. R. 1937. The Haugh unit for measuring egg quality. U.S. Egg Poult. Mag. *43*, 552–55, 572–573.

Heiman, V., and Carver, J. S. 1935. The yolk color index. U.S. Egg Poult. Mag. *41* (8), 40–41.

Heiman, V., and Carver, J. S. 1936. The albumen index as a physical measurement of observed egg quality. Poult. Sci. *15*, 141–148.

Helbacka, N. V. L., and Swanson, M. H. 1958. Studies on blood and meat spots in the hen's egg. 2. Some chemical and histological characteristics of blood and meat spots. Poult. Sci. *37*, 877–885.

Holder, D. P., Newell, G. W., Berry, J. G., and Morrison, R. D. 1968. The effect of rations on vitelline membrane strength. Poult. Sci. *47*, 326–329.

James, P. E., and Retzer, H. J. 1967. Measuring egg shell strength by beta backscatter technique. Poult. Sci. *46*, 1200–1203.

Jensen, L. S., Sauter, E. A., and Stadelman, W. J. 1952. The detection and disintegration of blood spots as related to age of eggs. Poult. Sci. *31*, 381–387.

Karunajeewa, H., Hughes, R. J., McDonald, M. W., and Shenstone, F. S. 1984. A review of factors influencing pigmentation of egg yolks. World's Poult. Sci. J. *40*, 52–65.

Nalbandov, A. V., and Card, L. E. 1944. The problem of blood clots and meat spots in chicken eggs. Poult. Sci. *23*, 170–180.

Parsons, C. H., and Mink, I. D. 1937. Correlation of methods for measuring the interior quality of eggs. U.S. Egg Poult. Mag. *43*, 484–491, 509–512.

Petersen, C. F. 1965. Factors influencing egg shell quality—a review. World's Poult. Sci. J. *21*, 110–138.

Polin, D. 1957. Biochemical and weight changes of mottled yolks in eggs from hens fed Nicarbazin. Poult. Sci. *36*, 831–835.

Romanoff, A. I., and Romanoff, A. J. 1949. The Avian Egg. John Wiley & Sons Co., New York.

Sauter, E. A., Harns, J. V., Stadelman, W. J., and McLaren, B. A. 1953. Relationship of candled quality of eggs to other quality measurements. Poult. Sci. *32*, 850–854.

Sauter, E. A., Stadelman, W. J., Harns, V., and McLaren, B. A. 1951. Methods for measuring yolk index. Poult. Sci. *30*, 629–630.

Sharp, P. F., and Powell, C. K. 1930. Decrease in internal quality of hen's eggs during storage as indicated by the yolk. Ind. Eng. Chem. *22*, 909–910.

Tyler, C., and Coundon, J. R. 1965. Apparatus for measuring shell strength by crushing, piercing or snapping. Br. Poult. Sci. *6*, 327–330.

U.S. Department of Agriculture (USDA) 1978. Egg Grading Manual, Agric. Handb. No. 75. U.S. Government Printing Office, Washington, DC.

U.S. Department of Agriculture (USDA) 1981. Revision of shell egg standards and grades. Fed. Regis. *46*(149), 39566–39573.

Van Wagenen, A., and Wilgus, H. W. 1935. The determination and importance of the condition of the firm albumen in studies of egg white quality. J. Agric. Res. *51*, 1129–1137.

Vuillemier, J. P. 1969. The Roche Yolk Colour Fan—an instrument for measuring yolk colour. Poult. Sci. *48*, 767–779.

Wesley, R. L., and Stadelman, W. J. 1959. Measurements of interior egg quality. Poult. Sci. *38*, 474–481.

Wilhelm, L. A., and Heiman, V. 1936. Albumen index determination by nomogram. U.S. Egg Poult. Mag. *42*, 426–429.

The Preservation of Quality in Shell Eggs

W. J. Stadelman

In considering the preservation of egg quality, it is assumed that a chicken egg is at its highest quality at the time of laying. Anything done to the egg later results in some quality loss. The definition of quality given in Chapter 3 must be remembered. Egg quality consists of those characteristics of an egg that affect its acceptability to the consumer. As far as the consumer is concerned, the eggshell is the first point for evaluation. Of course, since the packaging of any product is a major consideration, this receives special attention in Chapter 8.

SHELL QUALITY

Shell quality characteristics that must be considered are cleanliness, soundness, smoothness, and shape. It is essential that eggs reach the consumer with a sound, clean shell if satisfaction is to be achieved. As indicated in Chapter 2, it is not possible to produce entirely clean eggs, so methods for cleaning must be employed.

The two most desirable shell qualities—cleanliness and soundness—are largely controlled by production and egg-handling practices. In marketing, eggs with weak and unsound shells, odd-shaped eggs, and rough-shelled eggs should be removed from eggs for retail trade. All stained or dirty shells should be cleaned, either by dry abrasives or by washing. Dry cleaning of egg shells was the recommended practice before the development of sanitizing compounds capable of maintaining washing water and equipment in a condition

EGG SCIENCE & TECHNOLOGY,
3rd Edition

so that wet cleaning methods could be used. The simplest dry cleaners were hand abrasive blocks to scratch the dirt or stain from the shell surface. Abrasives were put on mechanically rotated wheels so that eggs could be rapidly cleaned by holding individual eggs against the rotating abrasive surface. Although mechanical equipment for dry cleaning of shell eggs was developed, it has been made obsolete by wet-cleaning procedures.

From the viewpoint of appearance of eggs, washing is the most effective and simplest method for removing dirt and stains from the shell surface. With washing, however, there are possibilities of bacterial penetration of the shells, with resultant rotten eggs. A series of reports (Alford et al. 1950; Gillespie and Scott, 1950; Gillespie et al. 1950A,B,C; Salton et al. 1951) thoroughly explore the problems of washing eggs. A report by Kahlenberg et al. (1952) concluded that washed dirty eggs should be marketed promptly, since storage of such eggs leads to economic loss. Based on such data as well as the extensive review by Romanoff and Romanoff (1949), it was decided that improved egg cleaning equipment was needed.

Egg wash-water temperature needs to be at least 11°C higher than egg temperature, and the concentration of bacterial spoilage organisms in the wash water is critical (Brant et al. 1966). The temperature differential between the eggs and wash water is also critical, as a difference of more than 10°C results in an increased number of thermal cracks due to expansion of the contents of the egg (Wesley and Beane 1967). In older eggs with enlarged air cells, the thermal checking is not a serious problem. Also, a water temperature of at least 35°C is needed for adequate cleaning. By not recycling the water and by giving careful attention to the temperature difference between eggs and wash water, eggs can be washed with a minimum of hazard to quality (Brant et al. 1966).

In 1966, Brant and coworkers designed and built an experimental egg-washing machine incorporating a number of design features not found on existing commercial washers at that time. These were: egg wetting before washing; fogger-type nozzles; low water volume; high-pressure cutting sprays; rotating brushes with the drive shaft parallel to the conveyor spools so that the bristles are perpendicular to the direction of egg travel; abrasive-impregnated brushes; an egg-spinning device to increase contact between the egg and the brush; and a separate conveyor in the drying unit. Other features were shaped brushes, a detergent metering device and nonrecirculated wash water. In the experimental unit, a savings in egg breakage of over 90¢ per 30-dozen was achieved on field trials as compared with commercial washers in use. They found that eggs with sound shells could be cleaned effectively with minimal spoilage hazard and shell damage without affecting internal egg quality (Walters *et al.* 1966).

Most eggs are now cleaned by washing with a detergent sanitizer solution in an egg washer at the egg-packing plant. The equipment used is similar to that shown in Fig. 4.1. By washing in the plant rather than at the farm, abuse of egg washing can be controlled. As previously mentioned, porous egg shells and the expansion and contraction of egg contents during washing can cause problems. Cold eggs subjected to hot water will expand too fast and crack the shell (expansion "checks"). On the other hand, the interior contents of eggs that have been washed in hot water and then rinsed by immersion in cool water contract and aspirate bacteria through the shell. The organisms multiply and the eggs spoil.

Modern egg washers are designed to minimize the above conditions by spraying the eggs with water rather than immersing them, using a sanitizer in the water along with a detergent for cleaning them, using rinse water warmer than the wash water, and finally drying the eggs

FIG. 4.1.   A modern on-line egg washer used to assure a high percentage of clean shells.
Courtesy of Seymour Foods, Topeka, Kansas.

with hot air. The scrubbing action of the washers is provided by rotating the eggs during washing and using pressure sprays and oscillating brushes.

The use of an adequate sanitizer in egg washing is an accepted practice. The most often used chemical is one of several chlorine compounds. Chlorine, when used at levels of over 50 ppm active chlorine on white-shelled eggs, can result in the development of a tinted egg shell. This is probably caused by a reaction between the chlorine and some of the amino acids of the cuticle on the egg shell. The problem is minimal when brushes are used in washing of eggs, as most of the cuticle is removed from the shell (Hamre and Stadelman 1964).

The other major shell quality problem is maintaining an unbroken shell. There is no sure way to achieve this goal, but attention to details to maintain gentle handling is essential. Care must be exercised not to fill gathering baskets too full, to use the right size of carton for packaging, and not to stack cases so high on pallets that the lower cases are crushed. A problem sometimes develops when high humidity in the egg-holding room results in a loss of strength of the cases, so that partial collapse results. Care must also be given to adjustment of egg-handling equipment so that breakage of eggs is minimized. The use of sponge-rubber backing on roll-ways is frequently of value.

Other shell-quality factors are coloring and mottling. Shell color is determined by the breed of hen, most chicken eggs being either white or brown. With brown eggs, there are many shades of color. For greatest consumer satisfaction it is best to package eggs into at least three color groupings—white, light brown, and dark brown. There is no relationship between shell mottling and interior egg quality. Likewise, other shell-quality measurements are not related to the mottled condition. The tendency to produce eggs with a particular degree of mottling appears to be an individual hen's personal characteristic.

INTERIOR EGG QUALITY

Various quality attributes of the albumen and yolk are lost as an egg ages. Factors associated with extent of quality loss are time, temperature, humidity, and handling. The rate of change in the albumen and yolk is a function of temperature and movement of carbon dioxide through the shell. The closer the temperature to the freezing point, the slower the rate of quality decline. Stadelman *et al.* (1954) compared eggs stored at 10° to 27°C (with a mean temperature of 16°C) to eggs stored at 8° to 13°C (with a mean temperature of 10°C). All eggs averaged 80 Haugh units at the start of the study. Sixteen days later the eggs stored at 16°C had an average Haugh unit value of 49, as compared to 69 for eggs stored at 10°C. The Haugh unit loss for the eggs stored at 16°C was about 2 units per day, as compared to 0.5 unit for the eggs stored at 10°C. A general recommendation is to handle shell eggs at less than 10°C at all times.

To reduce the rate of carbon dioxide and moisture loss various shell treatments have been used (Romanoff and Romanoff 1949). More recently spray oiling has come into common usage for egg shells. The oil used must be odorless, colorless, and free of fluorescent materials. Light mineral oils of food quality are normally used.

When treating eggshells with oil or other materials, it is essential that the treatment be applied to the eggshell within a few hours of production. The rate of carbon dioxide loss is very high during the first few hours, and weight loss during the first few days can be significant. The effectiveness of spray oiling in reducing weight loss is shown by the data in Table 4.1. The effectiveness of spray oiling 1, 6, 24, 48, 72, and 96 hours after lay was compared with no oiling with respect to albumen pH and Haugh units after 7, 14, and 21 days storage of the eggs at a temperature of about 10°C. In all instances, the eggs oiled within one hour of lay had higher Haugh units than eggs oiled at other times. This is good evidence that eggs should be oiled at the time of gathering to retain a maximum Haugh unit. Eggs oiled within 6 hours of lay did not retain as high an average Haugh unit as those oiled one hour following lay, but were superior to all the other oiling times (Goodwin et al. 1962).

Homler and Stadelman (1963) studied the effect of no oiling, oiling before washing, oiling after washing, and oiling both before and after washing on Haugh units and weight loss of eggs held for 3 weeks at 22°C, or 6 weeks at 13°C. Nonoiled eggs had lower average Haugh units

TABLE 4.1. Effect of Spray Oiling, Temperature and Humidity on Weight Loss in Shell Eggs[a]

	Held at 10°C, high humidity		Held at 24°C, low humidity	
Age of egg	Oiled (g)	Nonoiled (g)	Oiled (g)	Nonoiled (g)
2 hr	0.018	0.025	0.029	0.041
4 hr	0.032	0.048	0.060	0.085
6 hr	0.042	0.064	0.077	0.113
1 day	0.107	0.172	0.197	0.328
2 days	0.167	0.228	0.313	0.572
3 days	0.212	0.374	0.411	0.795
4 days	0.260	0.469	0.506	1.017
5 days	0.309	0.575	0.604	1.256

SOURCE: *Unpublished data by W.J. Stadelman.*
[a] Eggs oiled within 5 min after laying.

and a higher weight loss than any of the oiling treatments in both storage conditions. Oiling before compared favorably with oiling after washing with respect to Haugh units. On a percentage weight loss, oiling after washing was preferable. Oiling both before and after washing resulted in the highest Haugh units and lowest weight losses of any of the treatment groups.

An old method of preserving the quality of shell eggs especially for home use was storage in water glass or limewater. A 1917 USDA publication gave instructions for the use of sodium silicate or water glass (Anon. 1917). One quart (908 ml) of sodium silicate was mixed with 9 (8.1 liters) quarts of water in a 5-gal. container (approximately 21 liters). Fifteen dozen eggs were placed in the container with the solution covering all of the eggs. The container was covered to minimize evaporation and placed in a cool, dry place. Hall (1945) reported that temperature of storage of water-glass-preserved eggs markedly affected quality of eggs. When stored at 1° to 3°C, the water-glass-preserved eggs were comparable to fresh eggs in all attributes except odor and flavor. After 6-months storage at 13° to 15°C all eggs candled as grade B or better, with over half of the eggs being grade A or better. After storage for 6 months at room temperature, which varied from 5° to 35°C, all eggs were classified as inedible due to stuck yolks.

A quite different treatment for the preservation of shell eggs was reported by Funk in 1943. This involved immersion of the eggs in hot water or oil for sufficient time to kill the embryo in fertile eggs and to stabilize the thick albumen. The upper limit on the time and temperature for thermostabilizing eggs is the point at which coagulation of the albumen occurs. The times and temperatures suggested by Funk (1943) were immersion in oil at 60°C for 14 min or immersion in still water at 60°C for 10 min. Goresline et al. (1952) recommended a temperature of 56.7°C for 16 min for thermostabilization in oil. They emphasized that the temperature of the eggs just prior to thermostabilization influences results. For each 10°F difference they found a 1.13°F difference in internal temperature of the thermostabilized egg. A regional publication (Anon. 1955) compared oil dipping, thermostabilization, and a combination of the two with untreated control eggs held for 3 weeks at room temperature. In every instance, measurements of egg quality showed that the combined treatments yielded the highest quality eggs. Scott and Vickery (1954) confirmed the work of Funk (1943) and Goresline et al. (1952) in studies involving shipment of eggs by boat from Australia to England. Salton et al. (1951) reported success in reducing bacterial rotting of washed shell eggs by using thermostabilization.

Some of the problems of commercial application of thermostabilization were discussed by Barott and McNally (1943). They point out that the coagulation temperature of albumen varies with the pH of the egg white and that temperatures achieved in the eggs is influenced by egg size, starting temperature of the eggs, agitation of the heating medium, as well as just time and temperature.

A flash-heat treatment was introduced by Romanoff and Romanoff (1944) based on the immersion of eggs in boiling water for 5 sec. This resulted in eggs with superior storage characteristics at storage temperatures of either 5°C or 21°C. Feeney et al. (1954) reported that immersion for only 3 sec. in boiling water was effective in reducing bacterial spoilage in fresh eggs. The report by Romanoff and Romanoff (1944) was confirmed by Bartolome (1953) as part of a study on egg storage in tropical climates.

Changes in the poultry industry in the last 30 years have largely eliminated egg storage, and therefore the only currently used preservation treatment for shell eggs is spray oiling in combination with refrigeration. Swanson (1953) proposed the overwrapping of egg packages with a carbon dioxide-impermeable film and the incorporation of low levels of carbon dioxide into each package, but economics and the success of spray oiling have kept this method from general adoption. For maximum protection of interior quality Swanson et al. (1956) suggest treating eggs with carbon dioxide on the day laid followed by dip oiling. The carbon dioxide treatment consisted of a pressurized chamber filled with carbon dioxide gas in which the eggs were held. The carbon dioxide readily entered the eggs and was retained by the oiling treatment during storage for 12 days at 75° to 80°F or for 15 days at 60°F.

The humidity of the atmosphere surrounding the eggs can have a major effect on weight loss during storage. An indication of this is evident in the data of Table 4.1. Except for weight loss and the concurrent increase in air-cell size, humidity does not affect albumen or yolk quality. For holding eggs, a high relative humidity, 75 to 80%, is recommended. When eggs are shell-treated, humidity control to maintain small air cells is not as critical.

Handling is one of the most important factors affecting egg quality. The effects of rough handling are easily identified as shell breakage, but can also be associated with internal quality. Vibration of an egg will result in a thinning of the thick albumen, an increase in incidence of tremulous and free air cells, and yolks that are off-center when candled or out of the thick white when broken-out.

The possibility of using beta- or gamma-ray radiation to destroy microorganisms on or in eggs has been suggested. Shell eggs of known internal quality were irradiated with beta rays from a 3 MeV linear accelerator at levels ranging from 1,000 to 300,000 reps. Levels of 100,000 or 300,000 reps resulted in extreme thinning of the egg white with changes of over 50 Haugh units. In addition, the vitelline membrane on yolks of irradiated eggs was weakened, so that most yolks broke when the egg was opened on a flat glass tray. Damage by the relatively light dosages rules out the possible application of radiation sterilization in shell egg preservation (Parsons and Stadelman 1957).

Thus far, most of the discussion has centered on changes in the egg albumen. Egg-yolk quality, as determined by the Yolk Index of broken-out eggs or yolk centering and roundness by candling, changes at about the same rate as Haugh units or other measures of albumen condition. As the egg ages, water migrates from the albumen to the yolk, thus increasing the weight of the yolk. The rate of water transfer is a function of holding temperature, with faster migration occurring at higher temperatures. The migration of water to the yolk results in a stretching and weakening of the vitelline membrane.

A slow movement of lipid materials from the yolk to the albumen was reported by Pankey and Stadelman (1969). A change in percent of lipids in egg albumen from 0.0037 to 0.0058 during 16 weeks of storage at room temperature has been reported. These eggs were dipped in mineral oil before storage, so that moisture loss from the eggs was minimal. This movement of lipids to the albumen is of importance, as these materials in the albumen will modify the functional properties.

A final consideration in the preservation of egg quality is the mainte-nance of flavor. Eggs will pick up flavors from strong odors in the storage area, such as those from apples, oranges, decaying vegetable matter, oil, gasoline, or organic solvents. Eggs with such acquired off-odors become unusable, irrespective of other quality maintenance. In addition to acquiring foreign odors as flavor, eggs develop a stale or off-flavor of their own when held for several months. The intensity of stale flavor was greater in eggs that were oil-dipped prior to storage, as the oil seal prevented loss of the off-flavors which occurs naturally in nonoiled eggs (Harns et al. 1954; McLaren and Stadelman 1954). With spray-oiling the shell is not completely sealed as with dip-oiling, so the off-flavor does not become a problem any sooner than with nonshell-treated eggs.

Having outlined the conditions that affect egg quality and that nor-mally result in superior quality, how can this quality be most completely retained to time of consumption?

The most important factors in egg handling for quality preservation are listed below (adapted from Snyder 1961):

Frequent Gathering. Three to four gatherings per day are recom-mended, as this results in less breakage, fewer dirty eggs, and more rapid cooling.

Proper Cooling. Extensive experience has shown that it is next to impossible in summer, and difficult at other seasons, to market eggs of consistently superior quality through any "quality program" without adequate refrigeration on the farm. Best results are obtained by cleaning immediately after gathering all eggs requiring it, then cooling for 12 to 24 hours to at least 13°C (better 10°C) before packing in cases or cartons. It requires that length of time to bring warm eggs down to the recommended temperature. Lower temperatures than 10°C are impractical for farm cooling rooms and may cause "sweating" when eggs are removed.

Holding under Controlled Humidity. Relative humidity of holding rooms should not drop below 60% RH nor rise above 85% RH; 70 to 80% is optimum, as this degree of humidity will sufficiently retard evaporation to keep air cells small, without danger from molds.

Careful Handling. To avoid breakage and to prevent damage to the air cells as well as to the general interior egg structure, careful handling is necessary to conserve quality and avoid loss.

Proper Packing. It is important to pack all eggs large end uppermost, and to see that trays and flats fit the cases snugly. Improper packing results in poorer quality and lowered grades. It is also important to pack all eggs in precooled containers only—cases held in the cooler overnight or at least for several hours until thoroughly cooled.

Frequent Marketing. Twice per week marketing or oftener is essential to shorten the period between production and consumption. Frequency will be conditioned by size of flock, refrigeration on the farm, and method of marketing. Grading station pick-up trucks need to be insulated and probably refrigerated. Economy of pick-ups is conditioned by the number of cases of eggs per call, the number of calls required per load, and the distance travelled. Thus the collection cost per case may vary widely.

Moving Eggs Rapidly through the Marketing Channels
1. Rapid and efficient handling in grading stations with adequate refrigeration facilities and controlled humidity in storage room or rooms.

2. Speedy transportation by insulated or, preferably, refrigerated truck to retail or wholesale outlet is recommended.

3. Frequent delivery to retailers—at least twice and preferably three to five times per week is preferred.

Up-to-Date Merchandising (1) by adequate refrigerated holding space at retail outlets, (2) by sale of all eggs from properly refrigerated self-serve counters, and (3) by having all eggs cartoned attractively and marked as to grade; also brand if so desired. Carton colors for the several grades and brands should be so designed as to avoid confusing the customers. Attractive cartons containing clean eggs of uniform shell color, uniform yolk color, and dependably uniform high interior quality will undoubtedly result in increased sales and consumer satisfaction.

Proper Care of Eggs in the Home. This involves holding all eggs in the home refrigerator at 7° or 13°C until required for use. For best results, holding time should not be extended beyond one week.

It must be realized that eggs are a perishable product. Careless handling at any one point in the egg handling and marketing program will be reflected to a corresponding degree in lowered quality—and to some extent by less enthusiastic consumer acceptance.

SELECTED REFERENCES

Alford, L. R., Holmes, N. E., Scott, W. J., and Vickery, J. R. 1950. Studies in the preservation of shell eggs. 1. The nature of wastage in Australian export eggs. Aust. J. Appl. Sci. *1*, 208–214.

Anon. 1917. Preserving Eggs in Water-glass solution and Limewater, A.I. 30. Animal Husbandry Division, Bureau of Animal Industry, U.S. Department of Agriculture, Washington, DC.

Anon. 1955. Treating shell eggs to maintain quality. Mo., Agric. Exp. Stn., Bull. *659*, North Cent. Reg. Publ. No. 62.

Baker, R. C., and Curtiss, R. 1958. Strain differences in egg shell mottling, internal quality, shell thickness, specific gravity, and the interrelationships between these factors. Poult. Sci. *37*, 1086–1090.

Barott, H. G., and McNally, E. H. 1943. Opacity and infertility produced by thermostabilization process at 125°F and 144°F. U.S. Egg Poult. Mag. *49*, 320–322.

Bartolome, M. M. 1953. The influence of the "flash-heat" treatment on the keeping quality of eggs. Philipp. Agric. *37*, 188–202.

Brant, A. W., Starr, P. B., and Hamann, J. A. 1966. The bacteriological, chemical and physical requirements for commercial egg cleaning. U.S., Agric. Res. Serv., ARS, Market. Res. Rep. *740*.

Feeney, R. E., MacDonnell, L. R., and Lorenz, F. W. 1954. High temperature treatment of shell eggs. Food Technol. *8*, 242–245.

Funk, E. M. 1943. Stabilizing quality in shell eggs. Mo., Agric. Exp. Stn., Res. Bull. *362*.

Gillespie, J. M., and Scott, W. J. 1950. Studies in the preservation of shell eggs. IV. Experiments in the mode of infection by bacteria. Aust. J. Appl. Sci. *1*, 514–530.

Gillespie, J. M., Scott, W. J., and Vickery, J. R. 1950A. Studies in the preservation of shell eggs. II. The incidence of bacterial rotting in unwashed eggs and in eggs washed by hand. Aust. J. Appl. Sci. *1*, 215–223.

Gillespie, J. M., Scott, W. J., and Vickery, J. R. 1950B. Studies in the preservation of shell eggs. III. The storage of machine-washed eggs. Aust. J. Appl. Sci. *1*, 313–329.

Gillespie, J. M., Salton, M. R. J., and Scott, W. J. 1950C. Studies in the preservation of shell eggs. V. The use of chemical disinfectants in cleaning machines. Aust. J. Appl. Sci. *1*, 531–538.

Goodwin, T. L., Wilson, M. L., and Stadelman, W.J. 1962. Effects of oiling time, storage position and storage time on the condition of shell eggs. Poult. Sci. *41*, 840–844.

Goresline, H. E., Hayes, K. M., and Otte, A. W. 1952. Thermostabilization of shell eggs: Quality retention in storage. U.S., Dep. Agric., Circ. *898*.

Hall, G. O. 1945. Preserving eggs in water glass. Poult. Sci. *24*, 451–458.

Hamre, M. L., and Stadelman, W. J. 1964. The effect of water sanitizing compounds on the discoloration of the egg shell. Poult. Sci. *43*, 595–599.

Harns, J. V., Sauter, E. A., McLaren, B. A., and Stadelman, W. J. 1954. The effect of season, age and storage conditions on the flavor of eggs and products made using eggs. Poult. Sci. *33*, 992–997.

Homler, B. E. and Stadelman, W. J. 1963. The effect of oiling before and after cleaning in maintaining the albumen condition of shell eggs. Poult. Sci. *42*, 190–194.

Kahlenberg. O. J. *et al.* 1952. A study of the washing and storage of dirty shell eggs. U.S., Dep. Agric., Circ. *911*.

McLaren, B. A., and Stadelman, W. J. 1954. Relationships among physical, functional and flavor properties of eggs. Tech. Bull.—Wash. Agric. Expt. Stn. *14*.

Mountney, G. J. 1976. Poultry Products Technology, 2nd Edition. AVI Publishing Co., Westport, CT.

Pankey, R. D., and Stadelman, W. J. 1969. Effect of dietary fats on some chemical and functional properties of eggs. J. Food Sci. *34*, 312–317.

Parsons, R. W., and Stadelman, W. J. 1957. Ionizing irradiation of fresh shell eggs. Poult. Sci. *36*, 319–322.

Romanoff, A. L., and Romanoff, A. J. 1944. A study of preservation of eggs by flash heat treatment. Food Res. *9*, 358–366.

Romanoff, A. L., and Romanoff, A. J. 1949. The Avian Egg. John Wiley & Sons Co., New York.

Salton, M. R. J., Scott, W. J., and Vickery, J. R. 1951. Studies in the preservation of shell eggs. VI. The effect of pasteurization on bacterial rotting. Aust. J. Appl. Sci. *2*, 205–222.

Scott, W. J., and Vickery, J. R. 1954. Studies in the preservation of shell eggs. VII. The effect of pasteurization on the maintenance of physical quality. Aust. J. Appl. Sci. *5*, 89–102.

Snyder, E. S. 1961. Eggs. The Production, Identification and Retention of Quality in Eggs, Publ. No. 446. Ont. Agric. Coll., Guelph, Ontario, Canada.

Stadelman, W. J., Baum, E. L., Darroch, J. G., and Walkup, H. G. 1954. A comparison of quality in eggs marketed with and without refrigeration. Food Technol. *8*, 488–490.

Swanson, M. H. 1953. A proposal for the use of carbon dioxide in retail egg cartons. Poult. Sci. *22*, 369–371.

Swanson, M. H., Benson, H. N., and Skala, J. H. 1956. Quality preservation of shell eggs by treating with carbon dioxide prior to oiling. Poult. Sci. *35*, 1175.

Walters, R. E., Robbins, R. O., Brant, A. W., and Hamann, J. A. 1966. Improved methods, techniques and equipment for cleaning eggs. U.S., Agric. Res. Serv., ARS, Market. Res. Rep. *757*

Wesley, R. L., and Beane, W. L. 1967. The effect of various wash water and internal egg temperatures on the number of thermal checks of oiled and unoiled eggs. Poult. Sci. *46*, 1336.

The Microbiology
of Eggs

R. G. Board
H. S. Tranter

INTRODUCTION

The avian egg has evolved so that an embryo can develop in an axenic environment having a need for limited interactions with the extrinsic environment. Indeed, a source of heat, a means of turning, and a supply of oxygen are the sum total of the latter, the embryo being provided at oviposition with sufficient nutrients and water to sustain its development and that of the chick for a day or so after hatching. From the viewpoint of the poultryman producing eggs for the table, it is the need of the egg for a facility to exchange respiratory gases with the environment that can endanger the well-being of his product during collection, distribution, and protracted storage. In other words, the pores in the shell that permit gaseous diffusion are also the cause of eggs losing water and the portals by which microorganisms infect the egg. Indeed, attention to detail so that the diffusion of water vapor across the shell is minimized (see Chapter 4) and the pores are not penetrated with microorganisms can go a long way toward ensuring the maintenance of egg quality. Of course the well-being of the developing embryo is not entirely dependent upon the structure and functions of the eggshell; the albumen contributes also, particularly before and during the first few days of incubation when it provides both a chemical and physical defense against microbial infection of the yolk. This chapter is concerned mainly with the anti-

EGG SCIENCE & TECHNOLOGY,
3rd Edition

microbial defense offered by the complementary functions of the shell, the shell membrane, and the albumen.

Before Laying

It was assumed in the Introduction that avian eggs are germ free at oviposition. Critical evidence in support of this contention is not available. The old literature indicates that antimicrobial agents in the oviduct (unspecified as to nature or extent) provide an effective barrier against the upward migration of organisms present in the cloaca and thereby prevent contamination of the yolk and white. Of course, the incidence of microbial contamination of eggs at the time of lay ought to be easily established by laboratory examination. In practice, however, technical difficulties associated with the sterilization of the shell and the aseptic handling of the very obvious albumen and yolk cast doubts on the results of many surveys that have sought to give a precise answer to the question: What is the incidence of microbial contamination of eggs at oviposition? Although essential detail is lacking, two separate lines of evidence suggest that very few eggs contain saprophytic microorganisms at the time of lay. Thus Harry (1963) had to use enrichment cultures to recover organisms from the ovaries removed surgically from both in and out of lay hens. This observation suggests that even when the ovaries are contaminated with saprophytic bacteria, such organisms are present in very low numbers only. Perhaps the best clue to the microbial status of eggs at oviposition comes from surveys of the incidence of rotting in clean eggs stored for long periods. From their review of many surveys, Brooks and Taylor (1955) concluded that well over 90% of hens eggs are microbiologically sterile at lay.

It is well known that pathogenic microorganisms can be transmitted on/in avian eggs. As such organisms do not pose problems to eggs intended for human consumption, the reader is referred to veterinary textbooks for further details.

After Laying

The shell probably receives its first load of microorganisms when passing through the cloaca. From that time and until the egg is used, the opportunities exist for the shell to acquire microorganisms from every surface with which it makes contact. General surveys have shown that there is a wide variation in the level of contamination of eggshells; the populations range from a few hundred to tens of millions of bacteria per shell, with an average of about 100,000. With the exception of heavily soiled shells, there is a poor correlation of the level of contamination and the appearance of the shell. Thus, lightly soiled eggs can harbor on their shells fewer organisms than nest clean eggs, and it is only with the averages obtained from a statistically acceptable number of eggs that it becomes obvious that the contamination of the shell increases with exposure to dirty conditions. As contamination is achieved by contact of the shell with nesting materials, Keyes trays, hands, etc. the flora is notably heterogenous. Bacteria from sixteen genera were recovered from eggshells in one survey (Table 5.1) and from a knowledge of their habitats, it can be deduced that, in order of importance, dust, soil, and feces are the principal sources of contaminants. There is little doubt that representatives of genera other than those listed in Table 5.1 would be isolated, if a comprehensive range of selection media were used to survey microbial contamination of eggshells. The available information shows that gram-positive bacteria, probably because of their tolerance of dry conditions, dominate the flora on eggshells. In contrast, gram-negative bacteria are the principal contaminants of rotten eggs. The conditions that favor these minor contaminants in the infection of the egg contents will be discussed subsequently.

Rotten Eggs

Rotten eggs normally contain a mixed infection of gram-negative bacteria and, on occasion, a few gram-positive organisms are present

TABLE 5.1. Types of Microorganisms Present on the Shell of the Hen's Egg

Type of organism	Frequency of occurrence[a]
Streptococcus	±
Staphylococcus	+
Micrococcus	†
Sarcina	±
Arthrobacter	+
Bacillus	+
Pseudomonas	+
Acinetobacter	+
Alcaligenes	+
Flavobacterium	+
Cytophaga	+
Escherichia	+
Aerobacter	+
Aeromonas	±
Proteus	±
Serratia	±

[a] Present: ± occasionally, + on most eggs but in small numbers, and † always present in large numbers.

also. The commonest contaminants are numbers of genera *Alcaligenes, Acinetobacter, Pseudomonas, Serratia, Cloaca, Hafnia, Citrobacter, Proteus,* and *Aeromonas* (Table 5.2). It is noteworthy that this range of organisms has been recovered from rotten eggs in all parts of the world and over a period during which the basis of the industry has changed from a craft to a technology. Thus, when allowances are made for changes in nomenclature and taxonomy, it is apparent that the range of organisms isolated from eggs by Haines in 1938 is essentially similar to those recovered 30 years later by Board in 1965, Board and Board in 1968, and Moats in 1980. This implies that properties intrinsic to the egg favor the selection of rot-producing bacteria, and that such features as breed, housing, methods of storage, marketing procedures, etc., are of minor importance. The identity of the selective factors will be discussed later; at this time it will suffice to note that they favor the growth of organisms that have a negative reaction to Gram's stain, relatively simple nutritional requirements, and, with some, the ability to develop at low temperatures. These properties are common also to the organisms that cause taints in eggs and so differ from the rot-producing bacteria in their failure to digest proteins, form H_2S, split lecithin, or produce a pigment (Table 5.3).

Molds appear to be less important than bacteria in the addling of eggs. Under humid storage conditions, the shell can be clothed in myce-

TABLE 5.2. Types of Bacteria Found in Rotten Eggs

Type of organisms	Frequency of occurrence[a]
Pseudomonas aeruginosa	±
Pseudomonas fluorescens	†
Pseudomonas putida	†
Pseudomonas maltophilia	+
Flavobacterium	±
Alcaligenes	†
Acinetobacter	±
Cloaca	±
Cytophaga	±
Aeromonas	+
Proteus	†
Escherichia	†
Hafnia	+
Citrobacter	+
Bacillus	±
Micrococcus	±
Serratia	†
Streptococcus	±
Arthrobacter	±

[a] Present: ± on rare occasions only, + infrequently, † commonly.

TABLE 5.3. Changes Occurring in Eggs Infected with Pure Cultures

Organism	Metabolic attribute						Change	Type of rot
	1	2	3	4	5	6[a]		
Proteus spp.	+	+	–	–	–	–	Dark brown mealy yolk; dark brown albumen	Black rot type 2
Aeromonas liquefaciens	+	+	+	–	–	–	Gelatinous yolk blackened throughout; gray watery white	Black rot type 1
Enterobacter spp.	(+)	(+)	+	–	–	–	Yolk encrusted with custardlike material and occasionally flecked with olive-green pigment	Custard rot
Serratia marcescens	(+)	–	+	+	¦	–	White stained red throughout; yolk surrounded with custardlike material	Red rot
Pseudomonas maltophilia	+	+	–	(+)	–	–	Gelatinous, amberlike yolk striped olive-green, almondlike odor	Green rot
Pseudomonas fluorescens	+	–	+	–	+	–	Fluorescent green changing to pink white; yolk surrounded with custard like material	Pink rot
Pseudomonas putida	–	–	–	–	+	–	Fluorescent green pigment throughout the albumen	Fluorescent green rot
Pseudomonas aeruginosa	–	–	–	–	+	–	Fluorescent blue pigment throughout the albumen	Fluorescent blue rot
Flavobacterium *Cytophaga*	–	–	–	+	–	–	Yellow pigment formed in membrane at site of microbial growth	Yellow rot
Other *Enterobacter* and *Alcaligenes* spp.	–	–	–	–	–	+	Although large populations develop, there are no microscopic changes of the yolk and albumen	Colorless rot

[a] Proteolysis, 1; H₂S production, 2; lecithinase, 3; pigment, soluble 4; pigment insoluble 5; and negative in all these tests, 6. Parentheses indicate that although the organism can possess the attribute, there is no evidence that it contributes to the rotting of an egg.

lium, a condition known as "whiskers" in the industry. Hyphae penetrate the pores, and their growth on the shell membrane is often associated with the gelling of albumen around patches of mycelia. These show up as dark or colored patches or rings in candled eggs, and "pin cushions" on the inner surface of the shell membranes of broken-out eggs. Further growth results in the complete gelling of the albumen and the breakdown of the yolk membrane.

ANTIMICROBIAL DEFENSE

As the fertilized cell in an egg is supplied with all the chemicals needed for the growth of a chick, it is not surprising that through selection the eggs have been endowed with means whereby the food reserves are protected from microbial attack. The defense is partly physical (the shell, its membranes, and the albuminous sac) and partly chemical (the shell membranes and albumen). Its efficiency is dependent upon the integrity of the egg. Thus, antimicrobial defense is partially destroyed by a fracture of the shell and completely destroyed by a puncture in the yolk membranes.

The Shell

The shell has many contributions to make to the well-being of the embryo. In addition to providing portals for the diffusion of respiratory gases, it contributes to the conservation of water and provides mechanical protection as well as scaffolding for the embryo [see Rahn and Paganelli (1981) for comprehensive reviews]. When these many demands on the shell are considered collectively, it is then obvious that many are in conflict and that evolution has resulted in compromises whereby nature has fashioned the shell so that it is *fitted* to an environment—the nest in the wild [this aspect of the breeding biology of birds has been reviewed by Board (1982)]. This concept that an egg is fitted to its environment needs to be adopted by those who are involved in every stage of egg production, collection, and distribution so that eggs are not exposed to abuses for which they are not equipped. In addition, geneticists might well consider this concept when attempting to increase fecundity so that they do not, through increasing egg numbers, inadvertently select for eggs that will be inadequately equipped for the rigors of the market.

Microbiologists who have studied eggs immediately following oviposition assert that the eggshell is easily breached by bacteria while the cuticle is still moist. As yet, however, there is no direct evidence to support such an assertion. Nevertheless, it is noteworthy that the moist appearance of the shell of an egg that has just been laid cannot be reproduced merely by adding water. This suggests that the cuticle at the time of lay may be different in fine structure from that on a shell that has dried out. Further work is required to test the validity of this surmise and the assertions of those who claim that the shell of a newly laid egg is vulnerable to microbial invasion.

If a shell having a dried-out cuticle is considered as a resistance network (Fig. 5.1), then the means whereby the biological conflicts noted above are resolved can be appreciated. For the purposes of this discussion, it has been assumed that the conservation of water is of primary importance. This has been achieved evolutionarily by the selection of shells having a pore area commensurate apparently with the minimum requirement of the embryo to obtain oxygen and to rid its immediate environment of CO_2. Some water has to be lost from the egg during incubation in order for the air space to become large enough to sustain the embryo for the short time that it breathes with its lungs and to provide space sufficient for the embryo to move during pipping and hatching (see Rahn and Paganelli 1981). Even the minimal pore area achieved by selective pressures causes problems for those who trade eggs in markets where weight rather than numbers determines the price. To minimize further the loss of water, two strategies are available in theory: (1) maintaining of a very high relative humidity around the

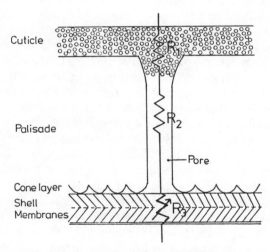

FIG. 5.1. Schematic drawing of radial section of hen's egg showing the major components of the resistance network. R_1 is cuticle, R_2 is the pore canal, and R_3 is the shell membranes which may change with time.

eggs so that a shallow diffusion gradient of RH exists across the shell, or (2) increasing the resistance that the shell offers to the flux of water vapor. As will be seen below, the former strategy cannot be exploited because it promotes the growth of molds. In the advanced stages of mold infection, the fluffy appearance of the shell makes an egg unmarketable. Even when mold growth is not readily apparent, there is always the danger that hyphal penetration of pores in the shell may have caused bacteria to be lodged in the shell membrane. Thus, even though storage conditions are changed sufficiently to impede mold growth, bacterial addling of the yolk and white may still ensue. With strategy 2, oils of various types have been applied to the shell so that evaporative water loss is reduced.

The final stage in the evolution of avian eggs can be considered to be the release from a need for exogenous water, a characteristic of all cleiodoic eggs. It has not been appreciated that through this achievement, the avian egg has become vulnerable to water. To permit the diffusion of respiratory gases, the shell of the avian egg contains upwards of 17,000 pores (Fig. 5.2), each one of which has dimensions that could be expected to support a column of water to a height far in excess of the thickness of the shell. If capillarity flooded pores each time that an egg in nature became wet, it could be assumed that death by asphyxiation would be a common fate of the embryo. In practice, the shell has the property of water repellency, which prevents the flooding of pores when eggs merely rest in water. In addition, the shells of all the domesticated birds—hen, guinea fowl, turkey, duck, goose, and quail—have a marked resistance to water due to the cuticle that covers the surface of the shell and plugs to varying extents the pore canals. The illustration of the cuticle (Figure 5.2) shows it to be formed of vesicles that contribute a bed of spheres though which gases and water vapor diffuse. Some hens lay eggs that are partially or completely lacking in cuticle; such eggs have the property of water repellency but not water resistance. There is an urgent need for studies of the physiology and genetics of cuticle synthesis and deposition. Indeed, should such studies indicate that the cuticle-deficiency state is a trait of particular hens only, well-established methods of selection would be expected to improve dramatically the overall quality of the cuticle on market eggs. Water resistance of the shell provides a barrier to the movement of bacteria; in the absence of water the bacteria cannot be translocated from the surface of the shell to the shell membranes. Energy has to be expended if the water resistance of the shell is to be overcome, a condition that is not appreciated sufficiently by those who design or use egg washing machines. Sufficient "work" can be done merely by causing the pressure within the egg to be less than that of the atmosphere, a situation that occurs when a warm egg is put into cold water. Those who seek to control the transmission of

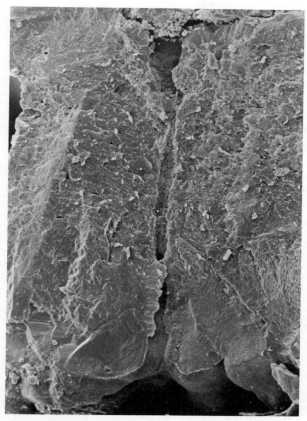

FIG. 5.2.   Radial section of hen's eggshell showing cuticular plug, pore canal, and cone layer. Magnification: × 400.

pleuropneumonia-like organisms (PPLO) via the egg have shown that therapeutic levels of antibiotics can be introduced into eggs submerged in a solution of antibiotics, if a vacuum is drawn and suddenly released or if a positive pressure is applied to the headspace above the liquid. In theory, the kinetic energy in a jet of water should be sufficient to overcome the water resistance of the shell. Since such resistance is due mainly to the cuticle, anything that damages the cuticle will contribute to a lessening of this resistance. There are many reports in the literature of increased levels of infection leading to rotting of eggs that were abraded before being challenged with rot-producing bacteria. Even in an egg having an undamaged cuticle, there are upwards of 10 to 20 pores that lack either an adequate cover or plug of cuticle; these uncovered pores provide the portals for bacteria to infect the contents of the egg. Indeed, those who have attempted to control PPLO-transmission in eggs have recognized that only a few pores (10–20) in the "average" egg permit the transfer of an antibiotic from the outer to the inner surface of the shell. Although the term "patent pore" has been coined, there is no evidence of the manner in which such pores differ from the majority. This area of egg microbiology has been reviewed by Board (1980), who stressed that the successful operation of egg-washing machines is dependent upon the temperature of the wash and rinse water being higher than the temperature of the egg.

The Shell Membranes

The exact role of shell membranes in the egg's defense is still uncertain. Each membrane is composed of anastomosing fibers (Fig. 5.3) in which a glycoprotein mantle (Fig. 5.4) envelopes a protein core notable for its content of the two amino acids, desmosine and isodesmosine. Although both of these are found in elastin, this protein and that in the shell membranes differ markedly in overall amino acid composition

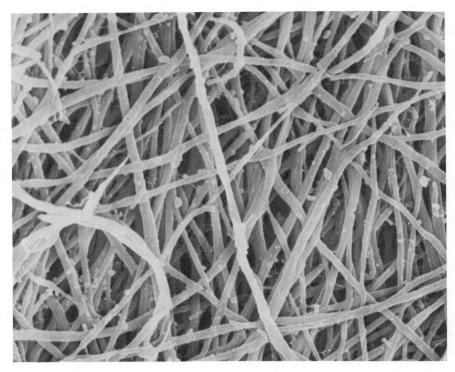

FIG. 5.3.   Interwoven fibers of hen's eggshell membranes. Magnification: × 900.

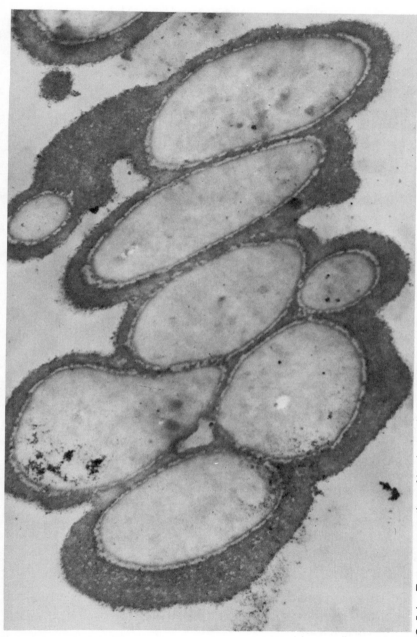

FIG. 5.4. Transverse section of hen's eggshell-membrane fibers showing protein cores surrounded by a glycoprotein mantle stained with ruthenium red. Magnification: × 10,000.

and susceptibility to digestion by elastinase (Leach *et al.* 1981). It is not surprising that the fibrous nature of the shell membranes has invoked the view that they may function as bacterial filters. Evidence in support of such an hypothesis comes from experiments of the type designed by Haines and Moran (1940), who replaced the yolk and white with a suspension of bacteria and then applied suction to the shell. The fluid drawn through the shell with an intact membrane did not contain microorganisms; in contrast, fluid drawn through a membrane-free shell contained microorganisms. This technique is open to criticism, since bacteria are made to traverse the shell and its membranes in a direction contrary to the normal direction for bacterial egg invasion. It is conceivable, for example, that pressure causes a compression of the shell membranes which may well force them into internal openings of the pores.

As an alternative method of proving or disproving this hypothesis, a technique was developed whereby the shell, with or without one or both of the shell membranes, is held in a suspension of bacteria and attempts are made to recover microorganisms from within the egg. When the time required for penetration is used as an index, it has been demonstrated that the shell with both of it membranes offers greater resistance to bacterial penetration than does the shell alone. Of the two membranes, the innermost offers the greater resistance, possibly because its inner surface is clothed with a limiting membrane (Fig. 5.5).

Although such evidence tends to support the theory that the outer structures of the egg act as bacterial filters, it should be remembered that the conditions of the test are highly artificial. Thus the results obtained are markedly different from those of other researchers using different methods. For example, when warm eggs were allowed to contract in chilled suspensions of bacteria, organisms were recovered from the inner surface of the shell membrane within a few hours. In another experiment, when the inner shell membrane of the intact eggs was challenged with *Serratia marcescens*, penetration was rapid in eggs held at 37°C but slow in those at 20° or 30°C. The evidence considered to date does not warrant an assumption that the outer structures of an egg are well adapted to resist bacterial penetration. There is no doubt that they retain microorganisms but, as will be seen in the section concerned with the course of microbial infection, this may aid rather than hinder the addling of an egg.

Organisms recovered from rotten eggs grow well when added to shell membranes suspended in a solution of mineral salts, providing the latter are not bactericidal. With *Pseudomonas maltophilia* and *Aeromonas liquefaciens*, growth is accompanied by the release of low molecular weight nitrogenous compounds from the membranes. There is no

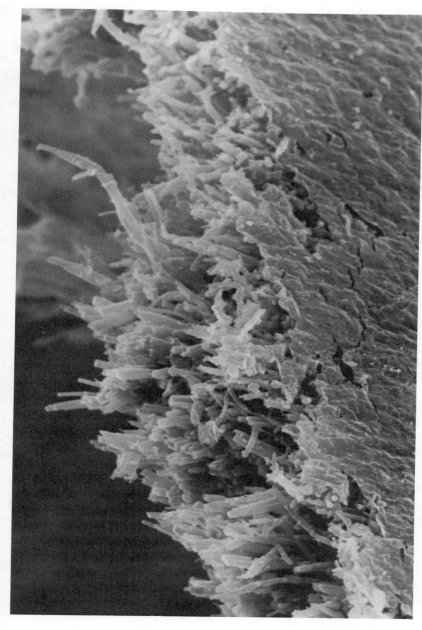

FIG. 5.5. Hen eggshell membranes prepared by critical-point drying showing inner limiting membrane. Magnification: × 900.

equivocal evidence, however, that such digestion aids in any way an organism's penetration of the membranes.

Although the shell membranes *in vitro* as well as those in the unincubated egg support microbial growth, incubation of fertile eggs causes changes whereby the membranes appear to offer an increasingly hostile environment to bacteria. Although it is now widely accepted (Tullett and Board 1976) that a reduction of the shell membranes' resistance to the diffusive flow of oxygen on the fifth to the eighth day of incubation of the hen's egg (Fig. 5.6) is a major event, there is no evidence to support an earlier assertion that this change in resistance is due to the membranes "drying out" and thereby imposing a water activity unfavorable to microbial growth.

The Albumen

The albumen makes two contributions to the antimicrobial defense of the egg—mechanical and chemical.

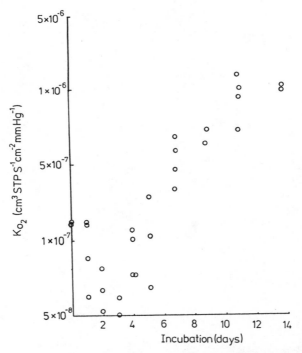

FIG. 5.6.   Changes in the oxygen flux across the integument of the hen's egg with incubation at 37.5° C.

SOURCE: *R. G. Board, H. S. Tranter, and Sparks, unpublished data (1982).*

With mechanical defense, there are two components: (1) the viscosity of the proteins, and (2) the organization of the albumen in the albuminous sac so that biological structure is conferred on the egg. Viscosity hampers the movement of bacteria that invade the shell membranes so that they do not have an unimpeded passage to the yolk. The albuminous sac of fresh eggs contributes to the central location of the yolk, thus maintaining it at the greatest distance from the contaminants restrained by the shell membranes. From their description of the thick white of the albuminous sac—"a weak gel interpenetrated by a system of microscopic, elastic fibres and not, as is usually believed, merely an entanglement network"—Brooks and Hale (1959) supported the many assertions that the fibers were the products of an interaction between lysozyme c and ovomucin. Further support for this concept has come from the studies of Professor D.S. Robinson (see Robinson and Monsey 1975).

Discussions of the chemical defense afforded to the yolk by the white have focussed on the latter's unsuitability as a medium for microbial growth. Such a concept is attractive when the constituents of the albumen are considered (Table 5.4). For, in addition to a toxic component, lysozyme, there are proteins that make certain cations and vitamins unavailable to organisms that require them. Moreover, the alkaline state (pH 9.5) of the albumen, apart from being itself an inimical factor, accentuates the chelating potential of ovotransferrin. The white contains only traces of nonprotein nitrogen. This deficiency has led to speculation about the possible role of the protease inhibitors in the albumen (Table 5.4); the argument implies that the inhibitors would prevent microbial contaminants from satisfying their nitrogen requirements at the expense of proteins. No one has yet shown that the inhibitors function in this way.

Although the lytic properties of lysozyme c can be demonstrated *in vitro*, no one has yet presented convincing evidence of this property being of paramount importance in the defense of the egg against microbial attack. It is tempting at this time to assume that its major role is in the mechanical defense system. As the lysozyme g of duck and goose eggs differs from lysozyme c of the hen's egg, the former should be investigated for its capacity to lyse gram-positive and gram-negative bacteria in the albumen in the egg. Of the components listed in Table 5.4, ovotransferrin appears to be the major contributor to the egg's defense against microbial infection and rotting. By depriving the microorganisms of Fe^{3+}, ovotransferrin prevents microbial multiplication over a temperature range of 0°–35°C. Above this temperature, many organisms, including strains of *Escherichia coli*, die as a consequence of iron deprivation. Thus, we are of the opinion that the defense offered by the albumen is optimal when the conditions are most alkaline and when

TABLE 5.4. Biological Properties of the Antimicrobial Proteins of Hen's Egg Albumen

Component	Activity	Source
Lysozyme	Hydrolysis of $\beta(1-4)$ glycosidic bonds in bacterial cell wall peptidoglycan	Geoffroy and Bailey (1975)
	Flocculation of bacterial cells	Friedberger and Hoder (1932)
	Formation of oligosaccharides from bacterial cell wall tetrasaccharides by transglycosylation	Chipman and Sharon (1969)
Ovotransferrin	Chelation of Fe^{3+}, Cu^{2+}, Mn^{2+}, Co^{2+}, Cd^{2+}, Zn^{2+}, Ni^{2+}	Tan and Woodworth (1969); Gelb and Harris (1980)
Avidin	Binding of biotin, rendering it unavailable to bacteria that require it	Green (1975); Chignell et al. (1975)
Ovoflavoprotein	Binding of riboflavin, rendering it unavailable to bacteria that require it	Clagett (1971); Miller et al. (1981)
Ovomucoid	Inhibition of bovine and porcine trypsin	Feeney and Allison (1969)
Ovoinhibitor	Inhibition of bovine and porcine trypsin	Liu et al. (1971);
	Inhibition of bovine and avian α-chymotrypsin	Zahnley (1980)
	Inhibition of subtilisin and fungal proteinase	
Ficin-papain inhibitor	Inhibition of ficin and papain	Sen and Whitaker (1973)

the temperature is at or near that of incubation. Claims have been made that ovotransferrin would induce microorganisms to produce chelating agents and thereby scavenge whatever iron is present in the albumen. Pyoverdine, a chelate produced by *Pseudomonas* spp., has been credited with such a role. In our work with Enterobacteriaceae, no evidence has been obtained of enterobactin, the major chelating agent of this family, being formed in albumen *in vitro*. Indeed, should it be formed, hydrolysis at the whites' pH of 9.5 might rapidly lead to a marked reduction of its activity, a situation that we have noted when incubating enterobactin in buffer of the same pH.

Since the rotting of eggs is due almost invariably to non-sporeforming bacteria, scant attention has been paid to bacterial endospores in albumen. We have demonstrated that ovotransferrin plays an important role in preventing normal growth of the nascent vegetative cells emerging from spores germinating in egg white, particularly at the high pH of the white (Tranter and Board 1982).

COURSE OF INFECTION

When considering the addling of eggs, most workers have tended to divide the process into quite arbitrary stages and to examine one or another of them in isolation. Thus we have much information concerning the probable function of the shell, the shell membranes and the albumen, but little direct evidence of how these are integrated in the antimicrobial defense of the egg. It was obvious in the discussion of this topic that the welfare of the embryo—and hence of the egg of commerce—cannot be assured by any one of the components. A plausible concept of their working in unison comes when extant data are discussed in terms of ecology. Thus the term "association" can be used for the flora of rotten eggs, and its genesis under commercial conditions can be considered to be the result of interplay of (1) the initial infection of the shell, (2) environmental conditions (extrinsic factors), (3) features inherent in the egg (intrinsic factors), and (4) physiological attributes of the organisms which make up the association (implicit factors).

Following oviposition, the shell acquires infection from all surfaces with which it makes contact, and the extent of infection is directly related to the cleanliness of these surfaces. Since nesting materials and Keyes trays, derive organisms from ubiquitous depots, the shell will harbor a flora of diverse composition (Table 5.1). Under normal conditions of handling and storage, few if any of the organisms on the shell multiply because of the low level of available water; in fact, there is evidence that, unless protected by fecal matter, soil, etc., many of these organisms die by desiccation (Fig. 5.7). The humid conditions in egg

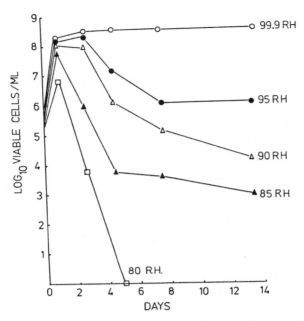

FIG. 5.7.   The effect of relative humidity on the growth of *Escherichia coli* at 37° C.

SOURCE: *S. Loseby, unpublished data (1977).*

stores can promote the growth of molds, and hyphae can thus penetrate the pores to infect the shell membranes and albumen. With storage under very humid conditions (RH > 98), the cuticle can be colonized and digested by *Pseudomonas* spp. (Board *et al.* 1979). Bacteria remain localized at or near the surface of the shell unless free water is present. They then penetrate the shell along with water drawn into the pores by capillary action or are sucked in when a warm egg contracts on cooling. The extent of the penetration can be increased merely by rubbing the shell with an abrasive. The presence of bactericides in wash water does not guarantee that viable bacteria will not be lodged within the pores; the chemicals are presumably rendered innocuous by reaction with the debris in the pore canal. Since the shell harbors a heterogeneous flora, and the organisms play only a passive role in infiltration of the shell, it is obvious that a range of organisms will be deposited on or near the shell membranes. It is suggested by some researchers that they must approach to with 15–20 μm of the membranes before the level of available water is favorable for growth.

A study of the colonization of the shell membranes by a mixed population of bacteria demonstrated that the selection of particular groups of organisms occurs. Thus the inoculum, obtained by washing dirty egg-shells, was dominated by gram-positive bacteria whose number declined gradually during incubation. In practice, it was the growth of gram-negative bacteria that caused the shell membranes to harbor a relatively small gram-positive flora after a period of incubation. The types of organisms selected was a function of the incubation temperature. Few microorganisms were recovered from the albumen until the onset of rotting. It was notable that only one strain of gram-negative bacteria made a major contribution to the contamination of the albumen, even though it and several other strains formed a consortium in the shell membranes. This information indicates that the initial colonization of the shell membranes is characterized by the selection of organisms best suited to the environment. Although scant, the available evidence suggests that selection favors organisms having simple nutritional requirements and that, among this group, growth rate will be the ultimate determinant of the viability of an organism retained by the shell membranes.

It is noteworthy that under commercial conditions of storage at ambient temperatures, there is a lag of 12–20 days following penetration of

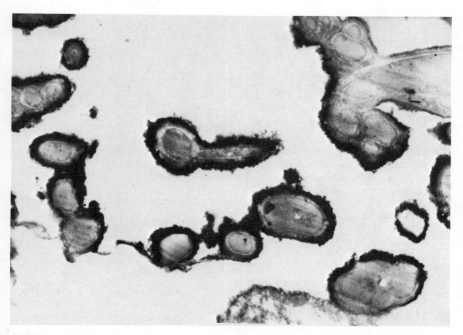

FIG. 5.8.   Transverse section of hen's eggshell membranes showing the binding of colloidal iron by mantle glycoprotein. Magnification: × 5000.

the shell and the recovery of large numbers of organisms from the albumen. Although the phase during which the infection is confined to the shell membranes is notable for the changing proportions of organisms, growth is of a limited extent normally and only a few organisms enter the albumen; these fail to multiply unless they make contact with the yolk or colonize the interface region between the yolk and shell membranes. Indeed, the lag period noted above can be attributed to the slow breakdown of the albuminous sac, which is a prerequisite of the union of the yolk and shell membranes. Rotting induced by organisms making contact with the yolk is a feature of eggs held at chill temperatures. Under such conditions, the contaminants of the albumen are not killed as quickly as those in albumen at ambient temperature, and thus the probability of a chance contact of the yolk with a contaminant is increased, because the latter occur in relatively large numbers. A similar position obtains when the shell membranes are contaminated with bacteria and iron. This element is bound avidly to the glycoprotein mantle on the membrane fibers (Fig. 5.8), and its transfer to the albumen is slow. The contaminants in the shell membranes grow quickly, because of the availability of iron and relatively large numbers of contaminants enter the albumen. The high incidence and rapid rotting of eggs washed in water containing iron can be attributed to (1) enhancement of microbial growth in the shell membranes, and (2) an increase in the probability of a contaminant of the albumen making contact with the yolk.

SELECTED REFERENCES

Board, P. A., and Board, R. G.   1968.   A diagnostic key for identifying organisms recovered from rotten eggs. Br. Poult. Sci. 9, 111-120.

Board, P. A., Henden, L. P., and Board, R. G.   1968.   The influence of iron on the course of bacterial infection of the hen's egg. Br. Poult. Sci. 9, 211-215.

Board, R. G.   1965.   The properties and classification of the predominant bacteria occurring in rotten eggs. J. Appl. Bacteriol. 28, 437-453.

Board, R. G.   1966.   Review article; the course of microbial infection of the hen's egg. J. Appl. Bacteriol. 29, 319-341.

Board, R. G.   1968.   Microbiology of the egg (a review). In Egg Quality: A Study of the Hen's Egg. T. C. Carter (Editor). Oliver and Boyd, Edinburgh, Scotland.

Board, R. G.   1969.   The microbiology of the hen's egg. Adv. Appl. Microbiol. 11, 245-281.

Board, R. G.   1980.   The avian eggshell: A resistance network. J. Appl. Bacteriol. 48, 303-313.

Board, R. G.   1982.   Properties of avian eggshells and their adaptive value. Biol. Rev. Cambridge Philos. Soc. 57, 1-28.

Board, R. G., and Ayres, J. C.   1965.   The influence of temperature on bacterial infection of the hen's egg. Appl. Microbiol. 13, 358-364.

Board, R. G., and Fuller, R.   1974.   Non-specific antimicrobial defenses of the avian egg embryo and neonates. Biol. Rev. Cambridge Philos. Soc. 49, 15-49.

Board, R. G., and Halls, N. A. 1973. The cuticle: A barrier to liquid and particle penetration of the shell of the hen's egg. Br. Poult. Sci. *14*, 69-97.

Board, R. G., Ayres, J. C., Kraft, A. A., and Forsythe, R. H. 1964. The microbiological contamination of eggshells and egg packing materials. Poult. Sci. *43*, 584-595.

Board, R. G., Loseby, S., and Miles, V. R. 1979. A note on microbial growth on egg-shells. Br. Poult. Sci. *20*, 413-420.

Brooks, J., and Hale, H. P. 1959. The mechanical properties of the thick white of the hen's egg. Biochim. Biophys. Acta *32*, 237-250.

Brooks, J., and Taylor, D. I. 1955. Eggs and egg products. G.B. Dep. Sci. Ind. Res. Board, Spec. Rep. Food Invest. *60*.

Buss, E. G. 1982. Genetic differences in avian egg shell formation. Poult. Sci. *61*, 2048-2055.

Chignell, C. F., Starkweather, D. K., and Sinha, B. K. 1975. A spin label study of egg white avidin. J. Biol. Chem. *250*, 5622-5630.

Chipman, D. M., and Sharon, N. 1969. Mechanisms of lysozyme action. Science *165*, 454-465.

Clagett, C. O. 1971. Genetic control of the riboflavin carrier protein. Fed. Proc., Fed. Am. Soc. Exp. Biol. *30*, 127-129.

Cooke, A. S., and Balch, D. A. 1970A. Studies of membrane, mammillary cover and cuticle of the hen eggshell. Br. Poult. Sci. *11*, 345-352.

Cooke, A. S., and Balch, D. A. 1970B. The distribution and carbohydrate composition of the organic matrix of the hen eggshell. Br. Poult. Sci. *11*, 353-365.

Edwards, N. A., Luttrell, V., and Nir, I. 1976. The secretion and synthesis of albumen in the magnum of the domestic fowl (*Gallus domesticus*). Comp. Biochem. Physiol. B *53B*, 183-186.

Feeney, R. E., and Allison, R. G. 1969. Evolutionary biochemistry of proteins. Homologous and analogous proteins from avian egg whites, blood sera, milk, and other substances. John Wiley and Sons, New York.

Friedberger, E., and Hoder, F., 1932. Lysozyme und Flockungsphanome bes Huhne-reiklars. Z. Immunitatsforsch. *74*, 445-447.

Gelb, M. H., and Harris, D. C. 1980. Correlation of proton release and ultraviolet difference spectra associated with metal binding by transferrin. Arch. Biochem. Biophys. *200*, 93-98.

Geoffroy, P., and Bailey, C. J. 1975. The action of hen and goose lysozyme on the cell wall peptidoglycan of *Micrococcus lysodeikticus*. Biochem. Soc. Trans. *3*, 1212-1214.

Green, N. M. 1975. Avidin. Adv. Protein Chem. *29*, 85-133.

Haines, R. B. 1938. Observations on the bacterial flora of the hen's egg, with a description of a new species of *Proteus* and *Pseudomonas* causing rot in eggs. J. Hyg. *38*, 338-355.

Haines, R. B., and Moran, T. 1940. Porosity of and bacterial invasion through the shell of the hen's egg. J. Hyg. *40*, 453-461.

Hamilton, R. M. G. 1982. Methods and factors that affect the measurement of egg shell quality. Poult. Sci. *61*, 2022-2039.

Harry, E. G. 1963. Some observations on the bacterial contents of the ovary and oviduct of the fowl. Br. Poult. Sci. *4*, 63-90.

Kutchai, H., and Steen, J. B. 1971. Permeability of the shell and shell membranes of hen's eggs during development. Respir. Physiol. *11*, 265-278.

Leach, R. M. 1982. Biochemistry of the organic matrix of the eggshell. Poult. Sci. *61*, 2040-2047.

Leach, R. M., Rucker, R. B., and Van Dyke, G. P. 1981. Eggshell membrane protein: a non-elastin desmosine/isodesmosine-containing protein. Arch. Biochem. Biophys. *207*, 353-359.

Liu, W.-H., Means, G. E., and Feeney, R. E.   1971.   The inhibitory properties of avian ovoinhibitors against proteolytic enzymes. Biochim. Biophys. Acta *229*, 176–185.

Miller, M. S., Buss, E. G., and Clagett, C. O.   1981.   Effect of carbohydrate modification on transport of chicken egg white riboflavin-binding protein. Comp. Biochem. Physiol. B *69B*, 681–686.

Moats, W. A.   1980.   Classification of bacteria from commercial egg washers and washed and unwashed eggs. Appl. Environ. Microbiol. *40*, 710–714.

Osuga, D. T., and Feeney, R. E.   1974.   Avian egg white proteins. *In* Toxic Constituents of Animal Foodstuffs. I. E. Liener (Editor), 2nd Edition. Academic Press, New York.

Parsons, A. H.   1982.   Structure of the egg shell. Poult. Sci. *61*, 2013–2021.

Rahn, H., and Paganelli, C. V.   1981.   Gas Exchange in Avian Eggs. State University of New York, Buffalo.

Robinson, D. S., and Monsey, J. B.   1975.   The composition and proposed sub unit structure of egg white-ovomucin. Biochem. J. *147*, 55–62.

Romanoff, A. L., and Romanoff, A. J.   1949.   The Avian Egg. John Wiley & Sons Co., New York.

Sen, L. C., and Whitaker, J. R.   1973.   Some properties of a ficin-papain inhibitor from avian egg white. Arch Biochem. Biophys. *158*, 623–632.

Seviour, E. M., and Board, R. G.   1972.   The behaviour of mixed bacterial infections in the shell membranes of the hen's egg. Br. Poult. Sci. *13*, 33–43.

Tan, A. T., and Woodworth, R. C.   1969.   Ultraviolet difference spectral studies of conalbumin complexes with transition metal ions. Biochemistry *8*, 3711–3716.

Tranter, H. S., and Board, R. G.   1982.   The inhibition of vegetative cell outgrowth and division from spores of *Bacillus cereus* T by hen egg albumen. J. Appl. Bacteriol. *52*, 67–74.

Tullett, S. G., and Board, R. G.   1976.   Oxygen flux across the integument of the avian egg during incubation. Br. Poult. Sci. *17*, 441–450.

Tullett, S. G., and Board, R. G.   1977.   Determinants of egg shell porosity. J. Zool. *183*, 203–211.

Zahnley, J. C.   1980.   Independent heat stabilization of proteases associated with multiheaded inhibitors. Complexes of chymotrypsin, subtilisin and trypsin with chicken ovoinhibitor and with lima bean protease inhibitor. Biochim. Biophys. Acta *613*, 178–190.

The Chemistry of Eggs and Egg Products

William D. Powrie
S. Nakai

INTRODUCTION

This chapter will be confined to the chemistry and structure of infertile eggs from hens classified as *Gallus domesticus*, since these eggs are used almost exclusively for human consumption. The average weight of eggs produced by selectively bred strains of hens is the vicinity of 60 g (Cotterill and Geiger 1977; Marion *et al.* 1964; Kline *et al.* 1965; Chung and Stadelman 1965). Shell eggs consist of about 9.5% shell, 63% albumen, and 27.5% yolk (Cotterill and Geiger 1977). The total solids of albumen, yolk, and whole egg are around 12%, 52% and 24%, respectively.

The shell shape and weight of hens' eggs is dependent on heredity, age of the bird, season of the year, and diet. To meet high quality standards of market eggs, commercial flocks have been selected through extensive breeding programs and, in addition, the nutritional adequacy of the birds has been established. Most hens' eggs destined for market have an oval shape. The shape of the egg can be expressed very roughly by the shape index (breadth/length × 100). According to Romanoff and Romanoff (1949), the standard egg of a hen has a shape index of 74 with blunt and pointed ends. A flock of 262 Leghorn hens laid eggs which had shape index values ranging from 63.1 to 81.7 and a median of 70 (Romanoff and Romanoff 1949).

Knowledge of the chemical composition and physicochemical properties of native albumen and yolk should be useful for interpreting the

97

changes that occur during shell egg storage and during pasteurizing, drying, and freezing. Alteration in the egg components will be reflected as a loss of functionality of albumen (foaming power) and yolk (emulsifying ability).

Reviews on the chemistry of egg components have been written by Bell and Freeman (1971); Brooks and Taylor (1955), Carter (1968), Feeney (1964), Osuga and Feeney (1974, 1977), Parkinson (1966), Powrie (1976), and Vadehra and Nath (1973).

STRUCTURE AND COMPOSITION OF THE SHELL AND SHELL MEMBRANES

Eggshell

The eggshell is made up for the most part of (1) a matrix consisting of interwoven protein fibers and spherical masses, and (2) interstitial calcite crystals. The proportion of matrix to crystalline material is about 1:50 (Romanoff and Romanoff 1949). In addition, the surface of the calcified shell is covered by a cuticle, a foamy layer of protein. A diagram of a radical section of an eggshell is shown in Fig. 6.1.

The matrix has been divided into two regions, the mammillary matrix and the spongy matrix, on the basis of differences in their staining capacity for polysaccharides, affinity for cations, and resistance to boiling the whole shell in 10% NaOH (Simkiss 1968). The mammillary matrix region is interconnected to protein fibers of the outer shell membrane. More specifically, the outer membrane fibers are associated with protein masses called mammillary cores, which are located about 20 μm inward from the inner front of the shell (Simkiss 1968). Calcite crystals are oriented randomly within each mammillary matrix to form a cone (Heyn 1963). The cones are cemented together at the top to form a cohesive mass (Fig. 6.1). According to Simons and Wiertz (1963) and El-Boushy et al. (1968), the spongy matrix has fine fibers (0.04 μm in diameter) that run parallel to the shell surface and are associated with numerous vesicles (0.4 μm in diameter). The crystals within the calcified spongy matrix (palisade layer) have their long axes oriented towards the shell surface.

Studies with light and electron microscopy indicate that the matrix fibers pass through calcite crystals rather than simply surrounding them (Simkiss 1968). Thus the matrix may have a significant influence on the shell strength. Simons et al. (1966), found a positive correlation between the nitrogen content of the shell and shell strength. Simkiss and Tyler (1957) found in their histochemical studies on decalcified shell that both the matrices consisted of protein-acid-mucopolysaccharide.

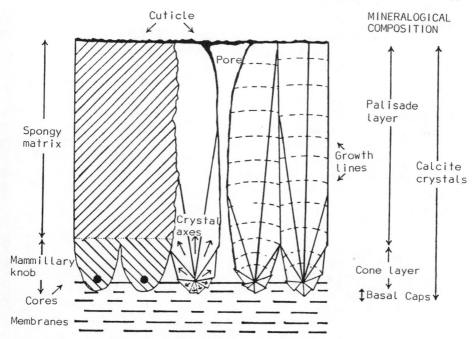

FIG. 6.1.   Diagram of a radial section of a hen's eggshell.

The polysaccharides contain chondroitin sulfates A and B, galactosamine, glucosamine, galactose, mannose, fucose, and sialic acid (Baker and Balch 1962; Frank *et al.* 1965).

The elemental composition of egg shell has been reported by Romanoff and Romanoff (1949) to be 98.2% calcium, 0.9% magnesium, and 0.9% phosphorus (present in shell as phosphate). Phosphate was detected by Terepka (1963) in the cone region, but most of the phosphate and magnesium has been found in the outer portion of the shell (Itoh and Hatano 1964). The inclusion of magnesium in the shell is probably limited by the properties of calcite. According to Brooks and Hale (1955), an increase in magnesium content of the shell is directly related to an increase in shell hardness. The calcite (calcium carbonate) crystals in egg shells have been studied by X-ray diffraction (Cain and Heyn 1964; Favejee *et al.* 1965), but the problem of orientation of crystal axes has not been resolved.

Numerous funnel-shaped (mouth at the outer surface) pore canals (7,000 to 17,000 per egg) are distributed at right angles to the shell surface and form connecting passages between the shell membrane and the cuticle (Romanoff and Romanoff 1949; Simkiss 1968). Pores are distributed unevenly over the shell surface. An abundance of pores exist at the blunt end, whereas the fewest number of pores per square centimeter are present at the center of the pointed end (Walden *et al.* 1956). The pore canals are filled with protein fibers (Romanoff and Romanoff 1949).

The water-insoluble cuticle forms a protective coating (about 10 to 30 μm thick) on the surface of the shell (covers the pores) and impedes microbial invasion of the egg contents (Board 1968). In the electron micrographs of Simons and Wiertz (1963), the cuticle appears to be constructed of two separate layers; one layer adjacent to the shell has a foamy appearance, whereas the outer layer is more compact. The cuticle is composed of about 90% protein which has a high content of glycine, glutamic acid, lysine, cystine, and tyrosine (Baker and Balch 1962). Hexosamines, galactose, mannose, fucose, glucose, and sialic acid are present as constituents of the polysaccharides (Cooke and Balch 1970).

Shell Membranes

Two membranes (consisting of protein fibers) reside between albumen and the inner surface of the shell. The total thickness of the two membranes have been found to range from 73 to 114 μm in eggs from White Leghorn and New Hampshire pullets (Tung and Richards 1972). According to Simkiss (1968), the outer membrane has a thickness of about 48 μm and the thickness of the inner membrane was estimated to be about 22 μm. Tung and Richards (1972) found that the thickness of the outer and inner membranes of eggs from White Leghorns ranged from 53.2 to 65.5 μm and from 19.5 to 24.3 μm, respectively. The outer shell membrane is attached firmly to the shell by numerous cones on the inner shell surface extending into the membrane and by fiber associations.

Electron micrographs of the shell membranes have provided evidence that each fiber of a membrane has an electron-dense core and a surrounding mantle layer having a high content of polysaccharide (Masshoff and Stolpmann 1961; Simons and Wiertz 1963; Tung and Richards 1972). The diameter of the fiber core ranges from 0.681 to 0.871 μm for the outer membrane and 0.481 to 0.592 μm for the inner membranes (Tung and Richards 1972). The mantle thickness in fibers of both types of membranes is about 0.1 μm for the White Leghorns. For eggs of New Hampshire pullets, the mantle thickness can be as high as 0.172 μm. The inner membrane has three layers of fibers, which are parallel to the shell and at right angles to each other. On the other hand, the outer membrane has six layers of fibers oriented alternately in different directions.

According to Britton and Hale (1977), the proteins of shell membranes have a high content of arginine, glutamic acid, methionine, histidine, cystine, and proline, but compared to cuticle or matrix proteins, the membrane proteins are relatively low in glycine (Baker and Balch 1962). The membrane proteins may be classed as keratins, since no hydroxyproline and high concentrations of sulfur-containing amino acids were found.

COMPOSITION OF ALBUMEN AND YOLK

Albumen

Albumen or egg white is made up of outer thin white, thick white, inner thin white and chalaziferous layers (inner thick white). The proportions and moisture contents of the various layers are presented in Table 6.1. The proportion of the layers has been found to vary widely (Almquist and Lorenz 1933). The variability is dependent on the breed, environmental conditions, size of the egg, and rate of production (Romanoff and Romanoff 1949). The proportion of thick white (dense albumen) to the other albumen layers in eggs from an individual hen is fairly constant.

Water is the major constituent of albumen. Romanoff and Romanoff (1949) reported that the moisture content decreased from the outer to inner albumen layers (Table 6.1). The total solids content of albumen as a whole ranges from 11 to 13%. Marion et al. (1964) reported mean total solids contents of 11.8 and 11.3% for two flocks over a 1-year production period. According to Rose et al. (1966), the mean solids content of albumen from eggs laid by hens in 20 flocks over one year was 12.11%. The total solids content of albumen is dependent on the strain and age of the hens (Cotterill et al. 1962; Kline et al. 1965; Marion et al. 1964; Rose et al. 1966).

The proximate analysis of albumen is presented in Table 6.2. Protein is the only major constituent. The age of the bird is highly significant in its effect on the protein content of albumen (Cunningham et al. 1960).

TABLE 6.1. Proportion and Moisture Content of Albumen Layers

| Layer | % Albumen | | % Moisture |
	Mean	Range	
Outer thin white	23.2	10–60	88.8
Thick white	57.3	30–80	87.6
Inner thin white	16.8	1–40	86.4
Chalaziferous (including chalazas)	2.7		84.3

SOURCE: Romanoff and Romanoff (1949).

TABLE 6.2. Composition of Albumen, Yolk, and Whole Egg

Egg component	% Protein	% Lipid	% Carbohydrate	% Ash
Albumen	9.7–10.6	0.03	0.4–0.9	0.5–0.6
Yolk	15.7–16.6	31.8–35.5	0.2–1.0	1.1
Whole egg	12.8–13.4	10.5–11.8	0.3–1.0	0.8–1.0

SOURCES: *Brooks and Taylor (1955), Chung and Stadelman (1965), Cunningham et al. (1960), Le Clerc and Bailey (1940), Marion et al. (1965), Romanoff and Romanoff (1949), Smith et al. (1954).*

Chung and Stadelman (1965) found that the protein content of albumen increased by 0.09 g for each gram increase in egg weight. The amount of lipid in yolk-free albumen is negligible.

The carbohydrates of the albumen exist in both free and combined forms with protein. According to Romanoff and Romanoff (1949), the amount of free carbohydrate, usually present as glucose, was 0.4% of the albumen and 0.5% was present in glycoproteins that contained mannose and galactose units. Tunmann and Silberzahn (1961) reported that 98% of uncombined carbohydrate in albumen was glucose and the total amount was 0.5%.

A wide variety of inorganic elements exist in the dissociated and bound forms. The amounts of some of the more important elements in albumen are given in Table 6.3. The elemental composition of albumen is extremely variable. Although the mineral content of the hen's diet is the most important factor influencing the amount of specific minerals in albumen, other factors such as environment, temperature, season, and age of bird are involved (Cunningham *et al.* 1960; Romanoff and Romanoff 1949).

Yolk

The vitelline membrane, surrounding yolk, is made up of two main layers; (1) the inner layer, which was formed in the ovary, and (2) the

TABLE 6.3. Elemental Composition of Albumen and Yolk

Element	% in Albumen[a,b]	% in Yolk[b,c,d]
Sulfur	0.195	0.016
Potassium	0.145–0.167	0.112–0.360
Sodium	0.161–0.169	0.070–0.093
Phosphorus	0.018	0.543–0.980
Calcium	0.008–0.02	0.121–0.262
Magnesium	0.009	0.032–0.128
Iron	0.0009	0.0053–0.011

SOURCES: [a]*Cunningham et al. (1960)*, [b]*Romanoff and Romanoff (1949)*, [c]*Chang (1969)*, [d]*Schmidt et al. (1956).*

outer layer, which was deposited in the oviduct (Bellairs *et al.* 1963). The two layers are separated by a thin continuous membrane. Fromm (1967) noted that the surface of the vitelline membrane in fresh eggs is composed of fibers connected to the chalaziferous layer. This fibrous network disappears during the storage of eggs at 95°F (35°C) for 3 days. The strength of the vitelline membrane decreases as the egg ages (Fromm 1964; Fromm and Martone 1962; Moran 1936). According to Heath (1976), the dry weight of the vitelline membrane increased when eggs were stored at a refrigerated temperature of 45°F (7°C) but decreased at an egg storage temperature of 72°F (22°C). Weight increase of the membrane may be attributed to fiber protein-thick albumen protein interaction. Although the increase in the membrane weight did not increase the strength of the vitelline membrane, the water movement into the yolk was reduced (Heath 1976). The strength of vitelline membrane can be assessed by measuring the yolk height (YH) and yolk width (YW) to obtain a Yolk Index, YH/YW (Fromm and Martone 1962).

The total solids content of yolk is generally about 50%. According to Marion *et al.* (1964) and Rose *et al.* (1966), yolk solids are influenced by age of the layer. Marion *et al.* (1964) indicated that, with fresh eggs from three strains of hens, the mean yolk solids content was 52.29%. A mean of 52.7% for yolk solids was reported by Kline *et al.* (1965) for fresh eggs from five commercial strains of birds. During the storage of shell eggs, water migrates from the albumen to the yolk, and as a consequence the yolk solids content decreases. Rose *et al.* (1966) calculated the solids in yolk from shell eggs stored at 39°F (4°C) for one week to be 50.09% throughout an 11-month period. Fromm (1966) found that yolk solids dropped from 53.5 to 49% when eggs were stored at 75°F (24°C) for 16 days. When eggs were stored at 34° to 39°F (1° to 4°C), the solids content of yolk dropped from about 52.8 to 50% (Meyer and Woodburn 1965).

Heath (1975) found that the yolk from eggs stored at 82°F (28°C) for 7 days had a moisture content of 47.3% compared to 46.1% for fresh yolk. When eggs were stored at 43°F (6°C) for 3 days, the moisture content of the yolk was significantly lower (2.8%) than that of fresh yolk, but after 7 days of storage, the moisture content rose to about the same level as that of fresh yolk.

The major constituents of the solid matter in yolk are proteins and lipids. The relative amounts of yolk constituents are presented in Table 6.2.

The protein content of egg yolk has been reported to be 15.7% by LeClerc and Bailey (1940) and 16.6% by Romanoff and Romanoff (1949). Using the %N data of Kline *et al.* (1965), the protein content of yolk with 50% solids would be 16.15%.

The lipid content of yolk from eggs of various strains of hens is between 32 and 36%. Variability has been attributed primarily to strain rather than diet (Marion *et al.* 1965; Romanoff and Romanoff 1949). The average lipid values obtained by Marion *et al.* (1965) for yolk from three strains of hens were 35.50, 32.67 and 31.95%. Using yolk from eggs collected over a 11-month period from White Leghorn hens of the same strain and age, Chung and Stadelman (1965) found lipid contents between 32 and 33.5%.

According to Privett *et al.* (1962), the composition of yolk lipid is 65.5% triglyceride, 28.3% phospholipid and 5.2% cholesterol. Rhodes and Lea (1957) calculated the composition of yolk phospholipid as: 73.0% phosphatidylcholine, 15.0% phosphatidylethanolamine, 5.8% lysophosphatidylcholine, 2.5% sphingomyelin, 2.1% lysophosphatidylethanolamine, 0.9% plasmalogen, and 0.6% inositol phospholipid.

The fatty acid composition of yolk lipid is influenced by the type of fat in the diets of the hens. Research has shown that the total amount of saturated fatty acids, mainly palmitic and stearic acids, does not change even with a large alteration in dietary fatty acid composition, but that the linoleic acid content of yolk increases with a concurrent decrease in oleic acid when the level of dietary polyunsaturated fatty acids is raised (Chen *et al.* 1965; Coppock and Daniels 1962; Evans *et al.* 1961; Fisher and Leveille 1957; Skellon and Windsor 1962). The total amount of palmitic and stearic acids in yolk lipid is generally between 30 and 38% (Table 6.4). According to Donaldson (1966), the concentration of total saturated fatty acids was 37% when the hens were on a fat-free diet.

The fatty acid composition of lipid fractions of yolk is presented in Table 6.4. It is of interest to note that the total amount of palmitic and stearic acids in the triglyceride fraction was about 30%, whereas these acids in phospholipids amounted to about 49% for lecithin and about 54% for cephalin. Oleic and linoleic acid contents of the triglyceride fraction were much higher than those for the phospholipids.

TABLE 6.4. Fatty Acid Composition of Lipid Fractions of Yolk and Dietary Lipid

Fatty acid	% of total fatty acids				
	Crude lipid	Triglyceride	Lecithin	Cephalin	Dietary lipid
16:0	23.5	22.5	37.0	21.6	14.0
16:1	3.8	7.3	0.6	trace	2.7
18:0	14.0	7.5	12.4	32.5	2.4
18:1	38.4	44.7	31.4	17.3	29.1
18:2	16.4	15.4	12.0	7.0	44.4
18:3	1.4	1.3	1.0	2.0	3.2
20:4	1.3	0.5	2.7	10.2	0.8
20:5					
22:5	0.4	0.2	0.8	3.0	0.8
22:6	0.8	0.6	2.1	6.4	1.3

SOURCE: *Privett et al. (1962).*

Becker *et al.* (1977) reported that the mean cholesterol value for yolks from eggs collected over 5 years from White Leghorns was 30.36 mg per gram of dry yolk with a standard error of 0.05. One yolk on the average contained about 226 mg of cholesterol. The cholesterol content of the yolk is influenced by the genotype of the birds. According to Harris and Wilcox (1963), yolk cholesterol per gram of wet yolk differed between dam families within sires in a random-bred White Leghorn strain. However, the differences were small. Turk and Barnett (1971) found that cholesterol concentration in the egg was not influenced by the age of the hen.

The amount of carbohydrate in yolk may be as high as 1.0% (Table 6.2). In the studies of Tunmann and Silberzahn (1961), the content of free carbohydrate was estimated to be 0.2% as glucose. Romanoff and Romanoff (1949) stated that free and combined carbohydrates were present in yolk at the 0.7 and 0.3% levels, respectively. Protein-bound carbohydrates in yolk are considered to be mannose-glucosamine polysaccharides.

The content of inorganic elements in yolk is about 1.1% as ash (Table 6.2). The major elements in yolk are phosphorus, calcium, and potassium (Table 6.3). A considerable amount of phosphorus (probably as phosphate) and potassium are in the dissociated form (Chang 1969).

PHYSICOCHEMICAL PROPERTIES OF ALBUMEN AND YOLK

Viscosity of Albumen and Yolk

The viscosity of egg albumen is dependent on the age, mixing treatment, temperature and rate of shear (Romanoff and Romanoff 1949; Tung *et al.* 1969, 1970). Using a narrow-gapped rotational viscometer, Tung *et al.* (1970) reported that unmixed albumen was pseudoplastic at 36°F (2°C) between shear rates of 8.1 to 147 sec^{-1}, and the shear stress-shear rate relationship is described by the power law and Casson models. With a constant shear rate, the albumen viscosity decreased with time and approached equilibrium in a few minutes. Tung *et al.* (1969) found that mixed albumen (forced through a #3 Buchner funnel six times) also displayed a pseudoplastic behavior at 50°F (10°C), 68°F (20°C) and 86°F (30°C). The plot of apparent viscosity in centipoises vs shear rate at 50°F (10°C) is shown in Fig. 6.2 for 12 samples of albumen with solids content of 12.92% and a pH of 8.4. With a shear rate of 24 sec^{-1}, mixed albumen had an apparent viscosity between about 15 and 27 centipoises, but a value of 18 centipoises is obtained from the curve in

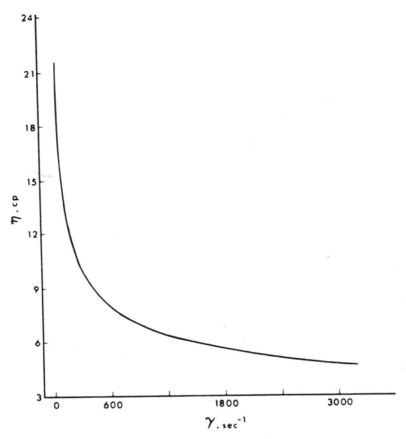

FIG. 6.2.   Apparent viscosity (centipoises) of an albumen at 50°F
(10°C) at various shear rates.
SOURCE: *Tung et al. (1969).*

Fig. 6.2. When the shear rate was increased to 3140 sec $^{-1}$, the apparent
viscosity of the mixed albumen dropped to about 5 centipoises.

Egg yolk is a pseudoplastic non-Newtonian fluid; Chang (1969) re-
ported that the shear stress (dynes/cm^2 or 10^{-1} Pascals) – shear rate (sec^{-1})
relationship was nonlinear. The particulate matter (granules) in the
yolk must be responsible for this nonlinearity, since plasma (yolk without the granules) is essentially a Newtonian fluid. Chang *et al.* (1970B),
using a Wells-Brookfield microviscometer (cone plate), demonstrated
that the apparent viscosity of native yolk (52.5% solids) at 77°F (25°C)
was dependent on the rate of shear. The apparent viscosity dropped
gradually from 23 to 18 poises with an increase in shear rate from 1.9 to

76.8 sec^{-1}. Kline *et al.* (1965), using a Brookfield spindle viscometer, reported that yolk at 77°F (25°C) with a solids content of 52.5 and 50.8% had viscosities of 1,462 and 782 centipoises, respectively. The spindle size and rate of spindle rotation were not reported. When Meyer and Woodburn (1965) used the Brookfield viscometer with a spindle No. 5 and a constant spindle speed of 20 rpm, the apparent viscosities of yolk with solids contents of about 52.7 and 50% were 27 and 9 poises, respectively. When thin albumen was added to yolk at 5, 10 and 20% levels, Chang *et al.* (1970B) noted a marked reduction in apparent viscosity. For example, the viscosity of yolk with 5% thin albumen (50.9% solids) at 77°F (25°C) was 9.5 poises with a shear rate of 1.9 sec^{-1}, as compared to a viscosity of 23 poises for native yolk. Yolk with 20% albumen (44.7% solids) had an apparent viscosity of 2 poises at 77°F (25°C).

Surface Activity of Albumen and Yolk

Since proteins and phospholipids are capable of lowering the surface and interfacial tensions, it is not surprising that albumen and yolk have surface tensions below 72 dynes/cm at 77°F (25°C), the surface tension of water. According to Peter and Bell (1930), the surface tension of a 12.5% solution of albumen at pH 7.8 was 49.9 dynes/cm (74°F, 24°C).

Vincent *et al.* (1966) reported that yolk, plasma and a 4.8% livetin solution (all at 77°F, 25°C) had surface tensions of about 44 dynes/cm and interfacial tensions (at water-cottonseed oil interface) of about 5 dynes/cm. Extensive dilution of yolk and plasma by either water or 10% NaCl caused a significant change in the surface activities. The low-density lipoprotein fraction of yolk lowered the surface tension to about 42 dynes/cm when the concentration was as low as 0.25%.

pH of Albumen and Yolk

The pH of albumen from a newly laid egg is between 7.6 and 8.5 (Brooks and Taylor 1955; Heath 1975, 1977; Romanoff and Romanoff 1949; Sharp and Powell 1931). During the storage of shell eggs, the pH of albumen increases at a temperature-dependent rate to a maximum value of about 9.7 (Heath 1977; Sharp and Powell 1931). After 3 days of storage at 37.4°F (3°C), Sharp and Powell (1931) found that the pH of albumen was 9.18. After 21 days of storage, the albumen had a pH close to 9.4, regardless of the storage temperature between 37.4°F (3°C) and 95°F (35°C).

Heath (1977) observed that when carbon dioxide loss was prevented by the oiling of egg shells, the albumen pH of 8.3 did not change over a 7-day period of egg storage at 72°F (22°C). With oiled eggs stored at 45°F

(7°C), albumen pH dropped from 8.3 to 8.1 in 7 days. Hickson *et al.* (1982) indicated that the pH of egg albumen is a major factor in controlling the rheological properties of gels formed during the heat treatment of albumen at 176°F (80°C).

The rise in the albumen pH is caused by a loss of carbon dioxide from the egg through the pores in the shell. The pH of albumen is dependent on the equilibrium between the dissolved carbon dioxide, bicarbonate ion, carbonate ion, and protein. The concentrations of the bicarbonate and carbonate ions are governed by the partial pressure of carbon dioxide in the external environment (Brooks and Pace 1938). Table 6.5 shows the influence of the percentage of carbon dioxide in the gaseous environment on the pH and concentrations of bicarbonate and carbonate ions. With an increase in the concentration of carbon dioxide in the environment, the concentration of bicarbonate ions increased as the carbonate concentration decreased.

The pH of yolk in freshly laid eggs is generally about 6.0, but during storage of eggs, the pH gradually increases to between 6.4 and 6.9 (Sharp and Powell 1931; Brooks and Taylor 1955). At storage temperatures of 36°F (2°C) and 99°F (37°C), the yolk reached a pH value of 6.4 in about 50 days and 18 days, respectively (Sharp and Powell 1931).

PROTEINS IN ALBUMEN

Albumen may be regarded as a protein system consisting of ovomucin fibers in an aqueous solution of numerous globular proteins. Several excellent review articles have been written on albumen proteins by Baker (1968), Feeney (1964), Osugo and Feeney (1974, 1977), Parkinson (1966), Vadehra and Nath (1973), and Warner (1954). The protein compositions of the thin and thick layers of albumen are different only in the ovomucin content (Forsythe and Foster 1949; Lanni *et al.* 1949).

TABLE 6.5. Influence of Carbon Dioxide Content in the Gaseous Environment on the pH and Concentrations of Bicarbonate and Carbonate Ions of Albumen at 25°C

% CO_2 in environment	pH of albumen	G-ions per liter	
		Bicarbonate	Carbonate
0.03 (air)	9.61	0.0205	0.0104
1	8.43	0.0448	0.0015
3	7.99	0.0490	0.0006
5	7.78	0.0505	0.0004
10	7.50	0.0528	0.0002
97	6.55	0.0580	—

SOURCE: *Brooks and Pace (1938).*

The principal protein fractions of albumen have been separated by the stepwise addition of ammonium sulfate (Warner 1954). The ion-exchange techniques using carboxymethylcellulose and diethylami-noethylcellulose have been used for the fractionation and purification of albumen proteins (Mandeles 1960; Rhodes *et al.* 1958). The albumen proteins and their characteristics are presented in Table 6.6. The major proteins are regarded as ovalbumin, conalbumin, ovomucoid, lysozyme, globulins and ovomucin.

Several electrophoretic methods have been used to separate the proteins of albumen and albumen fractions. According to Longsworth *et al.* (1940) and Forsythe and Foster (1949), the proteins of albumen were resolved by moving-boundary electrophoresis into seven major peaks, including ovalbumin A_1 and A_2, globulins G_2 and G_3, ovomucoid and conalbumin. Using paper electrophoresis, Evans and Bandemer (1956) obtained distinct bands of ovalbumin, ovomucoid plus ovoglobulin, conalbumin, and lysozyme. Using starch-gel electrophoresis, Lush (1961), Stevens (1961), and Feeney *et al.* (1963) separated the proteins of albumen into as many as 19 bands. Variation in the number of bands for albumen from various strains and inbred lines has been attributed to genetic polymorphism (Feeney *et al.* 1963). Globulins, in particular, were found to be genetically controlled. Polyacrylamide disc-gel electrophoresis was capable of resolving albumen proteins into 12 bands (Chang *et al.* 1970A; Galyean and Cotterill 1979; Matsuda *et al.* 1981).

Ovalbumin

Ovalbumin, the predominant protein in albumen, is classed as a phosphoglycoprotein, since carbohydrate and phosphate moieties are attached to the polypeptide. The complete amino acid sequence of hen's ovalbumin with 385 residues has been determined by Nisbet *et al.* (1981). The N-terminal amino acid is acetylated glycine and the C-terminal amino acid is proline. The molecular weight of the polypeptide chain is 42,699.

Ovalbumin can be obtained by fractionation with ammonium sulfate or sodium sulfate (Warner and Weber 1951, Warner 1954) and by chromatography (Rhodes *et al.* 1958; Mandeles 1960). Purified ovalbumin, having a molecular weight of about 45,000, is made up of 3 components, A_1, A_2, and A_3, which differ in phosphorus content (Cann 1949; Longsworth *et al.* 1940; Perlman 1952). Ovalbumin A_1 has two phosphates per molecule, ovalbumin A_2 has one, and ovalbumin A_3 has none. The relative proportion of the A_1, A_2 and A_3 components in the albumen fraction is about 85:12:3. The two phosphate groups are attached to serine residues, 68 and 344 (Nisbet *et al.* 1981).

Recent studies have indicated that ovalbumin molecule contains four

TABLE 6.6. Proteins in Egg Albumen

Protein	Amount of albumen (%)	pI[a]	Molecular weight	Amino acid residues[b]	T_d[c] (°C)	Characteristics
Ovalbumin	54	4.5	45,000	385	84.0	Phosphoglycoprotein
Ovotransferrin	12	6.1	76,000	—	61.0	Binds metallic ions
Ovomucoid	11	4.1	28,000	185	70.0	Inhibits trypsin
Ovomucin	3.5	4.5–5.0	$5.5–8.3 \times 10^6$	—	—	Sialoprotein, viscous
Lysozyme	3.4	10.7	14,300	129	75.0	Lyzes some bacteria
G_2 Globulin	4.0?	5.5	$3.0–4.5 \times 10^4$	—	92.5	—
G_3 Globulin	4.0?	4.8	—	—	—	—
Ovoinhibitor	1.5	5.1	49,000	—	—	Inhibits serine proteases
Ficin inhibitor	0.05	5.1	12,700	—	—	Inhibits thioproteases
Ovoglycoprotein	1.0	3.9	24,400	—	—	Sialoprotein
Ovoflavoprotein	0.8	4.0	32,000	—	—	Binds riboflavin
Ovomacroglobulin	0.5	4.5	$7.6–9.0 \times 10^5$	—	—	Strongly antigenic
Avidin	0.05	10	68,300	512	—	Binds biotin

[a] Isoelectric point.
[b] Number of residues in the published sequence.
[c] Denaturation temperature.

110

sulfhydryl groups, three of which are reactive to p-chloromercuribenzoate in native protein and the fourth in the denatured protein (Feeney 1964; Fernandez-Diez et al. 1964; MacDonnell et al. 1951). Fothergill and Fothergill (1970) found that, in addition to the four sulfhydryl groups, ovalbumin contained one disulfide group per molecule.

One carbohydrate moiety is attached to the polypeptide chain through asparagine residue 292 (Lee and Montgomery 1962; Nisbet et al. 1981). The carbohydrate moiety has a molecular weight between 1,560 and 1,580 (Lee et al. 1964; Montgomery et al. 1965) and consists of a core of two N-acetylglucosamine units and four mannose units with variable numbers of additional residues of the same sugars.

Smith (1964) and Smith and Back (1965) have shown that ovalbumin is converted to S-ovalbumin, a more heat-stable protein, during the storage of eggs. The denaturation temperature of native ovalbumin and S-ovalbumin are 84.5° and 92.5°C, respectively. An intermediate specie has a denaturation temperature of 88.5°C. Kurisaki et al. (1982) concluded that since there was no difference in electrophoretic and ion-exchange chromatographic behaviors for fresh and stored ovalbumin, the increased heat stability of S-ovalbumin could not be explained by an increase in net negative charge. Smith and Back (1965) suggested that sulfhydryl-disulfide interaction may be involved in the ovalbumin → S-ovalbumin transition. Nakamura and Ishimaru (1981) reported decreases in intrinsic viscosity and Stokes radius. They interpreted these changes to mean a greater compactness of ovalbumin molecules in stored albumen than in ovalbumin.

Ovalbumin in solution is readily denatured and coagulated by exposure to new surfaces (e.g., shaking) but is resistant to thermal denaturation (Warner 1954). According to Lineweaver et al. (1967), heating of albumen at pH 9 to 62°C for 3.5 min altered only 3 to 5% of the ovalbumin, whereas negligible amounts of this protein were changed by heating albumen with a pH of 7. Raman spectral shifts in the conformationally sensitive amide I and II lines of heated ovalbumin in solution indicated the development of intramolecular B-sheets in the thermal aggregation events (Painter and Koenig 1976). The heat sensitivity of ovalbumin in albumen has been discussed by Chang et al. (1970A), Ma and Holme (1982), and Matsuda et al. (1981).

Conalbumin (Ovotransferrin)

Conalbumin and ovotransferrin are synonymous terms. Conalbumin, a glycoprotein, is easily prepared by fractional precipitation with ammonium sulfate or separated from ovalbumin by shaking the protein solution until the ovalbumin coagulates (Azari and Baugh 1967; Warner 1954). Longsworth et al. (1940) considered conalbumin as a single protein, but when this fraction was examined by starch-gel electrophoresis, two forms were evident in the approximate ratio of 4:1 (Clark et al. 1963; Feeney et al. 1963). Rhodes et al. (1958) separated two conalbumin fractions from egg albumen by ion-exchange chromatography. Conalbumin with a molecular weight in the vicinity of 76,000 and a pI of 6 (Table 6.6) contains no phosphorus or free sulfhydryl groups.

Williams (1962A,B) has shown that the protein moieties of conalbumin and transferrin of chicken blood serum are identical, but the carbohydrate prosthetic groups are different. Most of the carbohydrate in ovotransferrin is in the form of a single oligosaccharide chain with four residues of mannose and eight residues of N-acetylglucosamines.

Conalbumin is more heat-sensitive than ovalbumin but less susceptible to surface denaturation (Warner 1954). Cunningham and Lineweaver (1965) showed that conalbumin in phosphate-bicarbonate buffer at the 1% level had minimum heat stability near pH 6. At pH 6, the conalbumin in buffer was altered to the extent of about 40% upon heating to 57°C for 10 min. When the conalbumin solution was adjusted to pH 9 and heated under the same conditions, the protein was not altered significantly. The heat stability of conalbumin in albumen was similar to that for purified conalbumin in buffer (Chang et al. 1970B; Cunningham and Lineweaver 1965; Seideman et al. 1963). However, the thermal stability of conalbumin is increased when albumen is mixed with yolk (Chang et al. 1970A; Torten and Eisenberg 1982; Woodward and Cotterill 1983).

Alderton et al. (1946) demonstrated that di- and trivalent metallic ions are bound firmly by conalbumin. Two atoms of Fe (III), Al (III), Cu (II) and Zn (II) per molecule of protein form stable complexes with conalbumin above pH 6 (Feeney 1964). These complexes are red, colorless, yellow, and colorless, respectively. When conalbumin is complexed with metallic ions, the complexes are resistant to thermal denaturation and proteolytic attack (Azari and Feeney 1958, 1961; Cunningham and Lineweaver 1965). Warner and Weber (1953) and Wishnia and Warner (1961) have suggested that iron is bound by three phenolic hydroxyl groups of protein as well as by one carbonate ion; two nitrogen groups may also be involved in the chelation (Windle et al. 1963).

Ovomucoid

Ovomucoid is a heat-resistant glycoprotein that can be prepared by first precipitating out the other albumen proteins with trichloroacetic acid at pH 3.5 and then precipitating ovomucoid with acetone (Lineweaver and Murray 1947). Ovomucoid can be precipitated out of solution upon saturation with ammonium sulfate (Warner 1954). Its molecular weight is approximately 28,000 and the isoelectric point is about 4.1 (Table 6.6). Ovomucoid was found by Lineweaver and Murray (1947) to be a trypsin inhibitor. Less than one molecule of ovomucoid is required to reduce the activity of one molecule of trypsin by 50%.

Two forms of ovomucoid were separated by moving-boundary electrophoresis (Longsworth et al. 1940; Fredericq and Deutsch 1949). According to Rhodes et al. (1960), three forms of ovomucoid with different contents of sialic acid were separated by column chromatography with buffers of different pH values.

With starch-gel electrophoresis, three discrete bands of ovomucoid were obtained by Wise et al. (1964) and Melamed (1967). In addition to the three forms of ovomucoid, the TCA-precipitated ovomucoid fraction contains ovoinhibitor which can be purified by column chromatography (Tomimatsu et al. 1966).

Montgomery (1970) has reported the carbohydrate composition of chicken ovomucoid: D-galactose, 1.0 to 1.5%; D-mannose, 4.3 to 4.7%; 2-amino-2-deoxy-D-glucose, 12.5 to 15.4%; sialic acid, 0.4 to 4% and total hexoses, 6 to 9%. The carbohydrate is present as three oligosaccharides, each joined to the polypeptide chain by an asparaginyl residue (Montgomery and Wu 1963). This glycoprotein is a single polypeptide chain consisting of three separate domains, each cross-linked by three disulfide bonds. The polypeptide chain consists of 26% α-helical structure, 46% β-structure, 10% β-turns and 18% random coil (Watanabe et al. 1981). In acidic solutions, ovomucoid is very resistant to heat denaturation, but in the alkaline region (pH 9) this protein is altered rapidly at 80°C (Lineweaver and Murray 1947). The criteria for thermal denaturation of ovomucoid were a loss of antitryptic activity and an increase in rate of hydrolysis by chymotrypsin. Presumably the helical structure is changed during heat treatment.

Lysozyme

Alderton et al. (1945) reported that the G_1 globulin of Longsworth et al. (1940) was lysozyme, the albumen enzyme that has a lytic action on bacterial cell walls (Johnson 1966). The lysozyme can be crystallized by the method of Alderton et al. (1945). Lysozyme may consist of two or three components that can be separated by cation exchange chromatography (Tallan and Stein 1951, 1953) and moving-boundary electrophoresis (Wetter and Deutsch 1951). The molecular weight of lysozyme is approximately 14,300, but aggregation of monomers occurs between pH 5 and 9 to form dimers (Osuga and Feeney (1974). The primary structure of lysozyme has been successfully elucidated. The enzyme contains 129 amino acid residues and four disulfide bonds (Canfield 1963; Canfield and Liu 1965; Jollès et al. 1964). No free sulfhydryl groups have been found. X-ray diffraction analysis has been used to obtain an insight into the secondary and tertiary structures (Blake et al. 1967).

Lysozyme is capable of hydrolyzing the β(1-4) linkages between N-acetylneuraminic acid and N-acetylglucosamine in bacterial cell walls (Board 1968).

The thermal inactivation of lysozyme as an enzyme is dependent on the pH and temperature (Cunningham and Lineweaver 1965). When lysozyme was dissolved in a phosphate buffer, no inactivation at 145°F (63°C) occurred for 10 min as the pH was raised to 9. However, at 149°F (65°C), the activity of lysozyme in buffer at pH 9 was reduced to about 70% during a 10-min period. Lysozyme is about 50 times more heat-sensitive in egg albumen than in phosphate buffer. In egg white heated to 145°F (63°C) for 10 min, the lysozyme is inactivated to a greater degree as the pH is increased above 7.

Ovomucin

Ovomucin, a glycoprotein, may contribute to the gel-like structure of thick white in the form of flexible, microscopic fibers (Brooks and Hale 1959; Forsythe and Berquist 1951; MacDonnell et al. 1951). The amount of ovomucin in the thick white is about four times greater than that in thin white (Brooks and Hale 1961; Feeney et al. 1952). Ovomucin has the ability to inhibit viral hemagglutination (Gottschalk and Lind 1949; Lanni et al. 1949).

Ovomucin can be precipitated from solution by diluting albumen with two or three volumes of water (pH 6 to 8), or by lowering the pH of albumen to pH 4 (Balls and Hoover 1940). The ovomucin precipitate is soluble in dilute salt solutions at about pH 7, or in alkaline solutions. To minimize the contamination of ovomucin by lysozyme, the lysozyme was removed from the albumen prior to ovomucin precipitation. Up to 85% of the original amount of lysozyme can be crystallized out by adjusting albumen to pH 9.5 and 5% NaCl, and then seeding with isoelectric lysozyme (Gottschalk and Lind 1949).

According to Sharp et al. (1951), the ovomucin fraction can be separated into three components by moving-boundary electrophoresis. Water-precipitated ovomucin is not resolved very effectively by starch-gel electrophoresis. Only one band, moving in the ovalbumin region, is formed in the starch gel (Baker and Manwell 1962; Oades and Brown 1965). When the solubility of ovomucin was increased by reduction with mercaptoethanol, the protein did not migrate on paper or starch gel, but did move as a single component on polyacrylamide gel during electrophoresis (Robinson and Monsey 1964). The carbohydrate content of ovomucin has been reported to be as high as 33%. The unfractionated ovomucin consists of 10 to 12% hexosamine, 15% hexose, and 2.6 to 8% sialic acid (Feeney *et al.* 1960; Gottschalk and Lind 1949; Odin 1951). The purified mercaptoethanol-reduced ovomucin prepared by the method of Robinson and Monsey (1964) had the following carbohydrate composition: 9.3% galactose, 1.4% mannose; 9.1% glucosamine; 4.8% galactosamine, and 8.7% sialic acid.

Two protein fractions with different carbohydrate contents have been separated from ovomucin. A carbohydrate-rich (50%) and a carbohydrate-poor (15%) fraction have been termed F and S, respectively, by Kato and Sato (1971) and β- and α-ovomucins, respectively, by Robinson and Monsey (1971). Smith *et al.* (1974) reported the presence of two types of carbohydrate moieties in ovomucin. Type A is O-glycosidically linked between *N*-acetylgalactosamine and serine or threonine, and type B is attached by an amino linkage between *N*-acetylglucosamine and asparagine. The composition of the oligosaccharides are discussed by Kato *et al.* (1978). Glycoproteins containing O-glycosidically linked carbohydrate moieties tend to form extensive hydrogen bonds with water and to give rise to gels (Dubois *et al.* 1974). Any alteration of the carbohydrate moieties may lead to a loss in gel structure.

The ovomucin in solution is resistant to heat alterations. According to Cunningham and Lineweaver (1965), the solutions of ovomucin between pH 7.1 and 9.4 did not change in viscosity or optical density during heating at about 194°F (90°C) for 2 hr. Ovomucin and lysozyme in solution can interact to form a water-insoluble complex (Cotterill and Winter 1955; Hawthorne 1950; Klotz and Walker 1948). It has been reported that the interaction is electrostatic in nature, involving the negative charges of the terminal sialic acid residues in ovomucin and the positive charges of the lysyl ε-amino groups in lysozyme (Kato *et al.* 1975). Over the pH range of 7.2 to 10.4, the interaction of these proteins decreases as the pH increases (Dam 1967). Cotterill and Winter (1955) noted that, in egg albumen, the amount of complex formation decreased as the pH approached the isoelectric point (10.7) of lysozyme.

Interactions between ovomucin and lysozyme may be involved in the maintenance of the gel structure of thick egg white and the process of egg-white thinning (Kato and Sato 1972; Kato *et al.* 1971, 1975; Robinson 1972; Robinson and Monsey 1972). However, contradictory hypotheses have been advanced.

Hawthorne (1950) has suggested that the thinning of thick albumen is caused, partly at least, by the interaction of ovomucin with lysozyme when the pH rises to around 9 during the normal loss of carbon dioxide from the egg upon storage. According to Feeney *et al.* (1952), the lysozyme activity loss of about 20 to 25% during storage of eggs at 36°F (2°C) for 45 days may be caused by the formation of an insoluble ovomucin-lysozyme complex. On the other hand, Cotterill and Winter (1955) have suggested that thinning is caused by a reduction in the amount of naturally occurring ovomucin-lysozyme complex at a pH of about 9 to 9.5. Miller *et al.* (1982) reported that at an ionic strength of 0.13 (ionic strength of albumen is about 0.1), the extent of lysozyme interaction with native ovomucin was only about 6%. Thus the ovomucin-lysozyme interaction is unlikely to be the cause of albumen thinning. Chemical scission of the *O*-glycosides from the polypeptide chains of ovomucin may bring about albumen thinning. Studies have shown that when eggs are stored at 86°F (30°C) for 20 days, the hexosamine and hexose contents of ovomucin decreased 50%, and the sialic acid dropped to 12% of the original level. The β-elimination of O-glycosidically linked carbohydrates in glycoproteins such as ovomucin can occur under alkaline conditions (Kato *et al.* 1979).

Avidin

Avidin is a glycoprotein that combines with biotin to form a stable complex incapable of absorption by the intestinal tract of animals (Green 1975; Osuga and Feeney 1974). This glycoprotein is composed of four polypeptide subunits each with 28 amino acid residues, and has a molecular weight of about 68,300. (Table 6.6). There is no appreciable α-helical structure in avidin (Green 1975). Huang and De Lange (1971) found four residues of glucosamine and five residues of mannose in each avidin subunit.

Purification procedures for avidin include ammonium sulfate precipitation (Pennington *et al.* 1942), absorption onto bentonite (Fraenkel-Conrat *et al.* 1952) and ion-exchange cellulose column chromatography (Melamed and Green 1963). Crystallization of avidin has been achieved (De Lange 1970; Green and Toms 1970) and its amino acid sequence analysis has been determined (De Lange and Huang 1971). The carbohydrate moiety is attached to asparagine residue number 17 (Green 1975).

Avidin binds four biotin molecules, one per subunit (Green 1964). The complex has a dissociation constant of 10^{-5} M and a free energy change of 20 kcal/mole of biotin bound (Green 1963). The avidin-biotin complex is resistant to denaturation and proteolysis. Pritchard *et al.* (1966) showed that avidin was irreversibly denatured at temperatures higher than 158°F (70°C) but that the complex was stable to 212°F (100°C).

Ovoglobulins

Longsworth *et al.* (1940) demonstrated by moving-boundary electrophoresis that the globulin fraction consisted of three proteins, G_1, G_2, and G_3. G_1 was considered to be lysozyme. The G_2 and G_3 components have been separated by starch-gel electrophoresis and by ion-exchange chromatography. Baker and Manwell (1962) identified the positions of bands G_2 and G_3 on starch-gel electrophoretograms. The G_2 component was isolated by Feeney et al. (1963) by chromatography on DEAE and CM cellulose, along with precipitation by ammonium sulfate. The G_2 isolate has a molecular weight of about 35,000 and an isoelectric point of 5.5. The ovoglobulins have been found by MacDonnell et al. (1955) to be excellent foaming agents.

Ovoinhibitor

According to Matsushima (1958), egg white contains a proteolytic enzyme inhibitor which is distinct from ovomucoid. Rhodes et al. (1960) demonstrated by paper electrophoresis that the ovoinhibitor is a homogeneous protein with negligible sialic acid. Purification of ovoinhibitor can be achieved by chromatography of TCA-precipitated ovomucoid (Tomimatsu et al. 1966). Ovoinhibitor is capable of inhibiting trypsin and chymotrypsin as well as fungal and bacterial proteases (Feeney et al. 1963; Matsushima 1958; Tomimatsu et al. 1966).

Flavoprotein

All the riboflavin in egg albumen is bound in the flavoprotein in a 1:1 ratio (Rhodes et al. 1959). The major function of the riboflavin-binding apoprotein is presumably to ensure transfer of the riboflavin from the blood serum to the albumen. Nishikimi and Yagi (1969) have indicated that the flavin nucleus is buried in the hydrophobic region of the apoprotein and the riboflavin-apoprotein was fixed in a constant configuration at pH levels between 4.5 and 9.0. With the dissociation constant for the complex of $10^{-6} M$, the flavin is bound tightly to the apoproteins (Winter et al. 1967). The apoprotein did not lose any binding ability after the protein solution (pH 7) was heated for 15 min at 212°F (100°C) (Rhodes et al. 1959). In paper and starch-gel electrophoresis the flavoprotein migrates slightly in front of the ovalbumin A band (Baker and Manwell 1962; Rhodes et al. 1959). The apoprotein migrated at the same rate as the flavoprotein in starch gel electrophoresis. Two fractions of apoprotein, separated on cellulose columns by eluting at pH 4.5 and 4.3, had seven and eight phosphate groups per mole, respectively (Rhodes et al. 1959). The molecular weight of the apoprotein is around 32,000 and the isoelectric point is about 4.0 (Table 6.6).

MICROSTRUCTURE OF YOLK PARTICLES

Yolk may be described as a complex system containing a variety of particles suspended in a protein (livetin) solution. The main types of particles are (1) yolk spheres, (2) free-floating drops or granules, (3) profiles or low-density lipo-protein, and (4) myelin figures.

Moran (1925) reported that, with the aid of an optical microscope, round masses with diameters ranging from 25 to 150 μm could be observed. Romanoff and Romanoff (1949) mentioned that the diameters of spheres in yellow yolk were much larger (25 to 150 μm) than the diameters of white yolk spheres (4 to 75 μm). According to Grodzinski (1946), yellow yolk spheres had numerous tightly packed droplets within them, whereas white yolk spheres had only a few droplets in a protein fluid. Bellairs (1961) used phase contrast and electron microscopy to confirm the presence of droplets in white and yellow yolk spheres. Numerous spheres (about 20 μm) were found by Chang (1969) in white yolk, but very few spheres (about 40 μm) were present in yellow yolk. Grodzinski (1946, 1951) considered spheres as osmometers, since each presumably has a superficial semipermeable membrane. A 0.16 M solution of NaCl was found to be isotonic with respect to the spheres. When spheres were immersed in hypotonic solutions of 0.10 to 0.13 M salt, they imbibed water and as a result the membranes broke. Bellairs (1961) noted three types of sphere surfaces in electron micrographs: lamellated capsule (several layers of membrane), unit membrane-like structure, and naked surface. The majority of spheres had naked surfaces.

Free-floating drops or granules in yolk are smaller and more numerous than yolk spheres (Bellairs 1961). In the electron micrographs of Chang *et al.* (1977), granules tended to be circular and ranged in diameter from 0.3 to 1.6 μm. Most of the granules had diameters in the vicinity of 0.5 μm. Bellairs (1961) reported that granules (lipid drops) had diameters of about 2 μm. Chang *et al.* (1977) used a Coulter Counter to estimate the size distribution of hydrated granules. The majority of the hydrated granules had diameters between 0 and 1.7 μm. Electron micrographs of Bellairs (1961) indicated that granules had high- and low-density patches with diameters between 30 and 60 Å. Chang *et al.* (1977) suggested that granules are made up of electron-dense subunits.

When Chang *et al.* (1977) diluted yolk with 0.34 M NaCl, the granules were partially disrupted and the electron-dense subunits could be easily seen. Complete disruption of the granules occurred when 1.71 M NaCl was added to yolk. When isolated granules were treated with 1.71 M NaCl, large and small globules (300 to 1000 Å) and myelin figures (560 to 1300 Å in length) as well as electron-dense masses were evident in the micrographs. Centrifugation of the dispersion of granules in 1.71 M

NaCl yielded 2 floating and 3 sedimenting fractions. As shown by electron microscopy, the floating fractions contained small and large globules as well as myelin figures. Fractions III and V consisted predominantly of fluffy masses (made up of entangled strands to which electron-dense microparticles were attached). The microparticles which were roughly circular had a mean diameter of about 100 Å. In Fraction IV, numerous electron-dense masses interspersed with small globules occurred. Chang *et al.* (1977) have hypothesized that the strands of the fluffy masses are phosvitin complexes and the attached microparticles are lipovitellin. Radomski and Cook (1964A,B) showed that lipovitellins and phosvitin have an affinity for each other and suggested that a phosvitin-lipovitellin complex is a basic unit of the granule.

Bellairs (1961) noted that small (diameters around 250 Å) round profiles were present in the electron micrographs of egg yolk. Chang *et al.* (1977) confirmed the presence of these particles in yolk. These profiles can be considered as low-density lipoprotein (LDL); Sugano and Watanabe (1961) calculated the diameters of hydrated LDL to be between 250 and 310 Å, whereas Martin *et al.* (1964) indicated the diameters of LDL range from 117 to 480 Å.

PROTEINS AND LIPOPROTEINS IN GRANULES AND PLASMA

Schmidt *et al.* (1956) reported that yolk can be separated by high-speed centrifugation into sedimented granules and a clear fluid supernatant called plasma. The granules make up about 19 to 23% of the solids in yolk and about 11.5% of the liquid yolk (Burley and Cook 1961; Saari *et al.* 1964). The moisture content of unwashed granules is about 44% (Burley and Cook 1961). Granules on a moisture-free basis contain about 34% lipid, 60% protein and 6% ash, including 0.5% divalent cations such as calcium (Saari *et al.* 1964). Phospholipid fractions, which represent about 37% of the total lipid, consist essentially of phosphatidyl-choline (82%) and phosphatidylethanolamine (15%).

The major portion of yolk is the plasma (about 78% of the total liquid yolk). The moisture content of plasma is about 49% (Saari *et al.* 1964). On a dry weight basis, plasma consists of 77 to 81% lipid, 2.2% ash and about 18% nonlipid residue which is mostly protein (Saari *et al.* 1964; Schmidt *et al.* 1956).

Separation of yolk proteins and lipoproteins have been achieved by using paper and gel electrophoresis (Evans and Bandemer 1957; McCully *et al.* 1962; Powrie *et al.* 1963). With paper electrophoresis, yolk lipovitellins remained at the origin, whereas low-density lipoproteins, livetins and phosvitin migrated. When disc-gel electrophoresis was used by Chang *et al.* (1970A), 19 bands including livetins, phosvitin (when EDTA was added to yolk), and lipovitellins were formed, but no low-density lipoproteins.

Proteins and Lipoproteins in Granules

According to Burley and Cook (1961), granules are composed of 70% α- and β-lipovitellins, 16% phosvitin and 12% low-density lipoprotein. Phosvitin is a nonlipid phosphoprotein, and the lipovitellins and low-density lipoproteins are lipid-protein complexes. Myelin figures were found in granules as a minor component (Chang *et al.* 1977).

Phosvitin

Mecham and Olcott (1949) isolated phosvitin by diluting yolk with a magnesium sulfate solution to form an insoluble complex. The phosvitin contains 12 to 13% nitrogen and about 10% phosphorus, which represents about 80% of the yolk protein phosphorus (Joubert and Cook 1958B; Mecham and Olcott 1949; Mok *et al.* 1961). This phosphoprotein can be isolated by chromatography (Burley and Cook 1961; Radomski and Cook 1964B; Wallace *et al.* 1966). Connelly and Taborsky (1961) separated phosvitin into two components by chromatography on a DEAE cellulose column and Abe *et al.* (1982) fractionated phosvitin into α- and β-phosvitin by gel filtration on a Sephadex G-200 column.

Using moving-boundary electrophoresis, Bernardi and Cook (1960A) obtained three peaks with isolated phosvitin when the ionic strength of the buffer was 0.3, but only two components were detected when the buffer strength was 0.1. When isolated phosvitin was used for gel electrophoresis, Chang *et al.* (1970A) obtained two diffuse bands. Although phosvitin in yolk cannot be detected on electrophoretograms, one or two phosvitin bands can be observed when ethylenediamine tetraacetate was added to yolk prior to application (Chang *et al.* 1970A; McCully *et al.* 1962). Phosvitin is homogeneous when analyzed by ultracentrifugation.

Lewis *et al.* (1950) and Allerton and Perlmann (1965) analyzed the amino acids in phosvitin and found little or no cysteine, cystine, methionine, tryptophan, or tyrosine. About 31% of the total amino acid residues are serine (Mok *et al.* 1961). The phosphorus in phosvitin is presumably present as phosphoserine (Mecham and Olcott 1949). Williams and Sanger (1959) have shown that at least six phosphoserine residues are grouped together in the polypeptide.

Investigations indicate that phosvitin has a molecular weight between 36,000 and 40,000 (Allerton and Perlmann 1965; Cook 1968, Mok *et al.* 1961). Abe *et al.* (1982) calculated the molecular weights of α- and

β-phosvitin to be 160,000 and 190,000, respectively. A partial amino acid sequence of phosvitin has been reported by Clark (1973). When these phosvitin fractions were subjected to polyacrylamide gel electrophoresis with a buffer containing sodium dodecylsulfate (SDS), the proteins were dissociated into smaller units. The SDS-treated α-phosvitin fraction contained polypeptides with molecular weights of 37,500, 42,500, and 45,000. Dissociation of β-phosvitin by SDS treatment produced a polypeptide with a molecular weight of 45,000. Several minor proteins were detected in the SDS-treated phosvitin fractions. According to Shantz and Dawson (1974), a total of 18 different proteins with molecular weights ranging from 12,400 to 119,000 were separated with SDS-polyacrylamide gel electrophoresis for a phosvitin fraction.

Phosvitin is presumably an elongated molecule with dimensions of about 14 A by 280 Å (Joubert and Cook 1958B). This phosphoprotein forms monodisperse soluble complexes with calcium and magnesium ions at low ionic strengths. Precipitation of phosvitin occurs at a concentration of 0.1 M magnesium sulfate.

Taborsky (1963) reported that ferric ions are bound tightly to phosvitin, yet the phosvitin-iron complex is in solution except when the iron is in excess of the amount which can be maximally bound (P/Fe ratio is about 2 at maximum level). Greengard et al. (1964) consider that phosvitin is the iron-carrier in yolk. Macromolecules can also interact with phosvitin. Radomski and Cook (1964A,B) demonstrated the formation of a lipovitellin-phosvitin complex. Chang et al. (1977) noted by electron microscopy that electron-dense lipovitellin micelles were attached to threadlike phosvitin complexes.

Low-Density Lipoproteins and Myelin Figures

Upon ultracentrifugation of a granule dispersion in 1.71 M NaCl solution, two floating and three sedimenting fractions were formed (Chang et al. 1977). The uppermost pellicle floating fraction (F I) consisted of small and large globules with diameters ranging from 30 to 100 Å and of myelin figures. The subpellicle floating fraction (F II) was made up of small globules (LDLg) with fairly uniform diameters about 27 nm and of myelin figures (45–130 Å). Garland and Powrie (1978A) found that LDLg (low-density lipoproteins from granules) and myelin figures (MF) in F II could be separated effectively by gel filtration on Sepharose 2B. According to Garland and Powrie (1978B), the LDLg fraction from yolk of eggs from 39-week-old hens contained about 84% lipid which consisted of about 3.7% cholesterol, 31% phospholipid, and 65% triglyceride. The total lipid content of LDLg was similar to that for LDL from yolk plasma. The lipid contents of plasma LDL_1 and LDL_2 fractions of Saari et al. (1964) were 89 and 86%, respectively. The cholesterol content of LDL from plasma was reported by Saari *et al.* (1964) to be 3.4% of the total lipid, a value close to that for LDLg.

Chang *et al.* (1977) have indicated that myelin figures (MF) from F II are composed of one or more electron-dense lamellae (fixed with osmium tetroxide) and a less electron-dense core (presumably triglycerides). The lamellae are equidistant from one another with a repeat period of around 45 Å. The yolk MF are similar in structure to the phospholipid myelin figures. Garland and Powrie (1978B) determined the amount of total lipid in yolk MF to be about 86%. About 35% phospholipid and 11.5% cholesterol are present in the lipid fraction. Concanavalin A affinity chromatography of the lipidless MF so-produced retained and unretained fractions with 1 and 4.8% total carbohydrate, respectively (Kocal *et al.* 1980).

Lipovitellins

Alderton and Fevold (1945) prepared a crude lipovitellin fraction from yolk by centrifuging water-diluted yolk, whereupon a high-density fraction was sedimented. Although these authors considered the high-density mass to be lipovitellin, in reality such a fraction is made up of granules (Burley and Cook 1961). Upon the addition of 10% or 2 M NaCl to yolk, granules are dispersed. When the NaCl-yolk dispersion is centrifuged, two floating fractions and two subnatant layers are formed (Bernardi and Cook 1960B; Joubert and Cook 1958A). The floating fractions consist of low-density lipoprotein (LDL), whereas the subnatant layers (high-density fraction, HDL) at the bottom of the tube contain lipovitellins, phosvitin, and livetins (Bernardi and Cook 1960B).

Joubert and Cook (1958A) developed a method of removing most of the other yolk proteins from the lipovitellin fraction. Basically the method consisted of diluting yolk with 0.4 M MgSO$_4$ to disperse granules, adjusting the molarity to 0.4 with MgSO$_4$, centrifuging to remove the floating LDL, diluting with water to 0.2 M MgSO$_4$ to precipitate the phosvitin, and diluting further to precipitate the lipovitellin. The high-density fraction is formed during centrifugation of the 0.4 M MgSO$_4$-yolk dispersion. Bernardi and Cook (1960B) found that lipoprotein isolated from the high-density fraction by a modified method of Joubert and Cook (1958A) could be resolved into two components, α- and β-lipovitellin, by moving-boundary electrophoresis. Chromatography has been used successfully to separate the two components. Bernardi and Cook (1960B) used hydroxyapatite columns, and Burley and Cook (1961) used Dowex-1 columns to remove phosvitin prior to hydroxyapatite chromatography, whereas Radomski and Cook (1964B) used gradient elution from triethylaminoethyl (TEAE)-cellulose columns. The lipid and protein phosphorus contents of α- and β-lipovitellin, isolated by various chromatographic procedures, are presented in Table 6.7. In the three most recent studies mentioned in Table 6.7, the protein phosphorus values for α- and β-lipoproteins were about 0.50 and 0.27%, respectively. Burley and Cook (1962A,B) concluded that, at pH values below 7.0, the lipoproteins existed in the dimer form, but monomers were created when the pH was raised. They found that α- and β-lipovitellins dissociated reversibly about 50% at pH 10.5 and 7.8, respectively. The molecular weights of the dimers of α- and β-lipovitellins are in the vicinity of 400,000 (Bernardi and Cook 1960B).

The lipid contents of the lipovitellins are quite variable and depend on the method of preparation. However, it may be stated that the lipid content of both α- and β-lipovitellins is in the vicinity of 20%, which includes 40% neutral lipid and 60% phospholipid (Martin et al. 1963). The neutral lipid includes about 4.1% free cholesterol and 0.14 to 0.2% of the ester. Each phospholipid fraction consists of about 75% phosphatidylcholine, 18% phosphatidylethanolamine, and 7% sphingomyelin and lysophospholipids.

Proteins and Lipoproteins in Plasma

Plasma is composed of livetins which are lipid-free globular proteins and low-density lipoproteins (LDL) (McCully et al. 1962). The livetins and low-density lipoproteins represent about 10.6 and 66%, respectively, of the total yolk solids.

Livetins.   The livetin fraction was first isolated in the undenatured state by Kay and Marshall (1928) with the procedure of removing lipoprotein, precipitating the livetins by half-saturation with ammonium sulfate, and extracting with an alcohol-ether mixture of 5°F (-15°C). Shepard and Hottle (1949) showed that three components of the livetin fraction could be resolved by moving-boundary electrophoresis. Martin et al. (1957) confirmed the presence of α-, β- and γ-livetins in egg yolk. McCully et al. (1962) noted the presence of three livetin bands on paper electrophoretograms. According to Chang et al. (1970A), 15

TABLE 6.7.   Lipid and Protein Phosphorus Contents of α- and β-Lipovitellins (LV) Isolated by Column Chromatography

Chromatographic column	Lipid content (%)		Protein P (%)	
	α-LV	β-LV	α-LV	β-LV
Hydroxyapatite (HA)	20	20	1.20	0.45
Dowex 1, then HA	22	22	0.50	0.27
TEAE cellulose	14.6	16.5	0.54	0.28

SOURCES: *Bernardi and Cook (1960B), Burley and Cook (1961), Radomski and Cook (1964B).*

bands were present in the disc-gel pattern formed during electrophoresis of the α-, β-livetin fraction. Under the same electrophoretic conditions, only one band was formed with the γ-livetin fraction. Mandeles (1960) obtained nine peaks during chromatography of livetin on a DEAE-cellulose column.

Martin et al. (1957) removed γ-livetin by precipitation from a solution containing 20% isopropanol at 32°F (0°C) or 37% ammonium sulfate. Martin and Cook (1958) purified the γ-livetin by repeated precipitation. The α- and β- livetins were separated by electrophoresis. The analyses of the livetins are reported in Table 6.8. The mean molecular weights of α-, β-, and γ-livetins were reported to be 80,000, 45,000 and 150,000, respectively (Martin et al. 1957; Martin and Cook 1958). The isoelectric point of livetin is about 4.8 to 5.0. The relative amounts of α-, β-, and γ-livetins in yolk have been reported as 2:3:5 by Shepard and Hottle (1949) and as 2:5:3 by Bernardi and Cook (1960A). Using immunological techniques, Mok and Common (1964) and Williams (1962A) indicated that livetins were indeed blood serum proteins of the chicken. Williams (1962A) identified α-livetin as serum albumin, β-livetin as $α_s$-glycoprotein and γ-livetin as γ-globulin.

Low-Density Lipoprotein.   Since the density of low-density lipoprotein is 0.98 (Martin et al. 1959), it is not surprising that this fraction has been successfully isolated by flotation. The supernatant, obtained by Alderton and Fevold (1945) when they centrifuged water-diluted yolk at a high speed to remove granules, contained a low-density lipoprotein fraction. However, the lipoproteins were altered during the isolation by ether extraction of the supernatant. The insoluble lipoprotein (about 40% lipid) between the ether and aqueous layers was termed lipovitellenin (Fevold and Lausten 1946). Several investigators (Weinman 1956; Evans and Bandemer 1957; Turner and Cook 1958) have stressed that lipids are associated with proteins in the native yolk and thus lipovitellenin must be regarded as an artifact. Nichols et al. (1954) isolated low-density lipoprotein (LDL) from egg yolk by a flotation procedure and found the fraction to have a lipid content of 89%.

Turner and Cook (1958) described a flotation procedure for the isolation of the low-density lipoprotein fraction (LDL) from egg yolk. This fraction contained about 80% lipid. Martin et al. (1963) and Sugano and

TABLE 6.8.   Composition of Livetins

	α-Livetin	β-Livetin	γ-Livetin
Nitrogen, %	14.3	14.3	15.6
Hexose, %	—	7	2.6
Hexosamine, %	—	—	1.8

Watanabe (1961) reported that their purified LDL contained about 89 and 84% lipid, respectively. According to Martin *et al.* (1963), the lipid in LDL consists of 74% neutral lipid (including about 4% free cholesterol and 0.2 of the ester) and 26% phospholipid (71 to 76% phosphatidylcholine, 16 to 20% phosphatidylethanolamine, and 8 to 9% sphingomyelin and lysophospholipids).

Ultracentrifugal flotation patterns of the low-density lipoprotein fraction show two peaks (Martin *et al.* 1959; Sugano and Watanabe 1961). Using repeated differential flotation, Martin *et al.* (1964) and Saari *et al.* (1964) isolated two ultracentrifugally distinguishable fractions, LDL_1 and LDL_2. The compositions of these LDL fractions were similar (Table 6.9). The total lipid for the LDL_1 was about 87 to 89%, whereas the LDL_2 had a lipid content of about 83 to 86%. The lipid from the LDL fractions consisted of 25 to 28% phospholipid, 3.3% cholesterol, 5% free fatty acid and the remainder as triglycerides. The average molecular weights of LDL_1 and LDL_2 have been estimated by Martin *et al.* (1964) to be 10 million and 3 million, respectively.

The diameters of LDL spheres range from about 170 to 600 Å with an average of about 300 Å (Chang *et al.* 1977, Kamat *et al.* 1972; Sugano and Watanabe 1961). Kamat *et al.* (1972) found that the polar head groups of lecithin in LDL are hindered in molecular motion because of an interaction presumably with LDL polypeptides. Chang *et al.* (1977) suggested that the proteins and phospholipids of LDL are grouped together in small particles (45 Å) which are adsorbed to the surfaces of triglyceride cores. These small particles (ultraparticles) can be released from the triglyceride cores by treatment with pepsin.

Lipidless plasma LDL (apo LDL or vitellenin) has been investigated by several researchers (Evans *et al.* 1973; Hillyard *et al.* 1972; Martin *et al.* 1959, 1963; Yamauchi *et al.* 1976). According to Yamauchi *et al.* (1976), who used SDS-polyacrylamide gel electrophoresis, 18 polypep-

TABLE 6.9. Composition of Low-Density Lipoproteins

	LDL_1	LDL_2
Total lipid, %	89	86
Lipid phosphorus	1.0	1.14
Lipid molar N/P ratio	1.04	1.14
Cholesterol, % of lipid	3.3	3.4
Free fatty acid, % of lipid	5	4.5
Phospholipid composition, %		
Phosphatidyl choline	84	83
Phosphatidyl ethanolamine	13	14
Nonlipid residue, %	11	14
Nitrogen, % of nonlipid residue	14.8	13.7
Phosphorus, % of nonlipid residue	0.15	0.16

SOURCE: *Saari et al. (1964).*

tides were present in the apo-LDL fraction. Two major components were glycoproteins with molecular weights around 71,000 and 135,000. The molecular weights of the other major polypeptides were about 16,000, 62,000, and 82,000. Kamat *et al.* (1972) pointed out that the protein in LDL appears to be predominantly in unordered and antiparallel β-conformations.

CAROTENOIDS IN YOLK

The yellow-orange color of yolk has been attributed to the presence of fat-soluble carotenoids in the lipid portion of lipoproteins (Shenstone 1968). The majority of carotenoids in yolk are hydroxy compounds called xanthophylls with minor amounts of carotenes (Ganguly *et al.* 1953; Smith and Perdue 1966). The types and amounts of carotenoids in yolk are diet dependent, since yolk pigmentation involves intestinal absorption and biotranslocation of feed carotenoids to the ovary (Brown 1938; Ganguly *et al.* 1953; Nelson and Baptist 1968; Scott *et al.* 1968). The major xanthophylls in yolk of commercial eggs are lutein, zeaxanthin, and cryptoxanthin, all of which are derived from commonly used pigmented feed ingredients such as yellow corn, alfalfa meal, and corn-gluten meal (Brockman and Volker 1934; Gillam and Heilbron 1935; Johnson *et al.* 1980; Smith and Perdue 1966). The percentages of various xanthophylls in yellow corn and alfalfa meal are presented in Table 6.10. Lutein is the principal xanthophyll in both plant materials (Bickoff *et al.* 1954) and has also been found to be the dominant carotenoid (about 62% of the total carotenoids) in yolk from eggs of hens fed yellow-corn and alfalfa-meal diets (Smith and Perdue 1966). Zeaxanthin was second highest with a level of 12.9% of the total carotenoids. Gillam and Heilbron (1935) reported that the quantity of carotene, cryptoxanthin, and zeaxanthin in yolk of hens on a yellow-corn diet were 0.015, 0.19, and 1.79 mg per 100 g of yolk, respectively.

The factors that influence the degree of yolk pigmentation include the chemical structure of the xanthophylls, the presence of antioxidants in

TABLE 6.10.   Carotenoids in Alfalfa, Yellow Corn and Dried Algae Meal

Carotenoid	Alfalfa	Yellow Corn	Algae
Lutein	46	54	86
Zeaxanthin	4	23	2
Violaxanthin	16	—	4
Neoxanthin	14	—	—
Cryptoxanthin	7	8	—
Others	13	15	8

SOURCE: *Smith and Perdue (1966).*

feed, and the fat content of feed. In addition, Scott *et al.* (1968) showed that the genetic capability to absorb and deposit xanthophylls in yolk varies among individual hens within a single strain.

The level of deposition of each plant pigment is dependent on the number of hydroxy or keto groups in the molecule. Generally, dihydroxy xanthophylls (e.g., lutein and zeaxanthin) and diketo xanthophylls (e.g., canthaxanthin) are transported to the yolk more efficiently than monohydroxy xanthophylls (e.g., cryptoxanthin) and monoketo xanthophylls (Braeunlich 1978). Peterson *et al.* (1939) noted the following efficiencies of xanthophyll transfer to yolk when carotenoid-low diets were supplemented with purified carotenoids: zeaxanthin, 24.35%; lutein, 11.45%; cryptoxanthin, 0%; and carotene, 0%. However, the deposition of cryptoxanthin in egg yolk was observed when rations high in yellow corn were fed. Brockman and Volker (1934) have shown that lutein and zeaxanthin are deposited in the yolk to a greater extent than cryptoxanthin and carotene. When Ganguly *et al.* (1953) fed purified cryptoxanthin and zeaxanthin to carotenoid-depleted hens, considerable amounts of these xanthophylls were present in the blood, liver, and ovaries. The pigmenting ability of a yellow pigment such as lutein is a straight-line function of the logarithm of the amount in the feed (Morehouse 1961). Very little provitamin A carotenoids are transferred to yolk from feed, since they are converted efficiently to vitamin A.

Fry and Harms (1975) evaluated yolk pigmenting potential of natural feed ingredients by estimating bioavailability as the total amount of xanthophyll deposited in yolk compared to the amount deposited by a standard (β-apo-8'-carotenal). The results indicated that the xanthophyll from alfalfa meal had an average bioavailability of 57.43% as compared to 75.51% and 107.88% for yellow-corn and corn-gluten meal, respectively. Scott *et al.* (1968) indicated that bioavailability of carotenoids in plant material may be restricted by binding in cells. Williams *et al.* (1963) estimated that the total carotenoid content of yolk from eggs of hens on a ration containing either yellow corn, 4% alfalfa meal or 3% corn gluten was about 14 μg per gram. Wheat, barley, and oats as ingredients in the hen's diet do not have any appreciable amounts of depositable xanthophylls.

Several studies have shown that the addition of antioxidants to poultry rations enhances pigmentation of yolk (Russo 1966; Bartov and Bornstein 1966). Biely *et al.* (1962) showed that high levels of dietary vitamin A caused a drop in yellow coloration of yolk, but the addition of 3% beef tallow to the diet eliminated the negative influence of the vitamin. Presumably vitamin A and xanthophyll compete for absorption in the intestinal mucosa under low-fat conditions.

Highly pigmented yolks are required for the manufacture of bakery products, pasta, and mayonnaise. Carlson *et al.* (1964) indicated that to achieve acceptable yolk pigment scores for the egg-breaking industry, 15 to 20% alfalfa meal would have to be included in the diet to supply 25 to 30 mg of xanthophyll per pound of feed. Pigment concentrates, such as dried algae meal and marigold-petal meal, have been developed to reduce the use of the alfalfa meal in the hen's diet (Scott *et al.* 1968. Synthetic red carotenoids, β-apo-8'-carotenal and canthaxanthin, have been studied as egg-yolk pigmenters (Fletcher *et al.* 1978; Marusich *et al.* 1960; Nelson and Baptist 1968). When each is used alone as a pigmenter, the yolk possess an orange to orange-red hue. Proper mixing of the yellow and red xanthophylls in the diet can result in acceptable yellow coloration of yolk.

The measurement of egg-yolk color has been carried out by visual comparison of yolk hue with colored standards, tristimulus reflectance colorimetry, and extraction of yolk with acetone to remove carotenoids for comparison with potassium dichromate standards or for spectrophotometric analysis (Francis and Clydesdale 1972; Fletcher 1980; Fry and Damron 1971; Hinton *et al.* 1973; Vuilleumier 1969).

REFERENCES

Abe, Y., Itoh, T., and Adachi, S. 1982. Fractionation and characterization of hen's egg yolk phosvitin. J. Food Sci. *47*, 1903–1907.

Alderton, G., and Fevold, H. L. 1945. Preparation of the egg yolk lipoprotein, lipovitellin. Arch. Biochem. *8*, 415–419.

Alderton, G., Ward, W. H., and Fevold, H. L. 1945. Isolation of lysozyme from egg white. J. Biol. Chem. *157*, 43–58.

Alderton, G., Ward, W. H., and Fevold, H. L. 1946. Identification of the bacteria-inhibiting iron-binding protein of egg white as conalbumin. Arch. Biochem. *11*, 9–13.

Allerton, S. E., and Perlmann, G. E. 1965. Chemical characterization of the phosphoprotein, phosvitin. J. Biol. Chem. *240*, 3892–3898.

Almquist, H. J., and Lorenz, F. W. 1933. The solids content of egg white. Poult. Sci. *12*, 83–89.

Azari, P., and Baugh, R. F. 1967. A simple and rapid procedure for preparation of large quantities of pure ovotransferrin. Arch. Biochem. Biophys. *118*, 138–144.

Azari, P. R., and Feeney, R. E. 1958. Resistance of metal complexes of conalbumin and transferrin to proteolysis and to thermal denaturation. J. Biol. Chem. *232*, 293–302.

Azari, P. R., and Feeney, R. E. 1961. The resistances of conalbumin and its iron complex to physical and chemical treatments. Arch. Biochem. Biophys. *92*, 44–52.

Baker, C. M. A. 1968. The proteins of egg white. *In* Egg Quality: A Study of the Hen's Egg. T. C. Carter (Editor). Oliver & Boyd, Edinburgh, Scotland.

Baker, C. M. A., and Manwell, C. 1962. Molecular genetics of avian proteins. I. The egg white proteins of the domestic fowl. Br. Poult. Sci. *3*, 161–174.

Baker, J. R., and Balch, D. A. 1962. A study of the organic material of hen's egg shell. Biochem. J. *82*, 352–361.

Balls, A. K., and Hoover, S. R. 1940. Behavior of ovomucin in the liquefaction of egg white. Ind. Eng. Chem. *32*, 594–596.

Bartov, I. and Bornstein, S.   1966.   Studies on egg yolk pigmentation. 2. Effect of ethoxy-guin on xanthophyll utilization. Poult. Sci. *45*, 297–305.

Becker, W. A., Spencer, J. V., Verstrate, J. A., and Mirosh, L. W.   1977.   Genetic analysis of chicken egg yolk cholesterol. Poult. Sci. *56*, 895–901.

Bell, D. J., and Freeman, B. M. (Editors)   1971.   Physiology and Biochemistry of the Domestic Fowl, Vol. 3. Academic Press, New York.

Bellairs, R.   1961.   The structure of the yolk of the hen's egg as studied by electron microscopy. I. The yolk of the unincubated egg. J. Biophys. Biochem. Cytol. *11*, 207–225.

Bellairs, R., Harkness, M., and Harkness, R. D.   1963.   The vitelline membrane of the hen's egg: A chemical and electron microscopical study. J. Ultrastruct. Res. *8*, 339–359.

Bernardi, G., and Cook, W. H.   1960A.   An electrophoretic and ultracentrifugal study on the proteins of the high density fraction of egg yolk. Biochim. Biophys. Acta *44*, 86–96.

Bernardi, G., and Cook, W. H.   1960B.   Separation and characterization of the two high-density lipoproteins of egg yolk, α- and β-lipovitellin. Biochim. Biophys. Acta *44*, 96–105.

Bickoff, E. M., Livingston, A. L., Bailey, G. E., and Thompson, C. R.   1954.   Xanthophylls in fresh and dehydrated alfalfa. J. Agric. Food Chem. *2*, 563–566.

Biely, J., Wood, D., and Topliff, J. E.   1962.   The effect of excessive amounts of dietary vitamin A on egg production in White Leghorn hens. Poult. Sci. *41*, 1175–1177.

Blake, C. C. F., Mair, G. A., North, A. C. T., Phillips, D. C., and Sarma, V. R.   1967.   On the conformation of the hen egg-white lysozyme molecule. Proc. R. Soc. London, Ser. B *167*, 365–377.

Board, R. G.   1968.   Microbiology of the egg: a review. *In* Egg Quality: A Study of the Hen's Egg. T. C. Carter (Editor). Oliver & Boyd, Edinburgh, Scotland.

Braeunlich, K.   1978.   The chemistry and action of pigmenters in poultry diets. World's Poult. Congr., Proc., 15th, 1978 pp. 236–240.

Britton, W. M., and Hale, K. K.   1977.   Amino acid analysis of shell membranes of eggs from young and old hens varying in shell quality. Poult. Sci. *56*, 865–871.

Brockman, H., and Volker, O.   1934.   The yellow pigment of the canary and the occurrence of carotenoids in birds. Hoppe-Seyler's Z. Physiol. Chem. *224*, 193–215.

Brooks, J., and Hale, H. P.   1955.   Strength of the shell of the hen's egg. Nature (London) *175*, 848–849.

Brooks, J., and Hale, H. P.   1959.   The mechanical properties of the thick white of hen's egg. Biochim. Biophys. Acta *32*, 237–250.

Brooks, J., and Hale, H. P.   1961.   The mechanical properties of the thick white of hen's egg. II. The relation between rigidity and composition. Biochim. Biophys. Acta *46*, 289–301.

Brooks, J., and Pace, J.   1938.   The distribution of carbon dioxide in the hen's egg. Proc. R. Soc. London, Ser. B *126*, 196–210.

Brooks, J., and Taylor, D. J.   1955.   Eggs and egg products. G. B., Dep. Sci. Ind. Res., Food Invest. Board, Spec. Rep. *60*.

Brown, W. L.   1938.   The influence of pimento pigments in the color of egg yolk of fowls. J. Biol. Chem. *122*, 655–659.

Burley, R. W., and Cook, W. H.   1961.   Isolation and composition of avian egg yolk granules and their constituents α- and β-lipovitellins. Can. J. Biochem. Physiol. *39*, 1295–1307.

Burley, R. W., and Cook, W. H.   1962A.   The dissociation of α- and β-lipovitellin in aqueous solution. Part I. Effect of pH, temperature, and other factors. Can. J. Biochem. Physiol. *40*, 363–372.

Burley, R. W., and Cook, W. H.   1962B.   The dissociation of α- and β-lipovitellin in aqueous solution. Part II. Influence of protein phosphate groups, sulphydryl groups and related factors. Can. J. Biochem. Physiol. *40*, 373–379.

Cain, C. J., and Heyn, A. N. J.   1964.   X-ray diffraction studies of the crystalline structure of the avian egg shell. Biophys. J. *4*, 23–29.

Canfield, R. E.   1963.   The amino acid sequence of egg white lysozyme. J. Biol. Chem. *238*, 2698–2707.

Canfield, R. E., and Liu, A. K.   1965.   The disulfide bonds of egg white lysozyme (muramidase). J. Biol. Chem. *240*, 1997.

Cann, J. R.   1949.   Electrophoretic analysis of ovalbumin. J. Am. Chem. Soc. *71*, 907–909.

Carlson, C. W., Halverson, A. W., and Kohler, G. O.   1964.   Some effects of dietary pigmenters on egg yolks and mayonnaise. Poult. Sci. *43*, 654–662.

Carter, T. C. (Editor)   1968.   Egg Quality: A Study of Hen's Eggs. Oliver & Boyd, Edinburgh, Scotland.

Chang, C. M.   1969.   Studies on egg yolk. Ph.D. Thesis, University of Wisconsin, Madison.

Chang, P. K., Powrie, W. D., and Fennema, O.   1970A.   Disc-gel electrophoresis of proteins in native and heat-treated albumen, yolk and centrifuged whole egg. J. Food Sci. *35*, 774–778.

Chang, P. K., Powrie, W. D., and Fennema, O.   1970B.   Effect of heat treatment on viscosity of yolk. J. Food Sci. *35*, 864–867.

Chang, C. M., Powrie, W. D., and Fennema, O.   1977.   Microstructure of egg yolk. J. Food Sci. *42*, 1193–1200.

Chen, P. H., Common, R. H., Nikolaiczuk, N., and Macrae, H. F.   1965.   Some effects of added dietary fats on the lipid composition of hen's egg yolk. J. Food Sci. *30*, 838–845.

Chung, R. A., and Stadelman, W. J.   1965.   A study of variations in the structure of the hen's egg. Br. Poult. Sci. *6*, 277–282.

Clark, J. R., Osuga, D. T., and Feeney, R. E.   1963.   Comparison of avian egg white conalbumins. J. Biol. Chem. *238*, 3621–3631.

Clark, R. C.   1973.   Amino acid sequence of a cyanogen bromide cleavage peptide from hen's egg phosvitin. Biochim. Biophys. Acta *310*, 174–187.

Connelly, C., and Taborsky, G.   1961.   Chromatographic fractionation of phosvitin. J. Biol. Chem. *236*, 1364–1368.

Cook, W. H.   1968.   Macromolecular components of egg yolk. *In* Egg Quality: A Study of the Hen's Egg. T. C. Carter (Editor). Oliver & Boyd, Edinburgh, Scotland.

Cooke, A. S., and Balch, D. A.   1970.   Studies of membrane, mammillary cores and cuticle of the hen egg shell. Br. Poult. Sci. *11*, 345–352.

Coppock, J. B. M., and Daniels, N. W. R.   1962.   Influence of diet and husbandry on the nutritional value of hen's eggs. J. Sci. Food Agric. *13*, 459–467.

Cotterill, O. J., and Geiger, G. S.   1977.   Egg product yield trends from shell eggs. Poult. Sci. *56*, 1027–1031.

Cotterill, O. J., and Winter, A. R.   1955.   Egg white lysozyme. 3. The effect of pH on the lysozyme-ovomucin interaction. Poult. Sci. *34*, 679–686.

Cotterill, O. J., Stephenson, A. B., and Funk, E. M.   1962.   Factors affecting the yield of egg products from shell eggs. Proc. World's Poult. Congr. 12, 1962 pp. 443–447.

Cunningham, F. E., and Lineweaver, H.   1965.   Stabilization of egg white proteins to pasteurization temperatures above 60°C. Food Technol. *19*, 136–141.

Cunningham, F. E., Cotterill, O. J., and Funk, E. M.   1960.   The effect of season and age of bird. 2. On the chemical composition of egg white. Poult. Sci. *39*, 300–308.

Dam, R.   1967.   A Review of the Chemistry and Functional Properties of Egg White Proteins, and the Characterization of Ovomucin, Proc. Egg Sci. Semin. Poultry and Egg Natl. Board, Lake Sunapee, New Hampshire.

DeLange, R. J.   1970.   Egg white avidin. 1. Amino acid composition; sequence of amino- and carboxy-terminal cyanogen bromide peptides. J. Biol. Chem. 245, 907–916.

DeLange, R. J., and Huang, T.-S.   1971.   Egg white avidin. 3. Sequence of the 78-residue middle cyanogen bromide peptide. Complete amino acid sequence of the protein subunit. J. Biol. Chem. 246, 698–709.

Donaldson, W. E.   1966.   Fatty acid interconversion by laying hens. Poult. Sci. 45, 473–478.

Dubois, M., Jouannet, P., Berge, P., and David, G.   1974.   Spermatozoa motility in human cervical mucus. Nature (London) 252, 711–713.

El-Boushy, A. R., Simons, P. C. M., and Wiertz, G.   1968.   Structure and ultrastructure of the hen's egg shell as influenced by environmental temperature, humidity and vitamin C additions. Poult. Sci. 47, 456–467.

Evans, R. J., and Bandemer, S. L.   1956.   Separation of egg white proteins by paper electrophoresis. J. Agric. Food Chem. 4, 802–810.

Evans, R. J., and Bandemer, S. L.   1957.   Separation of egg yolk proteins by paper electrophoresis. J. Agric. Food Chem. 5, 868–872.

Evans, R. J., Davidson, J. A., and Bandemer, S. L.   1961.   Fatty acid and lipide distribu- tion in egg yolks from hens fed cottonseed oil from Sterculia foetida seeds. J. Nutr. 73, 282–290.

Evans, R. J., Bauer, D. H., Bandemer, S. L., Vaghefi, S. B., and Flegal, C. J.   1973.   Struc- ture of egg yolk very low density lipoprotein: Polydispersity of the very low density lipoprotein and the role of lipovitellenin in the structure. Arch. Biochem. Biophys. 154 (2), 493–500.

Favejee, J. C., Van Der Plas, Schoorl, R., and Floor, F.   1965.   X-ray diffraction of the crystalline structure of the avian eggshell, some critical remarks. Biophys. J. 5, 359–361.

Feeney, R. E.   1964.   Egg proteins. In Symposium on Foods: Proteins and Their Reac- tions. H. W. Schultz and A. F. Anglemier (Editors). AVI Publishing Co., Westport, CT.

Feeney, R. E., Ducay, E. D., Silva, R. B., and MacDonnell, L. R.   1952.   Chemistry of shell egg deterioration: The egg white proteins. Poult. Sci. 31, 639–650.

Feeney, R. E., Rhodes, M. B., and Anderson, J. S.   1960.   The distribution and role of sialic acid in egg white. J. Biol. Chem. 235, 2633–2637.

Feeney, R. E., Abplanalp, H., Clary, J. J., Edwards, D. L., and Clark, J. R.   1963.   A genetically varying minor protein constituent of chicken egg white. J. Biol. Chem. 238, 1732–1736.

Fernandez-Diez, M. J., Osuga, D. T., and Feeney, R. E.   1964.   The sulfhydryls of avian ovalbumins, bovine β-lactoglobulin and bovine serum albumin. Arch. Biochem. Biophys. 107, 448–458.

Fevold, H. L., and Lausten, A.   1946.   Isolation of a new lipoprotein, lipovitellin, from egg yolk. Arch. Biochem. 11, 1–7.

Fisher, H., and Leveille, G. A.   1957.   Observations on the cholesterol, linoleic and linolenic acid content of eggs as influenced by dietary fats. J. Nutr. 63, 119–129.

Fletcher, D. L.   1980.   An evaluation of the AOAC method of yolk color analysis. Poult. Sci. 59, 1059–1066.

Fletcher, D. L., Harms, R. H., and Janky, D. M.   1978.   Yolk color characteristics, xanthophyll availability and a model system for predicting egg yolk color using beta-apo-8'-carotenal and canthaxanthin. Poult. Sci. 57, 624–629.

Forsythe, R. H., and Berquist, D. H.   1951.   The effect of physical treatments on some properties of egg white. Poult. Sci. 30, 302–311.

Forsythe, R. H., and Foster, J. F.   1949.   Note on the electrophoretic composition of egg white. Arch. Biochem. 20, 161–163.

Fothergill, L. A., and Fothergill, J. E.   1970.   Thiol and disulphide contents of hen ovalbumin: C-terminal sequence and location of disulphide bond. Biochem. J. *116*, 555–561.

Fraenkel-Conrat, H., Snell, N. S., and Ducay, E. D.   1952.   Avidin. I. Isolation and characterization of the protein and nucleic acid. Arch. Biochem. Biophys. *39*, 80–96.

Francis, F. J., and Clydesdale, F. M.   1972.   Miscellaneous. Part V. Egg yolks. Food Prod. Dev. August–September, pp. 47–55.

Frank, F. R., Burger, R. E., and Swanson, M. H.   1965.   The relationships among shell membrane, selected chemical properties and resistance to shell failure of *Gallus domesticus eggs.* Poult. Sci. *44*, 63–69.

Fredericq, E., and Deutsch, H. F.   1949.   Studies on ovomucoid. J. Biol. Chem. *181*, 499–510.

Fromm, D.   1964.   Strength, distribution, weight and some histological aspects of the vitelline membrane from the hen's egg yolk. Poult. Sci. *43*, 1240–1247.

Fromm, D.   1966.   The influence of ambient pH on moisture content and yolk index of the hen's yolk. Poult. Sci. *45*, 374–379.

Fromm, D.   1967.   Some physical and chemical changes in the vitelline membrane of the hen's egg during storage. J. Food Sci. *32*, 52–56.

Fromm, D., and Martone, G.   1962.   A rapid method of evaluating the strength of the vitelline membrane of the hen's egg yolk. Poult. Sci. *41*, 1516–1521.

Fry, J. L., and Damron, B. L.   1971.   Computer calculation of poultry and egg pigmentation data. Food Technol. *25*, 44–45.

Fry, J. L., and Harms, R. H.   1975.   Yolk color, candled grade and xanthophyll availability from dietary natural pigmenting ingredients. Poult. Sci. *54*, 1094–1101.

Galyean, R. D., and Cotterill, O. J.   1979.   Chromatography and electrophoresis of native and spray-dried egg white. J. Food Sci. *43*, 1345–1349.

Ganguly, J., Mehl, J. W., and Deuel, H. J.   1953.   Studies on carotenoid metabolism. XII. The effect of dietary carotenoids on the carotenoid distribution in the tissue of chickens. J. Nut. *50*, 59–72.

Garland, T. D., and Powrie, W. D.   1978A.   Isolation of myelin figures and low-density lipoproteins from egg yolk granules. J. Food Sci. *43*, 592–597.

Garland, T. D., and Powrie, W. D.   1978B.   Chemical characterization of egg yolk myelin figures and low-density lipoproteins isolated from egg yolk granules. J. Food Sci. *43*, 1210–1214.

Gillam, A. P., and Heilbron, I. M.   1935.   Vitamin A-active substances in egg yolk. Biochem. J. *29*, 1064–1067.

Gottschalk, A., and Lind, P. E.   1949.   Ovomucin, a substrate for the enzyme of influenza virus. I. Ovomucin as an inhibitor of haemagglutination by heated Lee virus. Br. J. Exp. Pathol. *30*, 85.

Green, N. M.   1963.   Avidin. 3.   The nature of the biotin-binding site. Biochem. J. *89*, 599–609.

Green, N. M.   1964.   The molecular weight of avidin. Biochem. J. *92*, 16c–17c.

Green, N. M.   1975.   Avidin. Adv. Protein Chem. *29*, 85–133.

Green, N. M., and Toms, E. J.   1970.   Purification and crystallization of avidin. Biochem. J. *118*, 67–70.

Greengard, O., Sentenac, A., and Mendelsohn, N.   1964.   Phosvitin, the iron carrier of egg yolk. Biochim. Biophys. Acta *90*, 406–407.

Grodzinski, Z.   1946.   Influence of the increase in the osmotic pressure upon the white yolk spheres of the hen's egg. Bull. Int. Acad. Sci. Cracovie *87*.

Grodzinski, Z.   1951.   The yolk spheres of the hen's egg as osmometers. Biol. Rev. Cambridge Philos. Soc. *26*, 253–264.

Harris, P. C., and Wilcox, F. H.   1963.   Studies on egg yolk cholesterol. 1. Genetic variation and some phenotypic correlations in a random bred population. Poult. Sci. *42*, 178–182.

Hawthorne, J. R.   1950.   The action of egg white lysozyme on ovomucoid and ovomucin. Biochim. Biophys. Acta 6, 28–35.

Heath, J. L.   1975.   Investigation of changes in yolk moisture. Poult. Sci. 54, 2007–2013.

Heath, J. L.   1976.   Factors affecting the vitelline membrane of the hen's egg. Poult. Sci. 55, 936–942.

Heath, J. L.   1977.   Chemical and related osmotic changes in egg albumen during storage. Poult. Sci. 56, 822–828.

Heyn, A. J. N.   1963.   The crystalline structure of calcium carbonate in the avian egg shell. J. Ultrastruct. Res. 8, 176–188.

Hickson, D. W., Alford, E. S., Gardner, F. A., Diehl, K., Sanders, J. O., and Dill, C. W.   1982.   Changes in heat-induced rheological properties during cold storage of egg albumen. J. Food Sci. 47, 1908–1911.

Hillyard, L. A., White, H. M., and Pangburn, S. A.   1972.   Characterization of apolipoproteins in chicken serum and egg yolk. Biochemistry 11, 511–518.

Hinton, C. F., Fry, J. L., and Harms, R. H.   1973.   The relationship of yolk pigmentation to candled grade. Poult. Sci. 52, 360–364.

Huang, T.-S., and De Lange, R. J.   1971.   Egg white avidin. 2. Isolation, composition and amino acid sequences of the tryptic peptides. J. Biol. Chem. 246, 686–697.

Itoh, H., and Hatano, I.   1964.   Variation of magnesium and phosphorus deposition rates during egg shell formation. Poult. Sci. 43, 77–80.

Johnson, E. A., Lewis, M. J., and Grau, C. R.   1980.   Pigmentation of egg yolks with astraxanthin from the yeast, Phaffia rhodozyma. Poult. Sci. 59, 1777–1782.

Johnson, L. N.   1966.   The structure and function of lysozyme. Sci. Prog. (Oxford) 54, 367–385.

Jollès, P., Jauregui-Adell, J., and Jollès, J.   1964.   Lysozyme of the white of hens eggs: distribution of disulfide bridges. C. R. Hebd. Seances Acad. Sci. 258, 3926.

Joubert, F. J., and Cook, W. H.   1958A.   Separation and characterization of lipovitellin from hen egg yolk. Can. J. Biochem. Physiol. 36, 389–398.

Joubert, F. J., and Cook, W. H.   1958B.   Preparation and characterization of phosvitin from hen egg yolk. Can. J. Biochem. Physiol. 36, 399–408.

Kamat, V. B., Lawrence, G. A., Barratt, M. D., Darke, A., Leslie, R. B., Skipley, G. G., and Stubbs, J. M.   1972.   Physical studies of egg yolk low-density lipoproteins. Chem. Phys. Lipids 9, 1–25.

Kato, A., and Sato, Y.   1971.   The separation and characterization of carbohydrate rich component from ovomucin in chicken eggs. Agric. Biol. Chem. 35, 439–440.

Kato, A., and Sato, Y.   1972.   The release of carbohydrate rich compound from ovomucin gel during storage. Agric. Biol. Chem. 36, 831–833.

Kato, A., Nakamura, R., and Sato, Y.   1971.   Studies on changes in stored shell eggs. 7. Changes in the physicochemical properties of ovomucin solubilized by treatment with mercaptoethanol during storage. Agric. Biol. Chem. 35, 351.

Kato, A., Imoto, T., and Yagishita, K.   1975.   The binding groups in ovomucin-lysozyme interaction. Agric. Biol. Chem. 39, 541–543.

Kato, A., Hirata, S., Sato, H., and Kobayashi, K.   1978.   Fractionation and characterization of the sulfated oligosaccharide chains of ovomucin. Agric. Biol. Chem. 42, 835–841.

Kato, A., Ogino, K., Kuramoto, Y., and Kobayashi, K.   1979.   Degradation of the O-glycosidically linked carbohydrate units of ovomucin during egg white thinning. J. Food Sci. 44, 1341–1344.

Kay, H. D., and Marshall, P. G.   1928.   The second protein (livetin) of egg yolk. Biochem. J. 22, 1264–1269.

Kline, L., Meehan, J. J., and Sugihara, T. F.   1965.   Relation between layer age and egg-product yields and quality. Food Technol. 19, 114–119.

Klotz, I. M., and Walker, F. M.   1948.   Complexes of lysozyme. Arch. Biochem. 18, 319–325.

Kocal, J. T., Nakai, S., and Powrie, W. D. 1980. Preparation of apolipoprotein of very low density lipoprotein from egg yolk granules. J. Food Sci. 45, 1761–1767.

Kurisakai, J., Murata, Y., Kaminogawa, S., and Yamauchi, K. 1982. Heterogeneity and properties of heat-stable ovalbumin from stored egg. J. Agric. Food Chem. 30, 349–353.

Lanni, F., Sharp, D. G., Beckert, E. A., Dillon, E. S., Beard, D., and Beard, J. W. 1949. The egg-white inhibitor of influenza virus hemagglutination. I. Prepartion and properties of semi-purified inhibitor. J. Biol. Chem. 179, 1275–1287.

LeClerc, J. A., and Bailey, L. H. 1940. Fresh, frozen and dried eggs and egg products (their uses in baking and for other purposes). Cereal Chem. 17, 279–312.

Lee, Y. C., and Montgomery, R. 1962. Glycoproteins from ovalbumin: The structure of the peptide chains. Arch. Biochem. Biophys. 97, 9–17.

Lee, Y. C., Wu, Y. C. and Montgomery, R. 1964. Modification of emulsion action on asparaginyl-carbohydrate from ovalbumin by dinitrophenylation. Biochem. J. 91, 9C–10C.

Lewis, J. C., Snell, N. S., Hirschmann, D. J., and Fraenkel-Conrat, H. 1950. Amino acid composition of egg proteins. J. Biol. Chem. 186, 23.

Lineweaver, H., and Murray, C. W. 1947. Identification of the trypsin inhibitor of egg white with ovomucoid. J. Biol. Chem. 171, 565–581.

Lineweaver, H., Cunningham, F. E., Garibaldi, J. A., and Ijichi, K. 1967. Heat stability of egg white proteins under minimal conditions that kill salmonellae. U.S., Agric. Res. Serv., ARS 74–39.

Longsworth, L. G., Cannan, R. K., and MacInnes, D. A. 1940. An electrophoretic study of the proteins of egg white. J. Am. Chem. Soc. 62, 2580–2590.

Lush, I. E. 1961. Genetic polymorphisms in the egg albumen proteins of the domestic fowl. Nature (London), 189, 981–984.

Ma, C.-Y., and Holme, J. 1982. Effect of chemical modifications on some physicochemical properties and heat coagulation of egg albumen. J. Food Sci. 47, 1454–1459.

MacDonnell, L. R., Silva, R. B., and Feeney, R. E. 1951. The sulfhydryl groups of ovalbumin. Arch. Biochem. Biophys. 32, 288–299.

MacDonnell, L. R., Feeney, R. E., Ganson, H. L., Campbell, A., and Sugihara, T. F. 1955. The functional properties of the egg white proteins. Food Technol. 9, 49–53.

Mandeles, S. 1960. Use of DEAE-cellulose in the separation of proteins from egg white and other biological materials. J. Chromatogr. 3, 256–264.

Marion, J. E., Woodroof, J. G., and Cook, R. E. 1965. Some physical and chemical properties of eggs from hens of five different stocks. Poult. Sci. 44, 529–534.

Marion, W. W., Nordskog, A. W., Tolman, H. S., and Forsythe, R. H. 1964. Egg composition as influenced by breeding, egg size, age and season. Poult. Sci. 43, 255–264.

Martin, W. G., and Cook, W. H. 1958. Preparation and molecular weight of γ-livetin from egg yolk. Can. J. Biochem. Physiol. 36, 153–160.

Martin, W. G., Vandegaer, J. E., and Cook, W. H. 1957. Fractionation of livetin and the molecular weights of the α- and β-components. Can. J. Biochem. Physiol. 35, 241–250.

Martin, W. G., Turner, K. J., and Cook, W. H. 1959. Macromolecular properties of vitellenin from egg yolk and its parent complex with lipids. Can. J. Biochem. Physiol. 37, 1197–1207.

Martin, W. G., Tattrie, W. G., and Cook, W. H. 1963. Lipid extraction and distribution studies of egg yolk lipoproteins. Can. J. Biochem. Physiol. 41, 657–666.

Martin, W. G., Augustyniak, J., and Cook, W. H. 1964. Fractionation and characterization of the low-density lipoproteins of hen's egg yolk. Biochim. Biophys. Acta *84*, 714–720.

Marusich, W., DeRitter, E., and Bauernfeind, J. C.   1960.   Evaluation of carotenoid pigments for coloring egg yolks. Poult. Sci. *39*, 1338–1345.

Masshoff, W., and Stolpmann, H. J.   1961.   Licht- und Electronenmikroskopische Untersuchungen an der Schalenhaut and Kalkschale des Huhnereies. Z. Zellforsch. Mikrosk. Anat. *55*, 818–832.

Matsuda, T., Watanabe, K., and Sato, Y.   1981.   Heat-induced aggregation of egg white proteins as studied by vertical flat-sheet polyacrylamide gel eletrophoresis. J. Food Sci. *46*, 1829–34.

Matsushima, K.   1958.   An undescribed trypsin inhibitor in egg white. Science *127*, 1178–1179.

McCully, K. A., Mok, C.-C., and Common, R. H.   1962.   Paper electrophoretic characterization of proteins and lipoproteins of hen's egg yolk. Can. J. Biochem. Physiol. *40*, 937–952.

Megham, D. K., and Olcott, H. S.   1949.   Phosvitin, the principal phosphoprotein of egg yolk. J. Am. Chem. Soc. *71*, 3670–3679.

Melamed, M. D.   1966.   Ovomucoid. *In* Glycoproteins. A. Gottschalk (Editor). Elsevier, Amsterdam.

Melamed, M. D. 1967. Electrophoretic properties of ovomucoid. Biochem. J. *103*, 805.

Melamed, M. D., and Green, N. M.   1963.   Avidin. 2. Purification and composition. Biochem. J. *89*, 591–599.

Meyer, D. D., and Woodburn, M.   1965.   Gelation of frozen-defrosted egg yolk as affected by selected additives: viscosity and electrophoretic findings. Poult. Sci. *44*, 437–446.

Miller, M. M., Kato, A., and Nakai, S.   1982.   Sedimentation equilibrium study of the interaction between egg white lysozyme and ovomucin. J. Agric. Food Chem. *30*, 1127–1132.

Mok, C. C., and Common, R. H.   1964.   Studies on the livetins of hen's egg yolk. II. Immunoelectrophoretic identification of livetins with serum proteins. Can. J. Biochem. *42*, 1119–1131.

Mok, C. C., Martin, W. G., and Common, R. H.   1961.   A comparison of phosvitins prepared from hen's serum and from hen's egg yolk. Can. J. Biochem. Physiol. *39*, 109–117.

Montgomery, R.   1970.   Glycoproteins. *In* The Carbohydrates. W. Pigman and D. Horton (Editors), Vol. 2B. Academic Press, New York.

Montgomery, R., and Wu, Y. C.   1963.   The carbohydrate-protein linkage in ovomucoid. Biochem. Biophys. Res. Commun. *11*, 249–254.

Montgomery, R., Lee, Y. C., and Wu, Y. C.   1965.   Glycopeptides from ovalbumin. Biochemistry *4*, 566.

Moran, T.   1925.   Effect of low temperature on hen's eggs. Proc. R. Soc. London, Ser. B *98*, 436–456.

Moran, T.   1936.   Physics of the hen's egg. II. The bursting strength of the vitelline membrane. J. Exp. Biol. *13*, 41–47.

Morehouse, A. L.   1961.   Dried algae meal as a source of xanthophyll for egg yolk pigmentation. Poult. Sci. *40*, 1432.

Nakamura, R., and Ishimaru, M.   1981.   Changes in the shape and surface hydrophobicity of ovalbumin during its transformation to S-ovalbumin. Agric. Biol. Chem. *45*, 2775–2780.

Nelson, T. S., and Baptist, J. N.   1968.   Feed pigments. 2. The influence of feeding single and combined sources of red and yellow pigments on egg yolk color. Poult. Sci. *47*, 924–931.

Nichols, A. V., Rubin, L., and Lindgren, F. T. 1954. Interaction of heparin active factor and egg-yolk lipoprotein. Proc. Soc. Exp. Biol. Med. *85*, 352–355.

Nisbet, A. D., Saundry, R. H., Moir, A. J. G., Fothergill, L. A., and Fothergill, J. E. 1981. The complete amino acid sequence of hen ovalbumin. Eur. J. Biochem. *115*, 335–345.

Nishikimi, M., and Yagi, K. 1969. Flavin-protein interaction in egg white flavoprotein. J. Biochem. (Tokyo) *66*, 427–429.

Oades, J. M., and Brown, W. O. 1965. A study of the water-soluble oviduct proteins of the laying hen and the female chick treated with gonadal hormones. Comp. Biochem. Physiol. *14*, 475–489.

Odin, L. 1951. On the hexosamine component of ovomucin. Acta Chem. Scand. *5*, 1420.

Osuga, D. T., and Feeney, R. E. 1974. Avian egg whites. *In* Toxic Constituents of Animal Foodstuffs. I. E. Liener (Editor). Academic Press, New York.

Osuga, D. T., and Feeney, R. E. 1977. Egg proteins. *In* Food Proteins. J. R. Whitaker and S. R. Tannenbaum (Editors). AVI Publishing Co., Westport, CT.

Painter, P. C., and Koenig, J. L. 1976. Raman spectroscopic study of the proteins of egg white. Biopolymers *15*, 2155–2166.

Parkinson, T. L. 1966. The chemical composition of eggs. J. Sci. Food Agric. *17*, 101–111.

Pennington, D., Snell, E. E., and Eakin, R. E. 1942. Crystalline avidin. J. Am. Chem. Soc. *64*, 469.

Perlman, G. E. 1952. Enzymic dephosphorylation of ovalbumin and plakalbumin. J. Gen. Physiol. *25*, 711–726.

Peter, P. N., and Bell, R. W. 1930. Normal and modified foaming properties of whey-protein and egg albumen solutions. Ind. Eng. Chem. *22*, 1124–1128.

Peterson, W. J., Hughes, J. S., and Payne, L. F. 1939. The carotenoid pigments. Kans. State Coll. Agric. Exp. Stn., Bull. *46*.

Powrie, W. D. 1976. Characteristics of edible fluids of animal origin: eggs. *In* Principles of Food Science. Part I. Food Chemistry. O. Fennema (Editor). Marcel Dekker, New York.

Powrie, W. D., Little, H., and Lopez, A. 1963. Gelation of egg yolk. J. Food Sci. *28*, 38–46.

Pritchard, A. B., McCormick, D. B., and Wright, L. D. 1966. Optical rotatory dispersion studies of the heat denaturation of avidin and the avidin-biotin complex. Biochem. Biophys. Res. Commun. *25*, 524–528.

Privett, O. S., Bland, M. L., and Schmidt, J. A. 1962. Studies on the composition of egg lipid. J. Food Sci. *27*, 463–468.

Radomski, M. W., and Cook, W. H. 1964A. Fractionation and dissociation of the avian lipovitellins and their interaction with phosvitin, Can. J. Biochem. *42*, 395–406.

Radomski, M. W., and Cook, W. H. 1964B. Chromatographic separation of phosvitin. α- and β-lipovitellin of egg yolk granules on TEAE-cellulose. Can. J. Biochem. *42*, 1203–1215.

Rhodes, D. N., and Lea, C. H. 1957. Phospholipides. 4. On the composition of hen's egg phospholipides. Biochem. J. *54*, 526–533.

Rhodes, M. B., Azari, P. R., and Feeney, R. E. 1958. Analysis, fractionation and purification of egg white proteins with cellulose action exchanger. J. Biol. Chem. *230*, 399–408.

Rhodes, M. B., Bennett, N., and Feeney, R. E. 1959. The flavoprotein-apoprotein system of egg white. J. Biol. Chem. *234*, 2054–2060.

Rhodes, M. B., Bennett, N., and Feeney, R. E. 1960. The trypsin and chymotrypsin inhibitors from avian egg whites. J. Biol. Chem. *235*, 1686–1693.

Robinson, D. S. 1972. Egg white glycoproteins and the physical properties of egg white. *In* Egg Formation and Production. B. M. Freeman and P. E. Lake (Editors). British Poultry Sci. Ltd., Edinburgh.

Robinson, D. S., and Monsey, J. B. 1964. Reduction of ovomucin by mercaptoethanol. Biochim. Biophys. Acta *83*, 368–370.

Robinson, D. S., and Monsey, J. B. 1971. Studies of the composition of egg white ovomucin. Biochem. J. *121*, 537–547.

Robinson, D. S., and Monsey, J. B. 1972. Changes in the composition of ovomucin during liquefaction of thick egg white. J. Sci. Food Agric. *23*, 29–38.

Romanoff, A. L., and Romanoff, A. 1949. The Avian Egg. John Wiley & Sons Co., New York.

Rose, D., Gridgeman, N. T., and Fletcher, D. A. 1966. Solids content of eggs. Poult. Sci. *45*, 221–226.

Russo, V. 1966. The influence of vitamin E on egg yolk pigmentation of hens fed different amounts of xanthophyll. Prod. Anim. *5* (1-2), 65–71.

Saari, A., Powrie, W. D., and Fennema, O. 1964. Isolation and characterization of low-density lipoproteins in native egg yolk plasma. J. Food Sci. *29*, 307–315.

Schmidt, G., Bessman, M. J., Hickey, M. D., and Thannhauser, S. J. 1956. The concentrations of some constituents of egg yolk in its soluble phase. J. Biol. Chem. *223*, 1027–1031.

Scott, M. L., Ascarelli, I., and Olson, G. 1968. Studies of egg yolk pigmentation. Poult. Sci. *47*, 863–872.

Seideman, W. E., Cotterill, O. J., and Funk, E. M. 1963. Factors affecting heat coagulation of egg white. Poult. Sci. *43*, 406–417.

Shantz, R. C., and Dawson, L. E. 1974. Electrophoretic examination of native and phosvitin fraction of avian egg yolk. Poult. Sci. *53*, 969–974.

Sharp, D. G., Lanni, F., Lanni, Y. T., Csaky, T. Z., and Bead, J. W. 1951. The egg white inhibitor of influenza virus haemagglutination. V. Electrophoretic studies. Arch. Biochem. *30*, 251–260.

Sharp, P. F., and Powell, C. K. 1931. Increase the pH of the white and yolk of hen's eggs. Ind. Eng. Chem. *23*, 196–199.

Shenstone, F. S. 1968. The gross composition, chemistry and physico-chemical basis of organization of the yolk and white. *In* Egg Quality: A Study of the Hen's Egg. T. C. Carter (Editor). Oliver & Boyd, Edinburgh, Scotland.

Shepard, C. C., and Hottle, G. A. 1949. Studies of the composition of the livetin fraction of the yolk of hen eggs with the use of electrophoretic analysis. J. Biol. Chem. *179*, 349–357.

Simkiss, K. 1968. The structure and formation of the shell and shell membranes. *In* Egg Quality: A Study of the Hen's Egg. T. C. Carter (Editor). Oliver & Boyd, Edinburgh, Scotland.

Simkiss, K., and Tyler, C. 1957. A histochemical study of the organic matrix of hen eggshells. Q. J. Microsc. Sci. *98*, 19–28.

Simons, P. C. M., and Wiertz, G. 1963. Notes on the structure of membranes and shell in hen's egg. An electron microscopical study. Z. Zellforsch. Mikrosk. Anat. *59*, 555–567.

Simons, P. C. M., Tyler, C., and Thomas, H. P. 1966. The effect of sodium hydroxide and sodium sulphide on the snapping strength of egg shells. Br. Poult. Sci. *7*, 309–314.

Skellon, J. H., and Windsor, D. A. 1962. The fatty acid composition of egg yolk lipids in relation to dietary fats. J. Sci. Food Agric. *13*, 300–303.

Smith, A. H., Wilson, W. O., and Brown, J. G. 1954. Composition of eggs from individual hens maintained under controlled environments. Poult. Sci. *33*, 898–908.

Smith, I. D., and Perdue, H. S. 1966. Isolation and tentative identification of the carotenoids present in chicken skin and egg yolks. Poult. Sci. *45*, 577–581.

Smith, M. B. 1964. Studies on ovalbumin. I. Denaturation by heat and the heterogeneity of ovalbumin. Aust. J. Biol. Sci. *17*, 261–270.

Smith, M. B., and Back, J. F. 1965. Studies on ovalbumin. II. The formation and properties of S-ovalbumin, a more stable form of ovalbumin. Aust. J. Biol. Sci. *18*, 365–377.

Smith, M. B., Reynolds, T. M., Buckingham, C. P., and Back, J. F. 1974. Studies on the carbohydrate of egg white ovomucin. Aust. J. Biol. Sci. *27*, 349–360.

Stevens, F. 1961. Starch gel electrophoresis of hen egg white, oviduct white, yolk, ova and serum proteins. Nature (London) *192*, 972.

Sugano, H., and Watanabe, I. 1961. Isolations and some properties of native lipoproteins from egg yolk. J. Biochem. (Tokyo) *50*, 473–480.

Taborsky, G. 1963. Interaction between phosvitin and iron and its effect on a rearrangement of phosvitin structure. Biochemistry *2*, 266–271.

Tallan, H. H., and Stein, W. H. 1951. Studies on lysozyme. J. Am. Chem. Soc. *73*, 2976.

Tallan, H. H., and Stein, W. H. 1953. Chromatographic studies on lysozyme. J. Biol. Chem. *200*, 507.

Terepka, A. R. 1963. Structure and calcification in avian eggshell. Exp. Cell Res. *30*, 171–182.

Tomimatsu, Y., Clary, J. J., and Bartulovitch, J. J. 1966. Physical characterization of ovoinhibitor, a trypsin and chymotrypsin inhibitor from chicken egg white. Arch. Biochem. Biophys. *115*, 536–544.

Torten, J., and Eisenberg, J. . 1982. Studies on colloidal properties of whole egg magma. J. Food Sci. *47*, 1423–1428.

Tung, M. A., and Richards, J. F. 1972. Ultrastructure of the hen's egg shell membranes by electron microscopy. J. Food Sci. *37*, 277–279.

Tung, M. A., Watson, E. L., and Richards, J. F. 1969. Rheology of egg albumen. Pap.—Am. Soc. Agric. Eng., Winter Meet. 69–874.

Tung, M. A., Watson, E. L., and Richards, J. F. 1969. Rheology of egg albumen. Pap.—Am. Soc. Agric. Eng., Winter Meet. 69.

Tung, M. A., Watson, E. L., and Richards, J. F. 1970. Rheology of fresh, aged and gamma-irradiated egg white. J. Food Sci. *35* (6), 872–874.

Tunmann, P., and Silberzahn, H. 1961. Uber die Kohlenhydrate im Huhnerei. I. Freie Kohlenhydrate. Z. Lebensm. Unters. Forsch. *115*, 121–128.

Turner, K. J., and Cook, W. H. 1958. Molecular weight and physical properties of a lipoprotein from the floating fraction of egg yolk. Can. J. Biochem. Physiol. *36*, 937–949.

Vadehra, D. V., and Nath, K. R. 1973. Eggs as a source of protein. *In* Critical Reviews in Food Technology. T. E. Furia (Editor). CRC Press Inc., Boca Raton, FL.

Vincent, R., Powrie, W. D., and Fennema, O. 1966. Surface activity of yolk, plasma and dispersions of yolk fractions. J. Food Sci. *31*, 643–648.

Vuilleumier, J. P. 1969. The "Roche yolk color fan"—an instrument for measuring yolk color. Poult. Sci. *48*, 767–779.

Walden, C. C., Allen, I. V. F., and Trussell, P. C. 1956. The role of the egg shell and shell membrane in restraining the entry of microorganisms. Poult. Sci. *35*, 1190–1196.

Wallace, R. A., Jared, D. W., and Eisen, A. Z. 1966. A general method for the isolation and purification of phosvitin from vertebrate eggs. Can. J. Biochem. *44*, 1647–1655.

Warner, R. C. 1954. Egg proteins. *In* The Proteins. H. Neurath and K. Bailey (Editors). Academic Press, New York.

Warner, R. C., and Weber, I. 1951. The preparation of crystalline conalbumin. J. Biol. Chem. *191*, 173–180.

Warner, R. C., and Weber, I. 1953. The metal combining properties of conalbumin. J. Am. Chem. Soc. *75*, 5094–5101.

Watanabe, K., Tsukasa, M., and Sato, Y. 1981. The secondary structure of ovomucoid and its domains as studies by circular dichroism. Biochim. Biophys. Acta *667*, 242–250.

Weinman, E. O. 1956. Lipide-protein complex in egg yolk. Fed. Proc., Fed. Am. Soc. Exp. Biol. *15*, 381.

Wetter, L. R., and Deutsch, H. F. 1951. Immunological studies on egg white proteins. IV. Immunochemical and physical studies of lysozyme. J. Biol. Chem. *192*, 237.

Williams, J. 1962A. Serum proteins and the livetins of hen's egg yolk. Biochem. J. *83*, 346–355.

Williams, J. 1962B. A comparison of conalbumin and transferrin in the domestic fowl. Biochem. J. *83*, 355–364.

Williams, J., and Sanger, F. 1959. The grouping of serine phosphate in phosvitin and casein. Biochim. Biophys. Acta *33*, 294–296.

Williams, W. P., Davies, R. E., and Couch, J. R. 1963. The utilization of carotenoids by the hen and chick. Poult. Sci. *42*, 691–699.

Windle, J. J., Wiersma, A. K., Clark, J. R., and Feeney, R. E. 1963. Investigation of the iron and copper complexes of avian conalbumins and human transferrins by electron paramagnetic resonance. Biochemistry, *2*, 1341–1345.

Winter, W. P., Buss, E. G., Claggett, C. O., and Boucher, R. V. 1967. The nature of the biochemical lesion in avian renal riboflavinuria. II. The inherited change of a riboflavin binding protein from blood and eggs. Comp. Biochem. Physiol. *22*, 897–906.

Wise, R. W., Ketterer, B., and Hansen, I. A. 1964. Prealbumins of embryonic chick plasma. Comp. Biochem. Physiol. *12*, 439–443.

Wishnia, A., and Warner, R. C. 1961. The kinetics of denaturation of conalbumin. J. Am. Chem. Soc. *83*, 2065–2071.

Woodward, S. A., and Cotterill, O. J. 1983. Electrophoresis and chromatography of heat-treated plain, sugared and salted whole egg. J. Food Sci. *48*, 501–506.

Yamauchi, K., Kurisaki, J., and Sasago, K. 1976. Polypeptide composition of hen's egg yolk very low density lipoprotein. Agric. Biol. Chem. *40*, 1581–1586.

The Nutritive Value of Eggs

Frances Cook
George M. Briggs

Eggs have been recognized as an important food from the time primitive men first snatched them from the nests of wild birds. Eggs were considered a delicacy in ancient times and where abundant, a staple of the diet. History associates the egg with science, superstition, and sports. It was also a symbol in pagan and religious customs. Many authorities credit Columbus with bringing chickens on his second voyage to the Western Hemisphere. Today, eggs remain a popular food in all countries of the world.

EGGS ARE NUTRITIOUS

Although eggs contain about 74% water, they are a rich source of such high-quality protein that experimental nutritionists often use them as a standard for measuring the quality of other food proteins. Eggs are also an important source of unsaturated fatty acids (mainly oleic), iron, phosphorus, trace minerals, vitamins A, E and K and the B vitamins, including B_{12}. As a natural source of vitamin D, eggs rank second only to fish-liver oils. Eggs are low in calcium (discarded in the shell) and contain very little, or no, vitamin C.

Eggs provide a unique, well-balanced source of nutrients for persons of all ages. Before the advent of present-day baby foods, hard-cooked egg yolk served as the major supplementary source of iron for young babies. Now iron-enriched, precooked baby cereals, strained meats, and

141

EGG SCIENCE & TECHNOLOGY,
3rd Edition

canned egg yolks offer alternative and more convenient sources of iron. However, home-prepared, hard-cooked egg yolks are still a practical and desirable supplement for infants. By the time the baby is a year old, he may also be fed the whites. Eggs contribute significantly to the body's nutrient needs during rapid growth, and are therefore an excellent food for young children and teen-agers.

Their high nutrient content, low caloric value, and ease of digestibility make eggs valuable in many therapeutic diets for adults. Most bland diets and the first light or soft diet during convalescence from surgery or other illness usually include eggs. They help patients return to good health and maintain it.

Eggs are valuable and readily acceptable in diets for older people, whose caloric needs are lower and who sometimes have difficulty in chewing certain types of food. Availability, modest cost, ease of preparation, popular taste appeal, and low caloric value give eggs a deserved place in geriatric diets.

Eggs are good for any meal or as snacks. They provide good nutrition, are satisfying, and have a comparatively low calorie count, important for persons with weight problems.

Most nutrition educators use the Daily Food Plan (or the so-called Basic-Four Food Plan) as a simple guide for persons of all ages in choosing their nutrient requirements from a wide variety of common and readily available foods. This guide, developed by nutritionists at the U.S. Department of Agriculture, divides foods into these four groups: milk, meat, vegetable-fruit, and bread-cereal.

Two or more servings per day are recommended from the meat group, which includes eggs, poultry, beef, veal, lamb, fish, and, as alternates, dry beans, dry peas, and nuts. (Most young children and older adults can meet their nutrient needs from the meat group with less than two servings.) Two eggs, 2 to 3 oz. lean, boneless, cooked meat, poultry or fish, 1 cup cooked dry beans, dry peas or lentils or 4 tbsp peanut butter count as a serving from the meat group.

EGG CONSUMPTION

The average per capita consumption of eggs in the United States declined from 1950 to a low of about 270 in 1977 (Table 7.1). This was the lowest annual per capita egg consumption since records were started in 1909. Since 1977 there has been an annual increase of about four eggs per year in per capita consumption. Each person who actually consumed the annual average of eggs ate, during the year, approximately 35 lb of edible portion, of which 26 lb were water and 9 lb were dry

TABLE 7.1. Average per Capita Civilian Consumption of Eggs in the United States for Selected Years

Year[a]	Number of eggs	Pounds
1950	377	49.0
1955	360	46.9
1960	324	42.5
1965	306	39.8
1970	304	39.5
1975	273	35.4
1980	281	36.5

SOURCE: *Adapted from U.S. Department of Agriculture (1968, 1972, 1982)*
[a] Data based on calendar year.

matter, including only 4 lb of fat. This amount of fat from egg yolks is very small when compared with the average per capita intake of 52.4 lb from fats and oils for the same period, plus the approximately 55 lb of fat in meats, fish, poultry, and dairy products. (The effect of egg yolk lipids on serum cholesterol levels is discussed later in this chapter.)

A motivation research study sponsored by the Poultry and Egg National Board (1958) indicated the attitudes toward eggs of a nationwide sample of slightly over 1,200 middle-income homemakers. Some of the more significant findings were:

1. Consumers associated the use of eggs most closely first with children and infants, then with men.
2. Homemakers 30 years of age or older used more eggs and considered them more economical than did younger homemakers.
3. Eggs were associated with breakfast in 91% of the homes, but were used as a breakfast food in only 59%. With the exception of coffee or tea and toast, eggs were associated with breakfast in more homes (59%) than was any other food. (Milk was not listed in the foods to be checked.)
4. Eggs were associated with baking by almost 80% of the homemakers. One-half of the homemakers considered eggs as a salad ingredient, whereas 27% thought of eggs as a main-dish ingredient.
5. Most homemakers had favorable opinions about eggs; 95% considered eggs a necessary food, 89% said eggs were high in protein, and 88% considered them a beneficial food.
6. Fifty-five percent of the homemakers said that eggs have no harmful effect on the heart, 7% thought eggs are harmful to the heart, and the remaining 38% expressed no opinion on this point.
7. Eighty-one percent of the homemakers did not think eggs were fattening, 11% thought they were, and 8% had no opinion.

A shift of consumer attitudes toward eggs since 1958 as well as other factors might explain the steady decline in egg consumption. Such factors might include:

- *Higher consumer incomes*, with accompanying greater consumption of most meats, as well as chicken, turkey, and convenience foods. With increased buying power, families tend to substitute more expensive and so-called "status" foods for less expensive ones, such as eggs, legumes, and cereals.
- *Changing family living and eating patterns.* Eggs have long been considered a basic breakfast food, but many persons now eat lighter breakfasts, or none at all, and thus fewer eggs are eaten for this meal. In many homes breakfast is no longer a family meal. Sometimes each member prepares and eats breakfast alone. This often means that the meal is simplified and includes nothing hot other than a beverage and possibly toast.
- *Concern about possible adverse effects of eggs on health.*

DISTRIBUTION OF NUTRIENTS IN EGGS

Most food composition tables are based on 100-g or 1-lb edible portions of foods. For many people, such tables, for eggs, are not meaningful because of differences in the weight of and nutrient distribution in whites and yolks. Table 7.2 shows the nutrient content of whole uncooked eggs, raw egg white, and raw egg yolk on a per egg basis. The composition of liquid or frozen and dehydrated egg products are based on 100 g of product. The composition of a large, whole raw egg and of a boiled egg is almost the same, of course, but that of a scrambled egg varies slightly due to the added milk and fat. The value listed in the table for a specific nutrient in a whole raw egg may vary from the combined values for a raw white and yolk because of differences in analytical techniques, sources of data, or how much the figures were rounded.

The fatty acid composition of the egg yolk is easily modified by adding fat to the hens diet. Vitamin and trace-mineral values also vary greatly when rations are heavily fortified with a particular vitamin or mineral. The average values in tables of food composition, however, are for eggs from hens receiving the usual commercial rations and for properly handled eggs passing through normal distribution channels, not stored under adverse conditions or for long periods of time.

The yolk comprises slightly over one-third of the edible portion, but it yields three-fourths of the calories and provides all or most of the fat, iron, vitamin A, thiamine and calcium, and almost half the protein and riboflavin of the whole egg. Protein and riboflavin are less concentrated in the white, but because there is almost twice as much white as yolk, more than half the total protein and riboflavin is in the white.

COMPOSITION OF EGGS

A review by Everson and Sanders (1957) summarized egg-composition data from many sources. These data and more recent studies were used by Cotterill and Glauert (1979) to generate the data in Table 7.2.

Tables of food composition evolve with changing times. Nutrient values reported from specific studies may be simple averages, but those listed in tables of food composition do not necessarily bring together the averages of available analyses. To prepare accurate tables of food composition, certain factors must be considered, such as: analytical equipment; methods and techniques; selection and number of samples included; soundness and significance of data; and whether values are representative for year-round and area-wide or nationwide use. A knowledge of which factors were considered in compiling tables of food composition can help in evaluating the accuracy of the values listed.

Many factors directly influence the concentration of nutrients in eggs. These include age, breed and strain of hens, differences in eggs produced by individual hens, rations and environmental temperatures for hens, egg-storage conditions and length of storage time, and processing, preparation, and cooking methods.

Cotterill and Glauert (1981) published nutrition-labeling panels for one and two medium, large and extra-large eggs according to Food and Drug Administration regulations (Table 7.3). Note that the values from Table 7.3 have been rounded off according to the rules. If used, this is the way a panel would look on an egg carton. The fatty acid composition of eggs and other selected foods is summarized in Table 7.2.

The composition of eggs of some domestic birds other than chickens were reported by Leung et al. (1952) and Leung and Flores (1961).

Effect of Breed, Strain, and Age of Hens

Size may influence the nutrient composition or, at least, the proportions of parts and solids of eggs. Some variations in size are due to the breed and/or strain of hen as well as to differences of individual hens within a strain or breed. But the age of hens has more effect on egg size; hens produce smaller eggs at the start of their laying cycle than later.

The proportion of yolk tends to be higher in small eggs than in larger ones, according to Forsythe (1963). As egg size increases, the relative amount of yolk decreases, but the total absolute amount increases. The

TABLE 7.2. Statistical Estimate Values for Nutrient Composition of Eggs Expressed on Shell (per Egg), Liquid/Frozen (per 100 g), and Dehydrated (per 100 g) Bases

Nutrients and units	Shell (per egg)[a]			Liquid/frozen (per 100 g)				Dehydrated (per 100 g)			S.D.[h]
	Whole	White	Yolk	Whole[b]	White[b]	Yolk Pure[c]	Yolk Commercial[d]	Plain whole[e]	Stab. white[f]	Plain yolk[g]	
Proximate											
Solids, g	13.47	4.6	8.81	24.5	12.1	51.8	44.0	96.8	93.6	97.2	—
Calories	84	19	64	152	50	377	313	600	388	692	4.98
Protein (N × 6.25), g	6.60	3.88	2.74	12.0	10.2	16.1	14.9	47.4	79.1	32.9	0.421
Total lipids, g	6.00	—	5.80	10.9	—	34.1	27.5	43.1	—	60.8	0.715
Ash, g	0.55	.26	0.29	1.00	0.68	1.69	1.49	4.0	5.3	3.3	0.081
Lipids											
Fatty acids, g											
Saturated, total	2.01	—	1.95	3.67	—	11.42	9.16	14.51	—	20.35	0.561
8:0	0.027	—	0.027	0.05	—	0.16	0.13	0.20	—	0.29	0.003
10:0	0.082	—	0.080	0.15	—	0.47	0.38	0.59	—	0.84	0.020
12:0	0.027	—	0.026	0.05	—	0.15	0.12	0.20	—	0.27	0.008
14:0	0.022	—	0.022	0.04	—	0.12	0.09	0.16	—	0.20	0.014
16:0	1.37	—	1.31	2.5	—	7.7	6.2	9.84	—	13.8	0.227
18:0	0.462	—	0.459	0.84	—	2.70	2.14	3.36	—	4.73	0.100
20:0	0.022	—	0.022	0.04	—	0.12	0.10	0.16	—	0.22	0.002
Monounsaturated, total	2.53	—	2.50	4.60	—	14.67	11.80	18.18	—	25.64	0.411
14:1	0.005	—	0.005	0.01	—	0.03	0.03	0.04	—	0.07	0.002
16:1	0.214	—	0.211	0.39	—	1.24	0.97	1.54	—	2.14	0.047
18:1	2.31	—	2.28	4.2	—	13.4	10.8	16.6	—	23.43	0.352
Polyunsaturated, total	0.73	—	0.72	1.32	—	4.20	3.37	5.22	—	7.45	0.133
18:2	0.660	—	0.650	1.20	—	3.82	3.07	4.74	—	6.79	0.216
18:3	0.011	—	0.014	0.02	—	0.08	0.06	0.08	—	0.13	0.020
20:4	0.055	—	0.051	0.10	—	0.30	0.24	0.40	—	0.53	0.029
Cholesterol, g	0.264	—	0.258	0.48	—	1.52	1.23	1.90	—	2.72	0.053
Lecithin, g	1.27	—	1.22	2.32	—	7.20	5.81	9.16	—	12.84	0.112
Cephalin, g	0.253	—	0.241	0.46	—	1.42	1.15	1.82	—	2.54	0.028

Vitamins											
A, IU	264	—	260	480	—	1527	1240	1896	—	2740	709
D, IU	27	—	27	50	—	161	129	198	—	285	11.6
E, mg	0.88	—	0.87	1.6	—	5.1	4.1	6.7	—	9.1	0.837
B$_{12}$, µg	0.48	2.58	0.48	.88	—	2.83	2.27	3.5	—	5.0	0.616
Biotin, µg	11.0	0.46	8.35	20.0	6.8	49.1	40.8	79	53	90	4.55
Choline, mg	237	238	238	430	1.2	1400	1130	1699	9	2497	172
Folic acid, mg	0.023	0.006	0.026	0.060	0.016	0.154	0.128	0.24	0.12	0.28	0.035
Inositol, mg	5.94	1.52	4.35	10.8	4.0	25.6	21.4	43	31	47	4.84
Niacin, mg	0.045	0.035	0.010	0.082	0.092	0.061	0.067	0.32	0.71	0.15	0.033
Pantothenic acid, mg	0.83	0.09	0.73	1.52	0.24	4.3	3.5	6.0	1.9	7.7	0.815
Pyridoxine, mg	0.065	0.008	0.057	0.119	0.021	0.334	0.273	0.47	0.16	0.60	0.042
Riboflavin, mg	0.18	0.11	0.07	0.33	0.28	0.44	0.41	1.30	2.2	0.91	0.028
Thiamine, mg	0.05	0.004	0.048	0.09	0.011	0.28	0.22	0.36	0.09	0.49	0.026
Minerals, mg											
Calcium	29.2	3.8	25.2	53	10	148	121	209	78	267	7.88
Chlorine	96.0	66.1	29.9	175	174	176	176	691	1349	389	6.58
Copper	0.033	0.009	0.024	0.061	0.023	0.145	0.121	0.24	0.16	0.27	0.672
Iodine	0.026	0.001	0.024	0.047	0.003	0.141	0.114	0.19	0.02	0.25	34.1
Iron	1.08	0.053	1.02	1.97	0.14	6.0	4.83	7.8	1.09	10.6	0.264
Magnesium	6.33	4.15	2.15	11.5	10.8	12.9	12.5	45	84	27.6	1.19
Manganese	0.021	0.002	0.019	0.038	0.007	0.11	0.09	0.15	0.05	0.19	0.008
Phosphorus	111	8	102	202	22	599	485	798	171	1072	43.6
Potassium	74	57	17	135	150	100	110	533	1163	243	9.28
Sodium	71	63	9	129	165	52	74	510	1279	164	9.88
Sulfur	90	62	28	164	163	165	165	648	1263	366	5.97
Zinc	0.72	0.05	0.66	1.30	0.12	3.89	3.15	5.1	0.93	7.0	0.293
Amino Acids, g											
Alanine	0.38	0.24	0.14	0.69	0.64	0.81	0.77	2.73	4.96	1.70	0.061
Arginine	0.42	0.23	0.19	0.77	0.60	1.14	1.03	3.04	4.65	2.28	0.044
Aspartic acid	0.65	0.40	0.25	1.18	1.06	1.44	1.37	4.66	8.22	3.03	0.237
Cystine	0.15	0.11	0.05	0.28	0.28	0.27	0.27	1.11	2.17	0.60	0.024
Glutamic acid	0.85	0.52	0.33	1.54	1.36	1.94	1.83	6.08	10.54	4.04	0.113
Glycine	0.22	0.14	0.08	0.40	0.36	0.49	0.47	1.58	2.79	1.04	0.025
Histidine	0.16	0.09	0.07	0.29	0.24	0.41	0.38	1.15	1.86	0.84	0.017
Isoleucine	0.36	0.21	0.15	0.66	0.56	0.87	0.81	2.61	4.34	1.79	0.069

(Continued)

TABLE 7.2. (Continued)

Nutrients and units	Shell (per egg)[a]			Liquid/frozen (per 100 g)				Dehydrated (per 100 g)			
						Yolk					
	Whole	White	Yolk	Whole[b]	White[b]	Pure[c]	Commercial[d]	Plain whole[e]	Stab. white[f]	Plain yolk[g]	S.D.[h]
Leucine	0.57	0.33	0.24	1.04	0.88	1.39	1.29	4.11	6.82	2.85	0.035
Lysine	0.45	0.25	0.20	0.82	0.66	1.17	1.07	3.24	5.12	2.37	0.046
Methionine	0.21	0.15	0.06	0.39	0.39	0.39	0.39	1.54	3.02	0.86	0.023
Phenylalanine	0.35	0.23	0.12	0.64	0.61	0.69	0.67	2.53	4.73	1.48	0.041
Proline	0.26	0.15	0.11	0.48	0.40	0.65	0.60	1.90	3.10	1.33	0.034
Serine	0.50	0.27	0.23	0.91	0.71	1.36	1.24	3.60	5.50	2.74	0.043
Threonine	0.32	0.18	0.14	0.59	0.47	0.85	0.78	2.33	3.64	1.72	0.036
Tryptophan	0.11	0.07	0.04	0.19	0.17	0.24	0.23	0.75	1.32	0.51	0.014
Tyrosine	0.28	0.16	0.12	0.51	0.41	0.73	0.67	2.02	3.18	1.48	0.023
Valine	0.43	0.27	0.16	0.79	0.72	0.96	0.91	3.12	5.58	2.01	0.080

SOURCE: Cotterill and Glauert (1979).

a Based on 60.9 g shell egg weight with 55.1 g total liquid whole egg, 38.4 g white, and a 16.7 g yolk; containing 24.1%, 12.1% and 51.8% solids, respectively (see Cotterill and Geiger 1977).
b Based on 24.5% and 12.1% solids, respectively, for whole and white liquid.
c Pure yolk containing 51.8% solids.
d Commercial yolk contains 44% egg solids, diluted with egg white only.
e Produced from whole egg containing 24.5% solids as in Footnote b.
f Produced from a bacteriologically fermented egg white.
g Produced from yolk containing 44% solids as in Footnote d.
h Standard deviation about the regression line for liquid and frozen data.

TABLE 7.3. Nutritive Value of Eggs per Serving Size[a]

	Eggs	
	One	Two
Calories	80	160
Protein	6 g	13 g
Carbohydrates	b	1 g
Fat[c] (68% of calories)	6 g	12 g
Polyunsaturated	1 g	2 g
Saturated	2 g	4 g
Cholesterol[c] (480 mg/100 g)	260 mg	520 mg
Sodium (130 mg/100 g)	70 mg	140 mg
Percentage of U.S. recommended daily allowances (U.S. RDA)		
Protein	15	30
Vitamin A	6	10
Vitamin C	d	d
Thiamine	4	6
Riboflavin	10	20
Niacin	d	d
Calcium	2	6
Iron	6	10
Vitamin D	6	15
Vitamin E	2	6
Vitamin B_6	4	6
Folic acid	8	15
Vitamin B_{12}	8	15
Phosphorus	10	20
Iodine	15	35
Magnesium	2	4
Zinc	4	10
Copper	2	4
Biotin	4	8
Pantothenic acid	8	15

SOURCE: *Cotterill and Glauert (1981).*
[a] Serving size = 1 or 2 U.S. Grade Large eggs (54 or 108 g edible portion; serving per container = 6 or 12.
[b] Contains less than 1 g.
[c] Information on fat and cholesterol content is provided for individuals who, on the advice of a physician, are modifying their total dietary intake of fat and/or cholesterol.
[d] Contains less than 2% of U.S. RDA of these nutrients.

author indicated that 88% of the variations in proportion of parts and/or solids in the yolks and whites results from differences in egg size rather than from the strain or age of the hens. Faster handling and shell-coating result in greater moisture retention and lower solids content (25% or less).

Eggs from two commercial strains of White Leghorn layers during their first year of production were analyzed by Kline *et al.* (1965). It was found that layers reaching or exceeding their first year of lay produced larger eggs with a higher proportion of yolks, solids in yolks, and edible contents, i.e., less shell. Age had no effect on the nitrogen and fat content of the yolks, but whites from eggs produced by older hens had a lower solids content.

In a study of the eight most common commercial strains of hens in this country, four consecutive eggs were collected from each of ten hens from each strain when the hens were 253 days old. Analyses of these eggs showed that strain differences significantly influenced the concentration of fat and cholesterol and the iodine number of the fat. The difference among strains in the dry-matter contents of eggs was not significant. Differences for hens within a strain were significant for the dry-matter and cholesterol content of fat (Edwards *et al.* 1960).

Egg, shell and contents weights, Haugh units and percentage of moisture and protein were determined on eggs from five strains of hens for two bimonthly periods and from four strains for three bimonthly periods; 20 to 30 hens of each strain were about 8 months old when the experiments began. This study showed that the strain of hens has a significant influence on the percentage of moisture and protein of the fresh eggs and the protein of the freeze-dried eggs. The age of the hens or the season significantly affects the weight of the egg contents, albumen height, Haugh units, and grams of protein per egg (May and Stadelman 1960B).

To determine how the chemical composition of egg whites is affected by season and the age of birds, four groups of hens were studied from September through the following August. The ages of these birds when the study began were 4, 7, 12, and 15 months. The season had a highly significant effect on the sodium, calcium, and chlorine content of the eggs, but little or no effect on the potassium, phosphorus, and protein content. The age of the birds showed a highly significant effect on the phosphorus, chlorine, and protein content, a lesser effect on calcium, and no effect on sodium and potassium content (Cunningham *et al.* 1960).

These few studies are examples of many others that might be cited to show that breed, strain, and age of hens directly influence the size and composition of eggs. In addition, eggs produced by one hen often vary in composition from those of any other hen from the same or a different strain or breed. These and other factors must be considered when the significance and validity of research findings are being weighed.

Effect of Environment

A study was carried out on two groups of hens (40 in the first trial and 75 in the second) to determine the effect of exposure to environmental temperatures of 8°, 19°, and 30°C. When held at 30°C, the hens produced smaller eggs than when held at the two lower temperatures. The highest temperature affected all components of the eggs proportionately, but the percentage of water was the same for all sizes of eggs. The authors suggested that environmental temperature might affect egg size by influencing the rate at which the component parts were deposited. If egg producers control environmental temperatures by various methods, however, excessive heat will have no significant effect on egg size (Carmon and Huston 1965).

Effect of Hen's Diet

It has been known for many years that diet can influence the fatty acid composition of the egg. Only examples of recent work will be cited here. To show the possible effect of certain fats on the nutrient composition of eggs, three groups of ten hens each were fed one of three diets for 8 weeks. The control diet included no fat, but the other diets were supplemented with 10% corn oil or 10% beef tallow. In eggs produced by the hens on the three diets, the ratio of protein, minerals, and vitamins to calories was the same. The iodine values of the lipids of the eggs from the control and beef-tallow groups did not increase, but those of the corn-oil group did (Jordan *et al.* 1962).

Five hens of the same strain were fed a diet for 1 year to deplete their body stores of linoleic acid. At the beginning of the experiment, linoleic acid in the form of corn oil was added to the diet. The linoleic acid content of all lipid fractions immediately increased. The addition of 8% corn oil to the diet produced a maximum of 20% linoleic acid in the egg-yolk fatty acids over a 14-day period. This increase was accompanied by a decrease in the oleic acid level. Minor adjustments in the amounts of the other egg-yolk fatty acids occurred along with these major changes (Balnave 1968).

Data from a group of hens over a 10-month period showed that hens on the same ration differ slightly in their ability to deposit vitamin A in their eggs. However, variations were much greater among individual hens than among individual laying cycles (Bandemer *et al.* 1958A).

Six pullets of the same strain were raised on rations low in vitamin A and provitamin A carotenoids. The hens produced eggs for about 10 weeks and then stopped quite abruptly. Analysis of the eggs laid during the last 5 weeks of production showed that the percentage of lipids in the yolks remained constant, but the total vitamin A content decreased (Plack *et al.* 1964).

Numerous other reports have been published showing a positive correlation of egg composition to nutrient-depleted diets or to diets with an excess of one or more nutrients. Such studies have little practical application, however, since commercial egg producers use nutritionally balanced rations.

Effect of Storage and Processing

The literature includes many studies on the effects of handling and of storage times and conditions on the nutritive value of shell eggs. Everson and Sanders (1957) included effects of storage and processing on nutritive value of shell eggs in their review. More recently, some additional studies have been reported.

Eggs with no shell treatment and those with shells coated with plastic (Hanning 1958) or oil (Hudspeth *et al.* 1965) were compared after storage at different times and temperatures. Both coatings significantly reduced moisture loss and increased pH during storage. Plastic-coated and untreated eggs stored at room temperature for 24 hours, 1 week, and 4 weeks varied slightly in thiamine content and showed an apparent increase in riboflavin. The untreated and oil-coated eggs stored at room and refrigerator temperatures for times ranging from 1 to 81 days showed some change in the relative proportions of the proteins.

The cold storage of eggs for a relatively long time may reduce the amounts of some nutrients. In tests comparing fresh eggs with those stored at about 2°C for 87 days, the total protein content remained constant, but threonine decreased significantly on both a percentage and weight basis (May and Stadelman 1960A). Eggs lost about 10% of vitamin A when stored 12 months at 0°C, most of this loss occurring between the fourth and eighth months. Under long storage, eggs with dirty shells lost no more of this vitamin than did those with clean shells (Bandemer *et al.* 1958B). A comparison of fresh eggs and those stored at 0°C for 6 and 12 months showed no consistent change in lipid distribution nor variation in the fatty acid content of lipids during storage (Evans *et al.* 1967).

The quality, flavor, composition, and functional properties of eggs are adversely affected more rapidly and to a greater extent by the speed and conditions of handling, uncoated shells, and storage times and temperatures than is the nutrient content. Current handling practices lessen the significance of earlier studies on storing fresh eggs for long periods at high or low temperatures.

The nutritive value of frozen and dried eggs is essentially the same as that of fresh eggs. The drying or freezing processes do not cause any significant loss of nutrients. Properly stored dried and frozen eggs show no subsequent nutrient loss.

Effect of Cooking

Most research on the effect of cooking on the nutritive value of eggs was done at least 20 years ago, and focused on the loss of various B vitamins. A comparison of hard-cooked and scrambled eggs showed that neither method had an advantage with respect to thiamine content. However, scrambled eggs had about 20% less riboflavin than did hard-cooked or raw eggs. (Hanning 1958). Poached and hard-cooked eggs had an average, but not significant, loss of 0.22% threonine. The loss of this amino acid was the same for both cooking methods (May and Stadelman 1960A).

FATTY ACID DISTRIBUTION AND CHOLESTEROL CONTENT OF EGGS AND SERUM CHOLESTEROL IN MAN

Fats act as carriers for vitamins A, D, E, and K. Linoleic acid, a fatty acid present in some fats, cannot be formed from other fats in the body by mammals, including man, but must be supplied by the daily diet. Therefore, it is an "essential fatty acid."

Foods and body fats contain about 18 common fatty acids. Each has different physical and chemical properties and different physiological functions. Based on their chemical structure, fatty acids are classified, on the one hand, as saturated, unsaturated, mono- and polyunsaturated, and, on the other, as short- or long-chain. Saturated fatty acids cannot take up any hydrogen or iodine. Unsaturated fatty acids have one or more double bonds and can thus take up hydrogen or iodine. Oleic acid is a long-chain, monounsaturated fatty acid, that is, each molecule has only one double bond that can be broken, to add hydrogen. Linoleic, linolenic, and arachidonic acids are more highly unsaturated (polyunsaturated) and contain two, three and four double bonds, respectively. Food fats contain chiefly long-chain fatty acids. Those in solid fats are mostly saturated and combinations of saturated and monounsaturated, and in oils, mainly monounsaturated or polyunsaturated. Coconut oil and butter are exceptions. They contain predominantly short-chain saturated fatty acids. One criterion for evaluating the nutritional value of fats and fatty acid distribution is based on the total fat and the saturated and unsaturated fatty acid content. A larger proportion of polyunsaturated fatty acid (linoleic) to monounsaturated fatty acid (oleic) is highly desirable. (Refer to Tables 7.2 and 7.4 for the total fat, fatty acid, and cholesterol content of eggs.)

Cholesterol is a sterol present in all animal tissues, including the human body. It is found in foods from animal sources, such as eggs, meat (especially brains, kidneys, and liver), poultry, fish, butter, and milk, but not in plants. The human body can synthesize and metabolize all the cholesterol it needs, but if some is supplied by food, the body tends to compensate by manufacturing a lesser amount. Cholesterol, like all

TABLE 7.4. Fatty Acid Composition of Selected Foods

	Weight (g)	Total saturated fatty acids (g)	Total unsaturated fatty acids (g)	Linoleic acid (g)
Bacon, 2 slices	16	2.93	5.43	0.76
Beef, round	120	5.22	7.32	0.20
Butter, 1 teaspoon	5	2.36	1.54	0.14
Chicken, leg	120	0.98	2.06	0.95
Eggs, 2 medium	108	3.42	6.46	1.26
Milk, skim, 1 cup	246	0.11	0.08	0.01
Milk, whole, 1 cup	244	5.24	3.79	0.55
Oil, corn, 1 tablespoon	14	1.56	11.74	5.63
Turkey, dark meat	120	0.98	2.06	0.95

SOURCE: *American Egg Board (1974).*

nutrients, is carried to the tissues by the blood. Because it has some chemical properties similar to those of fat, most of the cholesterol is transported throughout the body in combination with fatty acids.

Scientists are interested in the blood (serum) cholesterol, because it and fatty acids are found in thickened inner walls of the arteries of people with atherosclerotic heart diseases. Studies are still being made to determine whether there is a relationship between these deposits and serum cholesterol levels and to discover how the deposits form. Good evidence is accumulating which indicates that lowering the total fat content of the diet and raising the ratio of polyunsaturated to saturated fatty acids will decrease serum cholesterol.

There is no clear indication that lowering the serum cholesterol will reduce the incidence of heart attacks or strokes. In 1979 the American Medical Association Council on Scientific Affairs reported "It cannot be assumed that the proportions of saturated and unsaturated fat and the levels of cholesterol in the diet are of universal importance. For healthy people, moderation in fat intake should become the rule of thumb." In 1980 the Food and Nutrition Board of the National Academy of Sciences published a report, "Toward Healthful Diets," in which they state that "no significant correlation between cholesterol intake and serum cholesterol concentration has been shown in free-living persons in this country. . . . For these reasons, the Board makes no specific recommendations about dietary cholesterol for the healthy person."

The Dietary Guidelines for Americans as published by the U.S. Department of Agriculture and the U.S. Department of Health, Education, and Welfare in 1979 state, "Eating extra saturated fat and cholesterol will increase blood cholesterol levels in most people." They further state, "There is controversy about what recommendations are appropriate for healthy Americans. For the U.S. population as a whole, reduction in our current intake of fat, saturated fat and cholesterol is sensible." The discrepancies in the recommendations of the Food and Nutrition Board and the Dietary Guidelines is an example of the depth of the controversy relative to cholesterol levels in our diet. With such disagreement, the Food and Nutrition Board recommends

1. Selection of a nutritionally adequate diet from the foods available, by consuming each day appropriate servings of dairy products, meats or legumes, vegetables and fruits, and cereal and breads.
2. Selection of as wide a variety of foods in each of the major food groups as is practicable in order to insure a high probability of consuming adequate quantities of all essential nutrients.
3. Adjustment of dietary energy intake and energy expenditure so as to maintain appropriate weight for height; if overweight, achieve appropriate weight reduction by decreasing total food and fat intake and by increasing physical activity.
4. If the requirement for energy is low (e.g., reducing diet), reduction in consumption of such foods as alcohol, sugars, fats, and oils, which provide calories but few other essential nutrients.

Data supporting the absence of a recommendation relative to dietary cholesterol by the Food and Nutrition Board have since been published by O'Brien and Reiser (1980) and Flaim *et al.* (1981). Adult males fed up to four eggs daily (1400 mg cholesterol total daily intake) showed no evidence of increased plasma cholesterol levels.

The fatty acid distribution and cholesterol content of whole eggs and egg yolks have been studied extensively. Some of the reported results vary widely or actually conflict. Certain authors, in reporting their experiments, indicate some reasons for the apparent differences. For example: (1) To be valid, egg lipid assays must be made under strictly comparable conditions in the laying cycle (Coppock and Daniels 1962). (2) Present methods for lipid analysis may not be accurate enough to measure the order of change that occurs in some experiments. (Marion and Woodroof 1968). (3) Within a strain of hens, the size of the yolks is inversely related to the cholesterol concentration of the yolks (Nichols *et al.* 1963). (4) The methods used in many early studies to determine the effect of dietary fat on egg yolk cholesterol were inaccurate, since some of them reported increase as the result of the color produced by the unsaturated fatty acids rather than of increased cholesterol (Weiss *et al.* 1964).

Many studies have been done to determine the effect of diet on the fatty acid and/or cholesterol content of yolks. In one series of experiments, eggs produced by hens on a stock diet and a low-fat diet were compared with those of hens on diets that contained various percentages of safflower, linseed, cottonseed, corn, and soybean oils. The linoleic acid content of the eggs from hens fed cottonseed or safflower oil was approximately proportional to the amount of linoleic acid in the diet. Linoleic acid was deposited less efficiently in the eggs from hens fed diets containing linolenic acid (derived from linseed or soybean oil) than in eggs from hens fed cottonseed or safflower oil. The research suggests that linolenic acid might have an inhibitory or antagonistic effect on incorporation of linoleic acid in the yolk fat (Wheeler *et al.* 1959).

The fatty acid composition of the dietary fat also affects the cholesterol content of the egg. For instance, eggs produced by hens fed for 3 weeks on a diet that contained 30% safflower oil showed a 19% increase in cholesterol, as compared with eggs produced by the same hens on the basal diet. In contrast, a 30% linseed oil diet fed for 3 weeks produced eggs with 27% more cholesterol than those from hens on the basal diet (Weiss *et al.* 1964).

The effect of husbandry (battery-housing, deep-litter system, and semi-intensive free range with unlimited access to grass) has been compared with the effect of diet (basic control diet, and this diet plus 4% peanut oil or beef tallow) on the essential fatty acid (EFA) content of the body fat of the hens and of the eggs. Analysis of body fat showed an early response to the dietary inclusion of EFA (peanut oil). However, some time had to elapse before the level of EFA in body fat and in the eggs produced reached a constant value related to the dietary EFA. It was concluded that there was no significant difference between EFA content of eggs produced in the husbandry experiment and those produced on the special diet (unless hens had access to grass). Thus there was no proof that eggs produced by any modern system will increase human atherosclerosis. The desirability of increasing the normal level of EFA in the hens' rations by supplementation with vegetable oils, such as peanut or corn oil, was questioned by Coppock and Daniels (1962).

Changes in the composition of shell eggs during storage at 12.8°C for 28 to 40 days were studied. The expected physical changes occurred, but the differences between fresh and stored eggs were small in values for total lipids, individual lipid fractions, total fatty acids, or fatty acid composition of lipid fractions (Marion and Woodruff 1968).

The literature on the effects of egg lipids on serum lipids in man is extensive, and the results reported are often contradictory. For example, large amounts of egg yolk lipid (50 to 100 g per day) were fed to 13 human subjects; nine subjects showed little or no change in plasma lipids and three had moderate increases of phospholipids and plasma cholesterol. The other subject showed a prompt and maintained increase in phospholipids and cholesterol. It was felt that the effect was the result of the action of several lipid fractions and not of egg-yolk cholesterol (Splitter et al. 1965).

Since a specific weight of egg yolk lipid causes a higher and more rapid increase in the serum cholesterol level than does the same weight of any other dietary fat, experiments were carried out for 4 years on a 46-year-old man to determine which fractions of egg yolk lipid were responsible for the increase. The acetone-soluble fraction (cholesterol or other sterols) proved to be the cause of the rapid rise in serum cholesterol levels. If this fraction was saponified, the unsaponifiable and saponifiable fractions had to be fed together to increase the serum cholesterol levels of the patient. Further experiments showed that daily intake of between 40 and 500 mg cholesterol was required to affect the patient's serum level. Such amounts are below the usual daily intake with non-vegetarian diets (Wells and Bronte-Stewart 1963).

In a human feeding study, a group of men were equally divided with one group receiving no visible eggs and the other group getting two eggs per day per man. After several weeks on these diets, neither group showed any change in serum cholesterol levels (Flynn et al. 1979). Similarly, in one of many other studies, various fats and fat by-products were added to typical diets for laying hens (Summers et al. 1966). When highly saturated fats were added, the content of oleic acid in the egg yolks increased and the linoleic acid decreased. The more unsaturated fats, such as soybean and safflower oil, caused a marked increase in linoleic acid and a decrease in oleic acid as compared with the basic diet. The cholesterol level of the egg yolk increased with an increase in the degree of unsaturation of the dietary fat.

Although experiments have demonstrated that eggs with a 1:1 ratio of polyunsaturated and saturated acids can be produced, the implication of such a change should be considered. The normal egg and the 1:1 egg provide essentially the same amount of saturated fatty acids, the major change being in the relative amounts of oleic and linoleic acids. (Table 7.2 lists the total fat, saturated fatty acid and oleic and linoleic fatty acid content of an ordinary egg.) Since the average egg contains almost 6 g of fat, about 0.9 g of linoleic acid replaced the same weight of oleic acid in the 1:1 egg. Summers et al. (1966) believe that if the average person consumed about 100 g of fat daily, and ate two of the 1:1 eggs rather than two regular eggs per day, his increased consumption of polyunsaturated fatty acids would be about 2% at the expense of oleic acid. Whether producing 1:1 eggs, even if economically practical, would result in a significant change in the pattern of fatty acid intake by humans, is a matter of speculation.

On the other hand, ordinary and modified eggs containing more polyunsaturated fatty acids were compared with a diet known to be effective in reducing serum cholesterol in man (Brown and Page 1965). The ordinary eggs contained 13% polunsaturated fatty acids and the modified eggs, 45%. Both types had a saturated fatty acid content of about 35%. A series of 18-day dietary experiments were carried out with five normal men fed daily a vegetable oil diet with (1) no eggs, (2) two ordinary eggs, and (3) two modified eggs. The three diets were similar in total fat. The cholesterol contents of the diets were (1) 210, (2) 715, and (3) 845 mg, respectively. The diet with no eggs reduced the serum cholesterol by 19%, but the two diets containing eggs caused no change. It was concluded that eating two eggs daily completely inhibited the effectiveness of a diet that would otherwise reduce serum cholesterol. We might interpret these results to show that the relatively small beneficial changes resulting from the increase of polyunsaturated fatty acids in the modified eggs are balanced by the increased cholesterol content, or by other components that tend to raise serum cholesterol.

POSSIBLE ADVERSE EFFECTS OF EGGS IN CERTAIN DIETS

The first solid foods given to infants are often those with high iron content. The time at which the pediatrician recommends such foods usually depends on the baby's nutritional status rather than on its age. Hard-cooked egg yolk has frequently been one of the first such supplementary foods given to infants. Since the protein of raw or slightly cooked eggs is normally absorbed directly into the bloodstream, it may be antigenic, especially to infants. Thus some infants develop allergic reactions to eggs.

Infants with adequate stores of iron at birth do not need iron-rich foods until they are about 6 months old. If egg yolk is given as a source of iron, it should be hard-cooked, and fed with as much white removed as possible. Home-prepared egg yolk may retain enough white to cause allergic reactions, especially if the infant has a genetic background of allergy. Sterilized, canned yolks have less tendency to be allergic than do home-prepared yolks. Because of possible allergy, pediatricians may prescribe other iron-rich foods, or they may advise omitting egg white until the baby is 9 to 12 months of age. Of course, some infants, children, and adults are allergic to any and all forms of eggs.

In studies on the toxic effects of raw egg white, diets containing large amounts of raw egg white proved to be toxic to test animals and human subjects. Numerous studies on the composition of unheated whites showed that a protein, avidin, was the cause of such toxicity. Only about 0.05%, avidin is present in egg white. Heat inactivates avidin (Parkinson 1966).

Avidin can combine with biotin and render this essential vitamin unavailable, but the possibility of its causing a vitamin deficiency in humans is highly unlikely. Since biotin is widely distributed among common foods, including egg yolk, and avidin is easily inactivated, the ordinary use of raw or underheated egg whites would seldom, if ever, produce a shortage of biotin. An exception might occur with special hospital or emergency diets in which raw eggs are served often and food is limited, thus resulting in a low biotin intake.

Studies have been conducted to show the effect of feeding different levels of raw egg white powder as well as benzylpenicillin or caffeine to rats (Boyd et al. 1966). Greater and more pronounced toxic reactions occurred when the diets contained more raw egg-white powder. When either drug was also given, the combined toxicity was greater than expected. Another series of experiments with adult female and young male rats tested the direct toxic effects of diets that included from 20 to 100% raw egg white powder (Peters 1967). Including as little as 20% egg white produced measurable signs of toxicity, such as a 50% increase in the amount of urine excreted, high output of urinary protein, and an increased intake of water. The pH of urine of all rats receiving egg white increased to 9 as compared with 6.8 for the controls fed laboratory chow. Rats receiving 80% or more egg white exhibited acute toxic reactions, including significantly lower food intake and inhibited growth, loss of body weight, progressive decline in carcass weight, anorexia (lack of desire for food), glycosuria (sugar in urine), soft stools or actual diarrhea, and a six-fold increase in volume of urine. Toxicity was not prevented by a biotin supplement, but it was largely prevented by heat denaturation of the egg white powder. The rats tolerated 80% of denatured egg white well. The author concluded that the direct toxic effects were due to the large amounts of dietary raw egg white powder and not to a biotin deficiency.

Scientists have long recognized that ovomucoid, a protein in fresh egg white, is a factor responsible for the antitryptic activity of eggs. A more recent study has described the isolation of ovoinhibitor, another protein in egg white, which is a more effective and powerful trypsin inhibitor than ovomucoid. Trypsin is a protein-digesting enzyme secreted by the pancreas. Since large amounts of raw egg white must be ingested in order to significantly inhibit such action by this enzyme, adverse effects might occur in laboratory experiments but not with the usual human diets. (Matsushina 1958).

Experiments have shown that under exacting experimental conditions eggs can be carcinogenic to mice, rats, and chickens (Szepsenwol 1959, 1966; Szepsenwol et al. 1966), but there is no proof that eggs cause cancer in humans. One report (Hradec 1958) stated that the malignant tumor-enhancing factor seemed to be identical with the growth-promoting factor in egg yolks. However, another report indicated that material present in whites also might contain a factor having a carcinogenic effect (Szepsenwol 1959). In spite of these studies, eggs have not been proved to cause human malignancy. The diets fed to the animals contained proportions and amounts of eggs that would be far in excess of the number of eggs any person would normally eat daily.

SOME MISCONCEPTIONS ABOUT THE NUTRITIVE VALUE OF EGGS

Although certain factors, as previously discussed, have a positive influence on the nutrient composition of eggs, some persons erroneously believe that still others also affect nutritive value.

In some regions of the United States, consumers consider egg-shell color an indication of high quality and higher nutritive value. They are willing to pay more for eggs with the desired shell color. The color of the shells, whether dark or white, is directly related to the breed or strain of hens and has no effect on the nutrient composition of the eggs.

Many people think that all foods with a deep yellow color are higher in nutritive value than those of a lighter shade, and that darker yolks are therefore nutritionally better than lighter ones. Xanthophyll, the major substance causing yolks to have a deeper color, has no nutritive value. The idea that yolk color influences nutrient composition may have originated years ago when hens had the run of the farm and ate grass, which contains xanthophyll, as well as such essential vitamins as A and D. Present rations for commercial egg production are fortified with large amounts of synthetic vitamins, particularly the colorless, fat-soluble vitamins A and D. These rations include enough xanthophyll-containing ingredients to produce yolks with the desired medium-yellow color.

Some so-called nutrition "experts" state that fertile eggs are more nutritious than nonfertile eggs. No scientific proof exists to confirm such a recommendation, with the possible exception that the developing embryo might provide slightly more nutrients. Of course, if an embryo has developed enough to significantly increase the nutrient content, few people would wish to use the egg. Fertile eggs are more expensive to produce and deteriorate more rapidly than do nonfertile eggs.

"Organic" eggs may be promoted as much safer and more nutritious than those produced by hens on the usual commercial rations. Since pesticides, fungicides, herbicides, and commercial fertilizers must be proved safe before they can be used in the prescribed amounts at specific times in crop production, there is no reason for concern about eating eggs produced by hens on commercial rations. Organic eggs are no higher in nutritive value than regular eggs. If the ration for hens on an organic diet is not so well balanced as the usual commercial laying ration, the nutritive value of organic eggs tends to be lower. Too few organic eggs could be produced to meet the entire consumer demand, and such eggs are much more expensive.

Some persons consider raw eggs more digestible than cooked, perhaps because eggnogs are sometimes included in convalescent and geriatric diets. Cooked eggs are more readily digested than raw, but ultimately both are very completely digested and absorbed.

SELECTED REFERENCES

American Egg Board 1974. A Scientist Speaks About Eggs. American Egg Board, 1460 Renaissance Drive, Suite 301, Park Ridge, IL 60068.

Balnave, D. 1968. Influence of dietary linoleic acid on egg fatty acid composition in hens deficient in essential fatty acids. J. Sci. Food Agric. 19, 266–269.

Bandemer, S. L., Evans, R. J., and Davidson, J. A. 1958A. Seasonal variations in the vitamin A content of hens' eggs. J. Agric. Food Chem. 6, 549–552.

Bandemer, S. L., Evans, R. J., and Davidson, J. A. 1958B. The vitamin A content of fresh and stored shell eggs. Poult. Sci. 37, 538–543.

Boyd, E. M., Peters, J. M., and Krynen, C. J. 1966. The acute oral toxicity of reconstituted spray-dried egg white. Ind. Med. Surg. 35, 782–787.

Brown, H. B., and Page, I. H. 1965. Effect of polyunsaturated eggs on serum cholesterol. J. Am. Diet. Assoc. 46, 189–192.

Bunnell, R. H., Keating, J., Quaresimo, A., and Parman, G. K. 1965. Alpha-tocopherol content of foods. Am. J. Clin. Nutr. 17, 1–10.

Carmon, L. G., and Huston, T. M. 1965. The influence of environmental temperature upon egg components of domestic fowl. Poult. Sci. 44, 1237–1240.

Coppock, J. B. M., and Daniels, N. W. R. 1962. Influence of diet and husbandry on the nutritional value of the hen's egg. J. Sci. Food Agric. 13, 459–467.

Cotterill, O. J., and Geiger, G. S. 1977. Egg product yield trends from shell eggs. Poult. Sci. 56, 1027–1031.

Cotterill, O. J., and Glauert, J. L. 1979. Nutrient values for shell, liquid/frozen, and dehydrated eggs derived by linear regression analysis and conversion factors. Poult. Sci. 58, 131–134.

Cotterill, O. J., and Glauert, J. L. 1981. Shell egg nutrition labeling. Poult. Trib. 87(10), 16.

Cunningham, F. E., Cotterill, O. J., and Funk, E. M. 1960. The effect of season and age of bird. 2. On the chemical composition of egg white. Poult. Sci. 39, 300–307.

Edwards, H. M., Jr., Driggers, J. C., Dean, R., and Carmon, J. L. 1960. Studies on the cholesterol content of eggs from various breeds and/or strains of chickens. Poult. Sci. 39, 487–489.

Evans, R. J., Bandemer, S. L., and Davidson, J. A. 1967. Lipids and fatty acids in fresh and stored shell eggs. Poult. Sci. 46, 151–155.

Everson, G. J., and Sanders, H. J. 1957. Composition and nutritive importance of eggs. J. Am. Diet. Assoc. 33, 1244–1254.

Flaim, E., Ferreri, L. F., Thye, F. W., Hill, J. E., and Ritchey, S. J. 1981. Plasma lipid and lipoprotein cholesterol concentrations in adult males consuming normal and high cholesterol diets under controlled conditions. Am. J. Clin. Nutr. *34*, 1103–1108.

Flynn, M. H., Nolph, G. V., Flynn, T. C., Kakos, R., and Krause, G. 1979. Effect of dietary egg on human serum cholesterol and triglycerides. Am. J. Clin. Nutr. *32*, 1051–1057.

Food and Nutrition Board 1980. "Toward Healthful Diets." National Academy of Sciences, National Science Foundation, Washington, DC.

Forsythe, R. H. 1963. Chemical and physical properties of eggs and egg products. Cereal Sci. Today *8*, 309–312, 328.

Hanning, F. 1958. The effect of a plastic coating of shell eggs on changes which occur during storage and cooking. Poult. Sci. *36*, 1365–1369.

Hardinge, M. G., and Crooks, H. 1961. Lesser-known vitamins in foods. J. Am. Diet. Assoc. *38*, 240–245.

Hradec, J. 1958. Nature of the carcinogenic substance in egg yolks. Nature (London) *182*, 52–53.

Hudspeth, J. P., Newell, G. W., White, D. R., Berry, J. G., and Morrison, R. D. 1965. An electrophoretic study of chicken egg albumen. Poult. Sci. *44*, 149–158.

Jordan, R., Vail, G. E., Rogler, J. C., and Stadelman, W. J. 1962. Further studies on eggs from hens on diets differing in fat content. Food Technol. *16*, 118–120.

Kline, L., Meehan, J. J., and Sugihara, T. F. 1965. Relation between layer age and egg-product yields and quality. Food Technol. *19*, 114–119.

Leung, W. W., and Flores, M. 1961. Food Composition Table for Use in Latin America, pp. 80–81. The Institute of Nutrition of Central America and Panama, and the Interdepartmental Committee on Nutrition for National Defense, National Institutes of Health, Bethesda, MD.

Leung, W. W., Pecot, R. K., and Watt, B. K. 1952. Composition of foods used in Far Eastern countries. U.S. Dep. Agric., Agric. Handb. *34*, 38.

Marion, J. E., and Woodroof, J. G. 1968. Lipid changes in shell egg composition during storage. Food Technol. *22*, 333–335.

Matsushima, K. 1958. An undescribed trypsin inhibitor in egg white. Science *127*, 1178–1179.

May, K. N., and Stadelman, W. J. 1960A. Factors affecting protein and threonine in hens' eggs. J. Am. Diet. Assoc. *37*, 568–572.

May, K. N., and Stadelman, W. J. 1960B. Some factors affecting components of eggs from adult hens. Poult. Sci. *39*, 560–565.

Mountney, G. J. 1976. Poultry Products Technology, 2nd Edition. AVI Publishing Co., Westport, CT.

Nichols, E. L., Marion, W. W., and Balloun, S. L. 1963. Effect of egg yolk size on yolk cholesterol concentration. Proc. Soc. Exp. Biol. Med. *112*, 378–380.

O'Brien, B. C., and Reiser, R. 1980. Human plasma lipid responses to red meat, poultry, fish, and eggs. Am. J. Clin. Nutr. *33*, 2573–2580.

Orr, M. L., and Watt, B. K. 1957. Amino acid content in foods. U.S., Dep. Agric., Home Econ. Res. Rep. *4*, 48–49.

Parkinson, T. L. 1966. The chemical composition of eggs. J. Sci. Food Agric. *17*, 101–111.

Peters, J. M. 1967. A separation of the direct toxic effects of dietary raw egg white powder from its action in producing biotin deficiency. Br. J. Nutr. *21*, 801–809.

Plack, P. A., Miller, W. S., and Ward, C. M. 1964. Effect of vitamin A deficiency on the content of three forms of vitamin A in hens' eggs. Br. J. Nutr. *18*, 275–280.

Poultry and Egg National Board 1958. Consumer Attitudes toward Eggs, Chickens, Turkey. Report of a Motivation Research Study sponsored by PENB. PENB, Chicago, IL.

Splitter, S., Michaels, G., Schlierf, G., Wood, P., and Kinsell, L. 1965. Plasma lipid responses to the feeding of egg yolk lipids. Am. J. Clin. Nutr. *16*, 390.

Summers, J. D., Slinger, S. J., and Anderson, W. J. 1966. The effect of feeding various fats and fat by-products on the fatty acid and cholesterol composition of eggs. Br. Poult. Sci. *7*, 127–134.

Szepsenwol, J. 1959. Carcinogenic effect of hens' eggs as part of the diet in mice. Proc. Soc. Exp. Biol. Med. *102*, 748–751.

Szepsenwol, J. 1966. Carcinogenic effect of egg yolk in mice of the C57 B1 strain. Proc. Soc. Exp. Biol. Med. *122*, 981–986.

Szepsenwol, J., Roman, A., and Santiago, J. 1966. Changes in the pituitary of mice on diets supplemented with egg yolk, with extracts of eggs, or with cholesterol. Proc. Soc. Exp. Biol. Med. *122*, 1284–1288.

U.S. Department of Agriculture. 1968. National Food Situation, No. 124. Superintendent of Documents, U.S. Government Printing Office, Washington, DC.

U.S. Department of Agriculture. 1972. National Food Situation, No. 141. Superintendent of Documents, U.S. Government Printing Office, Washington, DC.

U.S. Department of Agriculture. 1982. National Food Review, Winter. Superintendent of Documents, U.S. Government Printing Office, Washington, DC.

U.S. Department of Agriculture. 1979. Dietary Guidelines for Americans. USDA/HEW, Washington, DC.

Watt, B. K., and Merrill, A. L. 1963. Composition of foods—raw, processed, prepared. U.S., Dep. Agric., Agric. Hand. *8*.

Weiss, J. F., Naber, E. C., and Johnson, R. M. 1964. Effect of dietary fat and other factors on egg yolk cholesterol. The "cholesterol" content of egg yolk as influenced by dietary unsaturated fat and the method of determination. Arch. Biochem. Biophys. *105*, 521–526.

Wells, V. M., and Bronte-Stewart, B. 1963. Egg yolk and serum-cholesterol levels: importance of dietary cholesterol intake. Br. Med. J. *1*, 577–581.

Wheeler, P., Peterson, D. W., and Michaels, G. D. 1959. Fatty acid distribution in egg yolk as influenced by type and level of dietary fat. J. Nutr. *69*, 253–260.

8

Merchandising Eggs in Supermarkets

Eric C. Oesterle

Eggs account for 1.3 to 1.6% of the total sales in supermarkets in the United States. They contribute a similar percentage to total supermarket gross margin dollars. Relatively few of the items handled in these supermarkets can boast of such a record. Clearly, eggs are a major contributor to supermarket sales and profits.

Eggs suppliers know these facts. The sales records, as well as the prices charged by their supermarket customers, are the only facts necessary for the calculation of such statistics.

Supermarket managers know these facts, too. They recognize the importance of eggs in their sales and profit picture. The egg industry has helped to promote egg sales in supermarkets with in-store studies of customer shopping habits, novel packaging, and special promotions.

Consequently, today's homemakers buy most of their eggs in a supermarket. Thirty years ago they patronized the farmer and the milkman, who delivered eggs to their homes. This trend toward supermarket sales of eggs means that the egg business in supermarkets exceeds 3 billion dollars annually. It will continue to grow.

This level of sales—and profits—can be improved, however. To do so, both the supermarket manager and the egg supplier need to clearly understand what it takes to make eggs move in the supermarket display case. In some supermarkets, eggs perform far better than industry averages; but in others, egg sales fail to measure up to so-called accepted levels. Reasons for these differences can partially be explained by a close examination of the concept of self-service selling in supermarkets.

EGG SCIENCE & TECHNOLOGY,
3rd Edition

Advanced studies in this method of retailing have helped to establish some basic principles or rules which, if properly applied, stimulate sales.

Application of tested supermarket self-service principles to eggs might well encourage more homemakers to buy eggs in their favorite supermarket. Even though it is true that eggs already hold their own in supermarket sales, a 5% increase in movement of an item as prominent as eggs could result in a substantial increase in supermarket gross margin dollars, as well as additional sales of eggs by the case for suppliers.

However, the so-called "credibility gap" between the egg supplier and his supermarket customer is wide indeed. Each needs to better understand the other's problems. This is probably more true for the egg supplier, however. To him, the supermarket might well be his only customer. On the other hand, the supermarket manager has many other suppliers besides the egg processor. He well realizes that despite the encouragement of the food manufacturer and the processor to buy his particular product and to give it the best location in the sales area, he must establish priorities for his limited shelf space. And high markups mean nothing, if the product does not move.

Egg suppliers—and other suppliers as well—stand to benefit by "thinking retail." An updating of the old Chinese proverb to "wearing the coat of the supermarket manager" might well apply here.

Eggs Are a Dairy Item in the Supermarket

To the supermarket manager, eggs are just one of 12,000 items that make up his product mix. To control this conglomerate of merchandise, both food and nonfood, the market manager departmentalizes his wares as grocery, meat, produce, dairy, frozen food, nonfoods, etc.

Eggs are classified as dairy products and are one of the 300 to 400 items that together are known as the dairy department. Consequently, supermarket operators do not tend to look at an individual item in terms of total store sales. They are concerned rather with the contribution of the different departments to the total sales and profit picture. Individual products are evaluated in terms of their role in a given departmental grouping.

The operating data for individual items in a department are time consuming and costly to generate. They are almost nonexistent at the supermarket level. Sales by department are recorded daily in the checkout; however, most supermarket operating statements detail sales by three departments—grocery, meat, and produce. But, the reporting of item movement and gross margin return is still in the infancy stage. Thus, eggs and the dairy department are recorded as part of grocery sales. Some supermarket statements departmentalize dairy sales. For the most part, however, supermarket managers do not know the specific role of their dairy department as far as its sales and gross margin are concerned. As a result, management thinking is geared to improving sales and gross margin—by departments. Eggs are a dairy item. How important a role do they play in dairy sales and margins? How can they help to increase dairy sales and gross margins?

These are the facts that food retailers can use in their planning. If egg suppliers would "wear the coat" of the supermarket operator and think retail, their approach would be directed toward the dairy department and its role in supermarket sales and gross profits.

Milk Industry Research Benefits the Egg Industry

In 1966, the Milk Industry Foundation provided funds for a continuing, detailed study of supermarket dairy departments by food distribution specialists at Purdue University. Results of this effort have proved most valuable in identifying the contribution of the dairy department to total supermarket sales and gross margin dollars. But of greater importance, these studies identified the role of specific dairy items and their contribution to total dairy sales and gross margin dollars. Space and inventory investment for each dairy item were also quantified. This research pointed the way for productive merchandising strategies which, when implemented, increased movement of dairy items with substantial benefit to both retailers and dairy product suppliers. The egg industry benefited from these studies. The majority of the meaningful operating statistics and merchandising strategies for eggs that are reported in this chapter are the direct result of this ongoing research.

Role of the Supermarket Dairy Department

The supermarket dairy department ranks third in sales and gross margin contribution (Table 8.1). Accounting for approximately 11 cents for every dollar of sales and a somewhat similar amount for every dollar of total gross margins, the importance of this department is unquestioned. Few supermarket operators have dairy department operating data, however.

ROLE OF EGGS IN THE SUPERMARKET DAIRY DEPARTMENT

Dairy department operating results, as summarized in Purdue University research studies, divide the dairy department into ten product

TABLE 8.1. Supermarket Department Operating Data, Independent Supermarkets, 1975

Department	Sales (per 100 dollars)	×	Gross margin ratio (%)	=	Contribution to total gross margin (dollars)	(%)
Grocery	$55.9	×	14.2	=	$7.94	45.76
Meat	20.7		18.5		3.82	22.02
Dairy	11.2		20.8		2.33	13.43
Frozen foods (includes ice cream)	6.4		25.6		1.64	9.45
Produce	5.8		28.0		1.62	9.34
Total	$100.0	×	17.35	=	$17.35	100.00

SOURCE: *Oesterle, 1975.*

groups (Table 8.2). Research has established the relative importance of eggs in terms of sales performance and return on inventory and space investment among other products sold in the dairy department. Some of the facts brought to light by this research are summarized as follows:

Sales: Eggs accounted for 11.5% of dairy department sales, ranking a strong fourth in the ten product classifications.

Gross Margin Contribution: Eggs contributed 11.4 cents of every gross margin dollar earned by the dairy department.

Inventory Turnover: The basic inventory of eggs turned over 3.5 times per week. Few other dairy or supermarket items in general can boast of such a record. Turnover for all dairy items averaged 1.3 times per week.

Display Space in the Dairy Case: Eggs were allocated 8.9% of the display space in the dairy department.

Implications from Dairy Department Operating Data

Examination of dairy department financial and operating data raises the question as to why eggs are not allocated shelf space commensurate with their sales and profit performance. A basic principle of self-service selling in supermarkets is to allocate space to a given product in relationship to the sales and profit performance of that product. Yet eggs, which account for 11.4% of dairy department gross profit dollars and which enjoy a high inventory turnover, and perform such feats despite the limited display space they are provided.

The limited display space for eggs can be dramatically illustrated by a study of the operating statistics for another dairy product group—biscuits, dinner rolls, cookies, and pastry (Table 8.2). These items are relatively new to the dairy department and account for approximately 2.2% of sales for this department. Their contribution to the dairy department gross margin dollar and inventory turnover are minimal, to

TABLE 8.2. Supermarket Dairy Department Financial and Operating Data in 150 Midwestern Supermarkets, 1973

Product group	Product performance		Inventory evaluation		Space evaluation: display shelving (%)
	Department sales (%)	Contribution to total dollar gross margin (%)	Inventory turnover	Gross margin dollar return of inventory investment	
Milk	36.2	34.0	4.1	83.0	18.5
Cheese	17.6	21.0	0.4	11.9	25.6
Margarine and butter	15.0	10.3	0.7	9.9	15.3
Eggs	12.3	11.4	2.2	43.9	8.4
Cultured products	7.3	8.9	1.5	43.8	9.5
Biscuits, dinner rolls, cookies, pastry	2.6	3.1	0.5	13.7	8.5
Cream and cream substitutes	2.1	2.9	1.6	52.0	2.2
Milk beverages	1.9	2.0	3.1	81.2	2.2
Juices (7 items)	1.0	2.0	2.0	2.0	3.0
Miscellaneous	5.0	6.4	0.8	20.3	9.8
Total	100.0	100.0	1.1	23.4	100.0

SOURCE: *Supermarket Dairy Department Financial and Operating Data, 141 Midwestern Firms Reporting, 1973; Department of Agricultural Economics, Purdue University, West Lafayette, IN.*

say the least. However, cookies and pastry enjoy almost the same amount of display space allocated to eggs.

Closer examination of operating data for other product groupings reveals similar disparities. One might well comment that generally the items in dairy departments are out of balance as far as sales and profit performance and allocated display space are concerned. Such misallocations of space can mean several things. First, the true profit potential of this department has yet to be fully realized in the supermarket. Second, imbalance of space and sales means frequent out-of-stock situations in the dairy display, and sales of fast-moving, profit-producing items suffer. Third, frequent restocking of fast-moving items means excessive labor costs.

A further breakdown of eggs into size classifications of medium, large, and extra large, as summarized in the study of 50 selected supermarkets, indicated that an imbalance existed between movement and space allocated for packages of eggs by size, as well as for the total egg display (Table 8.3). This was particularly true for extra-large eggs, which occupied approximately one-third of the egg display, yet generated only 18% of total egg sales. However, the sales-space balance among sizes of eggs in the egg display was not nearly so critical as the sales-space balance for eggs in the whole dairy case.

Why the Imbalance in Sales and Space for Dairy Items?

First, revealing as this information is, it must be remembered that such data are not typically generated by supermarket accounting or control systems. Current research using electronic data-processing equipment is paving the way for more operating statistics on individual item performance.

Second, the introduction of "new" items weekly on the supermarket scene is almost an accepted practice. With shelves stocked to capacity, the addition of a new item may be made either in the space freed by discontinuing a slow-moving product, or in space made available by reducing the space allotted to existing items.

TABLE 8.3. Egg Sales, Gross Margin Contribution, and Display Shelving Occupied by Size Classification in 150 Midwestern Supermarkets, 1973

Size classification of eggs	Dollar egg sales in dozens (%)	Contribution to total dollar gross margin for eggs (%)	Egg display shelving (%)
Medium	30.7	33.7	25.9
Large	53.6	50.0	44.5
Extra large	15.7	16.3	29.6
Average for 50 supermarkets	100.0	100.0	100.0

Since operating data on individual item movement are not often available, new-item introduction to the shelf is most often made by reducing facings of existing stock. More often than not, the fast-moving—and usually the most profitable—items suffer. These items, of necessity, are granted more facings than the item whose movement is average, as gauged by the number of times the display is restocked. The concept of "share the shelf" exists at the expense of sales and profits.

One need only examine the tremendous assortment of new "dairy" items in the pastry and dinner roll categories, in the dips and spreads, and in the cheese group to recognize that the introduction of this variety might well have come at the expense of eggs and milk (Table 8.2). A careful evaluation of the performance of new items is essential to preserve the balance of space and sales and profit.

Other Factors to Consider

Before one jumps to the conclusion that added display space for eggs in the supermarket dairy department could reverse the trend in egg sales and consumption, one must consider other facts. For an item to sell well in a supermarket, it must not only have display space, but a strategic location in the layout of the total sales area, as well as on the actual display shelf. Design of the package also plays a major role. The price of the item, in combination with these other factors, also helps decide the fate of a product in the self-service arena of merchandise known as the supermarket.

SCIENCE OF SELF-SERVICE—APPLIED TO EGGS

In supermarkets, customers wait on themselves. In their tour of the sales area, they spend on the average about 23 min and have available for their choice about 12,000 different items. Some of their choices are predetermined. These are the so-called shopping-list items that are necessities in the home—items which are purchased frequently. However, the majority of the purchases are made on impulse. Today fewer customers come to the supermarket with shopping lists (Table 8.4). They depend on the presentation of the items on display in the market to suggest new and different ways to satisfy their families at mealtime.

Consequently, for an item to sell in a self-service situation it must be readily observed by customers as they tour the displays. Specialists in self-service merchandising call this "exposure." In a well-merchandised supermarket a consumer is shown—is exposed—to many items, with little effort on his part. If customers decide that this product fits their menu plans, they may buy it.

Now how does one item in a myriad of 12,000 products catch the eye of

TABLE 8.4. Impact of Shopping Lists on Dairy
Purchases in Selected Lafayette Supermarkets[a]

	Customers (%)	Amount spent
Shopping list	69	$2.40
No shopping list	31	$1.80
	100	

[a] 540 observations, June 1982.

shoppers as they hurry up and down the aisles of the supermarket? Studies of thousands of customers' shopping tours have guided supermarket management in their development of a selling strategy in self-service situations. This strategy is based on the premise that certain items are frequently purchased by the majority of shoppers. Previous reference was made to the shopping-list items that are basic to the menu and to the maintenance of the home, for example, milk, soap, paper products, bread, hamburger, cereals. Supermarket strategists call these "power" items. Properly located throughout the selling area in a supermarket, these items literally "pull" or attract customers to them.

Thus by strategically locating the everyday shopping items in the layout, a supermarket manager can subtly encourage the customers to shop the majority of the displays. In so doing, they are exposed to buy by impulse the new, or seasonal, item which they had not preplanned to purchase.

Self-service selling in supermarkets depends on exposure of customer to item. This exposure depends on two factors. First, customers must see the item. Second, they should be assisted in seeing and discovering the item by attraction to a demand, or frequently purchased, item, so located that it encourages them to pass by other less frequently purchased items.

Eggs Are a Power Item

Today's family purchases approximately 1 dozen eggs each week of the year in the supermarket. Thus, eggs fall easily into the category of a shopping-list item. To the supermarket merchandiser they are a power item. When displayed in the same refrigerated case with milk, eggs will attract shoppers. Small wonder that 88 out of every 100 supermarket customers shop the dairy case.

Despite its natural attraction, the dairy department can receive even more attention if it is:

1. Placed early in the customer shopping tour,
2. Located directly across from another departmental grouping of power items,

3. Located in a corridor arrangement of display fixtures that contain no cross aisles to disrupt traffic flow and reduce exposure of customers to items on display in the dairy department,

4. Fronted by an aisle of medium width which discourages the installation of special display islands that detour traffic away from the main dairy display fixture.

This same strategy of locating power items throughout the entire sales area to expose customers to the 12,000 items on display works equally well for items grouped together in departments. Strategic locations of frequently shopped items like milk, eggs, and margarine in a given dairy department encourage customers to look over the majority of the 300 to 400 dairy products on display. When these three power items are grouped together, shoppers tend to pick up their milk, eggs, and margarine and fail to take full notice of the products in the remainder of the dairy department.

Studies of customer shopping patterns of multishelved self-service supermarket dairy displays revealed that the greatest amount of customer purchases, often as high as 78% of total purchases, were made from the bottom shelf. On the other hand, it was not unusual to observe customer purchases of only 6% from the top two shelves. Clearly, such a distribution of purchases by shelf indicated that dairy displays were poorly merchandised and lacked sales appeal (Fig. 8.1).

The reason for this imbalance of purchases was the location of fast-moving dairy items (milk, margarine, and eggs) on the lower shelf of the display case. These three product groupings account for better than two-thirds of total sales. The bottom shelf in conventional dairy display equipment is larger than the other shelves and thus virtually dictates that fast-moving items be placed there (Fig. 8.1). As a result, the eye of the customer is directed downward and across the bottom of the display. Merchandise bought by impulse on the upper three or four shelves often goes unnoticed. Dairy sales, overall, have suffered.

The Full-Vision Concept of Self-Service Displays

The majority of the conventional self-service dairy-display fixtures are designed with a spacious bottom shelf which has caused the fast-moving demand items to be placed on this lower shelf. Consequently, eggs are frequently displayed horizontally across the face of the display on the bottom shelf. Milk is usually also found in the lower shelf. It is not unusual for a customer to encounter a 5-ft lineup of eggs and a 20-ft lineup of half-gallons of milk in a supermarket dairy department. This conventional arrangement of products encourages shoppers to move along the display to purchase their everyday dairy items. But since these items are located on the lower shelf, their attention is directed

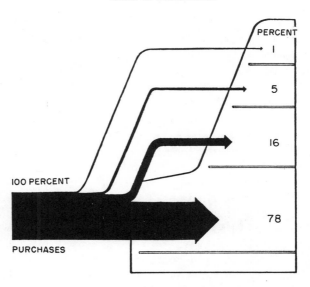

FIG. 8.1.   Conventional display arrangement of self-service supermarket dairy case.

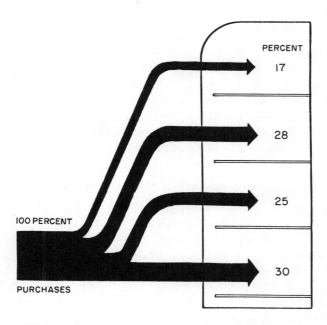

·FIG. 8.2.   Vertical or full-vision arrangement in a modi-fied air-curtain self-service supermarket dairy case.

downward, and they tend to miss the upper shelves of the display fixture. Previous reference has been made to shopping patterns for dairy cases which indicate that little strategy is being used for the placement of power items (Fig. 8.2).

The protruding lower shelf of the conventional dairy case limits accessibility to the top shelf of the display. In addition, customers must bend or stoop to obtain merchandise in the lower shelf area. Conventional dairy display equipment is not convenient for ease of shopping. The newer display equipment with no extending lower shelf but with all shelves of the same depth is much more conducive to customer shopping. In addition, it permits demand items to be displayed vertically, in full vision of the customers (Fig. 8.3).

The concept of full vision in self-service display was conceived by grocery merchandisers, who concluded from their extensive studies of customers' shopping habits that the optimum position for exposure in a self-service display is an arc extending a foot above and below eye level. Their recommendations thus placed high value on grocery shelves that were "eye level," or about 60 in. high. The shelves directly above and

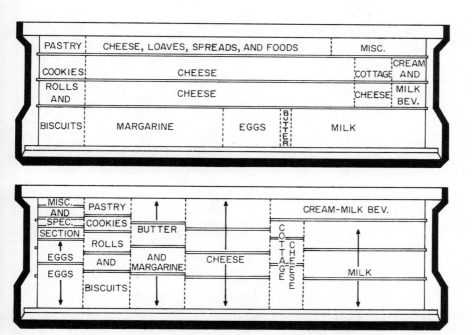

FIG. 8.3. Conventional (top) and revised product arrangement (bottom) in a self-service dairy case in the dairy department of a selected midwestern supermarket, 1968.

directly below the eye-level location were also prime locations, but ranked second to the eye-level area (Figs. 8.4A and 8.4B).

Consequently, merchandisers spaced as many of their items at eye level in the grocery section as possible. Additional stock of an item was displayed on shelves directly above and directly below. This vertical arrangement made possible the employment of power items to their fullest advantage. Result: substantial increases in grocery display gondolas of products formerly unnoticed by customers (Fig. 8.2).

This same principle can be effectively applied to self-service dairy-display equipment. Shopping patterns in dairy displays utilizing this method reflect a more uniform sale of items from the entire face of the dairy display. This is in striking contrast to the shopping patterns of the conventional arrangement where better than 95% of customer dairy purchases are made from the lower shelf (Fig. 8.1).

Strategy in Self-Service: An Application

The results of a recent study in a midwestern supermarket related misallocation of space and poor product arrangement to lack of sales and gross profit dollars in the dairy department. As a result, 168 slow-moving, unprofitable items from the original product group of 32 were discontinued. Display space was then reallocated among the remaining 155 dairy items in accordance with individual product movement and profit. In addition, products were displayed in family groupings—milk, eggs, margarine, and butter. A vertical or full-vision display of products was used when possible. Demand and impulse items were strategically dispersed throughout the display.

Results for the Dairy Department. The application of these basic self-service principles to the dairy department resulted in the following changes:

Sales/week	Increased	$700	+18.6%
Gross dollar profits/week	Increased	$100	+15.2%
Gross profit percentage	Decreased	1.1–1.77	− 0.3%
Inventory turns/week	Increased		+24%
Gross return on investment	Increased	23–29%	+25%
Wage costs/week	Decreased	$40.00	
Customer complaint	One during the entire 4 weeks		

Results for Eggs. Space allocated to eggs was increased from 8.8 to 12% of the total display shelving. A full-vision or vertical display of eggs by size was developed. Eggs were positioned at the end of the display to encourage customers to shop the entire dairy department. This position, combined with the vertical-display technique, made the egg display an impressive one indeed.

FIG. 8.4A. Vertical eye-level displays of eggs place the product in the greatest area of consumer exposure. Pilot studies of vertical displays in strategically merchandised supermarket dairy departments have produced 30% increases in egg sales.

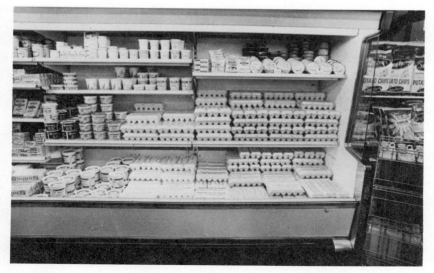

FIG. 8.4B. To be effective, vertical displays must be placed so as to appear fully stocked.

The results of these display changes were:

Egg sales/week	Increased	$156	+35.0%
Gross dollar profits/week	Increased	$30	+32.0%
Gross profit percentage	Decreased		− 1.0%
Inventory turns/week	Increased	1.4–2.0%	+30.0%
Gross return on investment	Increased	39.6–55.0%	+42.0%

In addition, similar results could be reported for most product groups; space and sales and profits were in better balance. The dairy department had been merchandised in accordance with basic self-service principles.

Summarizing, egg sales can be improved in supermarket dairy departments with the proper allocation of space and the application of basic self-service display techniques. However, before space can be reallocated in accordance with sales and profits, the individual performance of all items in the dairy department must be quantified.

When display space and product movement are in balance and the best self-service display techniques are employed, sales and gross margins for the total dairy department should be optimal. However, this level of sales is the direct result of the optimal performance of the individual products, the combined effect of which is far greater than if these products were sold individually without any influence from other items in the dairy case.

EGG PACKAGING

It is not the intent of this discussion to explore the vast area of egg packaging but rather to describe the tremendous impact of space and self-service practices on the sales of eggs in conventional cartons in the supermarket. In addition, the author feels that far more attention has been given to egg packaging and its affect on consumer purchases than has been given to the study of egg movement in a self-service environment, the supermarket. Even though the beneficial results of the latter approach have already been substantiated, implementing this approach presents a real challenge to the egg industry.

Certainly, an individual egg carton that appeals to the homemaker in color and label design and at the same time embodies the self-service principle of exposure, best complements its merchandising in the supermarket. For, once the strategy of location of the department and the placement of items within a department have exposed the customer to the actual egg display, her final decision might well be determined by the attractive label on the carton, a label that suggests a tempting dish based on the use of eggs as an ingredient.

Actually, customers respond more to pictures of eggs in the cooked form on cartons rather than to pictures of chickens and farms (Table 8.5). However, the majority of egg cartons still have illustrations of chickens, barns, and farms in their main label themes. Furthermore, many dairy departments are encased in a simulated barn, complete with hay fork and a bale of hay!

Recently, supermarket fixture manufacturers developed a display case especially designed to handle eggs in the wire master container. This equipment provides for rapid shelf stocking of the product and attractively displays the egg carton for the customer. In addition, this equipment is fully adaptable to eye-level or vertical merchandising (Fig. 8.5). However, many supermarket managers have installed these wire containers in an effort to gain more space for the many new refrigerated items that are now sold in the dairy case. In some instances, the author has observed eggs displayed in a section of the store far removed from the dairy department.

None of the dairy department studies conducted at Purdue University has allowed for the evaluation of the freestanding egg display as far as its ability to stimulate increased egg sales. However, when strategic merchandising concepts were implemented in two supermarket dairy departments equipped with both conventional dairy fixtures and freestanding egg merchandisers, increases in total dairy department sales were not marked, as compared to sales in two other supermarket dairy departments with conventional dairy display equipment containing the total mix of dairy items (Table 8.6).

Unquestionably, the installation of the separate egg-display cases in the two supermarkets had provided eggs with adequate space in relation to sales. However, once the egg display had been relocated to the

TABLE 8.5. Response of Homemakers to Situations Where Eggs Were Mentioned

Picture or situation thought of when eggs were mentioned	Homemakers responding (%)
Typical American breakfast	37
Food value	3
Baking cakes, etc.	4
Package, shopping, or display	3
Other miscellaneous; nonchicken or nonfarm	13
Big, fresh, white, clean eggs	15
Chickens, barns, and farms	25
	100

TABLE 8.6. Results of Strategic Merchandising in Dairy
Departments with and without Separate Egg-Display
Fixtures, 1969

| | Percent change | |
Operating standard	Separate egg case	Complete dairy case
Dairy sales/week	+13	+23.5
Egg display frontage	0	+68
Egg sales/week	+ 4	+25
Egg sales to total dairy sales	+12.4	+17

other departments, sales of eggs in these departments accounted for 17% of total dairy department sales, in contrast to the 13% of department sales generated by eggs in the egg-display fixture.

The efficient manner in which eggs can be stocked in freestanding egg displays cannot be questioned; nor can the fact that such display units often provide more space for eggs in relation to their movement and earned profits. The question remains, however, as to their effect on dairy sales—and egg sales as well—that is, when eggs are no longer merchandised as part of the product mix of the dairy department.

This question still remains, even though the egg-display fixture is located adjacent to or in between dairy-display fixtures. Such a combination of displays fixes the location of eggs and does not allow for the strategic location of eggs with regard to other items in the dairy-product mix. In addition, egg-display space is more or less determined by the capacity of the egg merchandiser. This space may or may not meet the needs of a dynamic, well-merchandised dairy department. Finally, the freestanding egg display that is not positioned near the dairy department may not receive the full attention of the dairy clerk responsible for the ordering, stocking, and merchandising of eggs.

Shelf Pricing Preferred over Carton Pricing

The adoption of the Unit Pricing Code (UPC) and front-end scanning by supermarkets is gaining momentum. And there is a noticeable trend in the industry to shelf price rather than unit price merchandise, especially fast-moving items. Consequently, in these situations eggs are often shelf priced rather than carton priced. Dairy department managers adopting this practice report a substantial reduction in egg breakage. Store managers report a decided reduction in labor costs when fast-moving items are not unit priced.

A well-documented study of the impact of shelf pricing on the sale of eggs in supermarkets revealed that egg sales increased 8% when this pricing technique was adopted. In-depth interviews with consumers revealed that shelf pricing served as point-of-purchase (POP) material, in addition to alerting the customer to the current price of eggs. Consumer response to shelf pricing was overwhelmingly preferred. The study also revealed that consumers favored eye-level positioning of price tags, which in turn clearly designated the size as well as the price of eggs as shown in Fig. 8.5.

Assembly Line to Display Packaging

The possibilities of a mobile cart designed both to transport as well as to store eggs offers a real challenge to the industry. Such a cart could be loaded at the packing plant with a prearranged assortment of egg sizes so positioned that the cart could be the basic vehicle to move the eggs onto the truck and directly into a specially designed dairy case. The size of the cart could be in accord with the space needed for the sales volume

FIG. 8.5.   Vertical display of eggs stocked into the display in tilted 15-dozen plastic cases. Point-of-sale pricing eliminates pricing of individual cartons and so reduces egg breakage and labor.

in a particular supermarket. The cart would demand full vision or vertical arrangement of the product. Such a device would not only reduce handling costs, but constitute the optimum as far as space and display techniques are concerned. Display carts of this type are also being studied by the milk industry for handling half and full-gallon cartons of milk.

Recent studies on the positioning of eggs in supermarket dairy fixtures in 15 dozen plastic or wire shipping containers have reflected labor savings at the supermarket level of 30–35% (Fig. 8.6). In addition, egg breakage was greatly reduced. The use of this container maximized the space and full-vision aspects of strategic merchandising.

Some egg producers objected to this display technique since egg cartons positioned on their sides rather than in the conventional end-up position. Research by food technologists at Purdue University indicated that the side positioning of eggs in no way reduced the quality of the egg (Cardetti *et al.* 1979). Tipping of the 15 dozen wire or plastic containers at the proper angle is assured in egg displays with construction as shown in Fig. 8.7.

Such modularization of the product is in line with the need for greater productivity in the food industry where supermarket labor costs account for more than 50% of total operating expenses.

FIG. 8.6. Stocking of a chest-type refrigerator with eggs in 15-dozen plastic cases. Vertical display and easily read prices are special features of this unit.

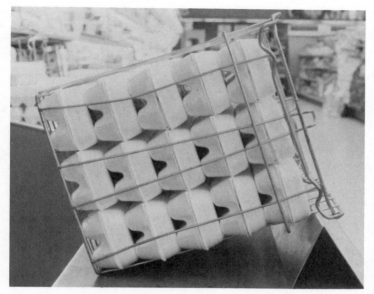

FIG. 8.7.   The angle of tilt is critical to provide for easy access and to prevent cartons from falling out.

CONCLUSION

Egg business in supermarkets is big business. It could be bigger. But the fact remains that in the supermarket world the egg is a dairy item; and to impress supermarket managers with the potential of their product, the egg processor must gear his thinking and the performance of his product to the dairy department.

Despite the fact that dairy products are no longer in the grocery sales classification, current research on the dairy department has repeatedly pointed up the misallocation of space in relation to product sales and movement. The conventional dairy display fixture is poorly designed and attracts customer shoppers to the large lower shelf where eggs, milk, and margarine—the three big sellers of the dairy—are located. Sales of products from other shelves are very low.

Future electronic data processing and optical scanners at the checkout will record product movement by individual items and thus alert the retailer to space, movement imbalance. But until this day arrives, the egg supplier who studies product sales and profits for the entire dairy case and who is aware of the potential of strategic self-service merchandising techniques and the equipment that best employs these, will benefit by counseling retailers not as to how more eggs can be sold, but as to how total dairy sales can be improved. Since eggs are one of the outstanding sales and profit performers in the department, both the egg supplier and the retailer stand to benefit.

REFERENCES

Branson, R. E., and Courtney, H. 1966. Market Performance of overwrapped egg cartons. Tex., Agric. Exp. Stn. [Bull.] *1062*.

Cardetti, M. M., Rhorer, A. R., and Stadelman, W. J. 1979. Effect of egg storage position on consumer quality attributes of shell eggs. Poult. Sci. *58*, 1403.

Coapstick, J., and Oesterle, E. C. 1978. Merchandising and packaging of eggs—a literature review. Purdue Univ. Agric. Exp. Stn., Bull. *190*.

Coapstick, J., and Oesterle, E. C. 1980. Update on consumer attitude about egg merchandising techniques. Poult. Trib. June, pp. 22–32.

Erickson, S. T. 1979. Merchandising in action: Eggs in baskets. Poult. Trib. (reprinted, July).

Goodale, R. S. 1978. Egg sales can be increased 50 percent. Poult. Trib. (reprinted, August).

Oesterle, E. C. 1975. Financial Performance of Selected Independent Supermarkets in Five Regions of the United States. Purdue University, West Lafayette, IN.

Oesterle, E. C. 1976. There's magic in strategic egg merchandising. Poult. Trib. *82*, (4), 14–18.

Oesterle, E. C. 1979. Performance Profiles of Eggs in Selected Supermarket Dairy Departments, EC 488. Cooperative Extension Service, West Lafayette, IN.

Oesterle, E. C. 1981. Operating Results: Independent Supermarkets; Five Regions of the U.S.A. Food Marketing Institute, Washington, DC.

Oesterle, E. C., and Coapstick, J. 1978A. Egg merchandising—an outdated science. Poult. Trib., July, pp. 22 and 28.

Oesterle, E. C., and Coapstick, J. 1978B. Green light for containerized eggs in supermarkets. Poult. Trib. September, pp. 14 and 16.

Oesterle, E. C., Cardetti, M. M., Rhorer, A. R., and Stadelman, W. J. 1979. Basket displays stimulate egg sales. Poult. Trib., April, pp. 14 and 18.

Progressive Grocer. April 1981. 49th Annual Report of the Grocery Industry.

Progressive Grocer. April 1982. 50th Annual Report of the Grocery Industry.

Single Service Institute (Undated). Eggs: How to Wake the Sleeping Giant. Egg Packaging Division, SSI.

Egg-Products Industry

Owen J. Cotterill

In 1981, The American Egg Board published a brochure entitled "A Scientist Speaks about Egg Products," in which egg products were defined as processed and convenient forms of eggs for commercial, food service, and home use (Cotterill 1981). About 341 million kg of liquid egg products are produced annually in the United States alone. The brochure lists 35 examples of products divided into four categories as follows:

Refrigerated Liquid Egg Products

Refrigerated liquid egg products are distributed by bulk tank-truck lots as well as in 30-lb cans and 4-, 5-, 8-, and 10-lb cartons.

Egg whites
Egg yolks
Whole eggs

Frozen Egg Products

Frozen egg products are generally packed in 30-lb cans and in 4-, 5-, 8-, and 10-lb pouches or waxed or plastic cartons.

Egg whites
Whole eggs
Whole eggs with yolks added
Egg yolks
Whole eggs with corn syrup
Whole eggs and yolks with corn syrup
Sugared egg yolks
Salted egg yolks
Salted whole eggs

EGG SCIENCE & TECHNOLOGY,
3rd Edition

Dried Egg Products

Dried eggs for foodservice use are sold in 6-oz pouches, Number 10 cans and 3- and 25-lb poly-packs. For commercial use, 25- and 50-lb boxes and 150-, 175-, and 200-lb drums are available.

Spray-dried egg-white solids
Flake albumen
Instant egg-white solids
Whole egg solids
Stabilized whole egg solids (glucose-free)
Free-flowing whole egg solids (with free-flow agent added)
Egg-yolk solids
Stabilized egg yolk solids (glucose-free)
Free-flowing egg-yolk solids (with free-flow agent added)
Blends of whole egg and/or yolk with carbohydrates (sugar or corn syrup) added

Specialty Egg Products

These products, which are designed for institutional and retail use, have mainly appeared on the market in the last 5 years. The packaging is more oriented toward individual portions or a limited number of servings per unit. Some products can be cooked or reheated in a microwave oven.

Whole hard-cooked, peeled eggs (plain or pickled)
Chopped hard-cooked, peeled eggs
Frozen hard-cooked egg rolls or long eggs
Egg-substitute products with egg white and other ingredients (dried, frozen, or refrigerated liquid)
Scrambled egg mixes (dried, frozen, or refrigerated liquid)
Frozen scrambled-egg mix in boilable pouch
Frozen scrambled eggs
Freeze-dried scrambled eggs
Frozen omelets
Frozen egg patties
Frozen quiches
Frozen quiche mix
Frozen fried eggs

HISTORY

The egg-drying industry developed before the frozen-egg industry. Patents were sought in the United States as early as 1865 on methods for drying eggs; however, it was not until 1878 that processing apparently started. A St. Louis, Missouri firm was transforming egg yolk and albumen, by a drying process, into a light-brown, meal-like substance. By the turn of the century, there was active competition from at least five companies: four in Missouri and one in Sioux City, Iowa. In 1984, there were 107 egg-breaking plants and 27 egg-dehydrating plants in the United States. A list of these plants operating under the Egg Products Inspection Program is published annually by the Poultry Division, Agricultural Marketing Service, U.S. Department of Agriculture, Washington, DC. 20250.

This industry began to disintegrate in 1915 and moved to China, following the opening of the Panama Canal, which aided the transport of Chinese exports to the Eastern Seaboard. The egg-products industry in the United States remained small until World War II.

The development of the frozen-egg industry followed the development of mechanical refrigeration. H. J. Keith, who is credited with the founding of the frozen-egg industry in the late 1890s, first packed frozen whole eggs without mixing the yolks and whites. Since freezing causes the yolk to gelatinize, the gummy lumps of yolk were not well accepted, whereupon Keith thoroughly mixed the whites and yolks to produce a more satisfactory product. In the early 1900s it was discovered that gelation of frozen yolk could be controlled by the addition of salt or sugar. In 1912, Harry A. Perry invented the hand separator, which improved the efficiency of egg breaking. Dr. Mary E. Pennington, a scientist with the USDA, improved the sanitation and physical facilities in frozen-egg plants to reduce bacterial counts.

Although the frozen-egg industry continued to expand during the 1920s and 1930s, few major technical advancements were made. Henningsen Bros., in 1938, was the first commercial company to pasteurize liquid eggs. A limited amount of pasteurization was done during the 1940s and 1950s, and some companies made good use of pasteurization during this period. However, during the early 1950s it was common to hear pasteurization referred to as too costly and impractical for liquid-egg processing. New regulations by the USDA and FDA regarding *Salmonella* levels made pasteurization virtually mandatory in 1966.

The production of dried eggs greatly expanded during World War II, before which there was little knowledge available on how to process these products. Although large quantities of dried whole eggs were produced for use by the United States Armed Forces or for shipment to allied countries, the general acceptance of these products was poor as the quality was far below present standards. Reportedly, 28% of 400 cartons of dried eggs sampled in Canada were positive for *Salmonella*— *Salmonella oranienburg* being the most common stereotype isolated. Today all products must be *Salmonella* negative. Also, the flavor and appearance of the egg products produced during the 1940s were undesirable. Consequently, when some improved products appeared on the market in the early 1950s, the industry adopted the term "Egg Solids" to avoid the adverse impressions associated with the "Dried Eggs" of the 1940s.

From 1900 to 1940, only China and the United States were processing major quantities of egg products. China was the main producer of dried eggs, and in 1925 they exported over 61 million lb of dried eggs, of which 36% were shipped to the United States. The liquid- and frozen-egg industry developed primarily in the midwest of the United States. The peak year for liquid-egg production in the United States, including all uses, was 1944, when over 680 million kg were broken. The current total liquid production in the United States is about 454 million kg per year. The development of the egg products industry in Canada paralleled that of the United States. More recently, an egg-products industry has developed in many other countries. The trend in the United States involves the bulk tank shipment of liquid products either for immediate consumption or for drying, thus avoiding freezing.

MAJOR DEVELOPMENTS

Two early achievements, freezing and dehydration, have made possible preserved eggs of high nutritional value. The egg-products industry would not have developed without these techniques of food preservation. However, these processes altered the product and created new problems which had to be solved. Some major developments are discussed briefly.

Egg-Breaking Equipment

The hand-breaking tray and separation clip were developed specifically for the egg-products industry (Fig. 9.1). A slotted trough is shown in Fig. 9.2. Until the 1950s most eggs were broken with these simple pieces of equipment. The evolution of the manually operated units to the sophisticated egg-breaking machines in use today is illustrated very well in the patent literature; several of these patents are listed in Chapter 18. At least three successful machines are currently on the market.

Freezing causes an irreversible gelation of egg yolk. In the early 1900s Keith is credited with adding sugar to yolk to inhibit gelation. Later, around 1926, salt was added for the same purpose. These ingredients are usually added at the 10% level.

FIG. 9.1.   Separator tray and clip for breaking eggs.

Pasteurization

The passage of the Egg Products Inspection in 1970 by Congress meant that all egg products in the United States must be pasteurized. The work by A. R. Winter (see references in Chapters 11 and 12) at Ohio State University in the 1940s was a major contribution toward understanding the problems of pasteurization. His work was the primary basis for using the conditions of 60°C for 3.5 min for pasteurizing whole eggs in the United States. Winter studied the pasteurization requirements of other products as well as the effects of pasteurization on the functional properties of egg products. He published the first paper on organisms that survive pasteurization.

The conditions for pasteurization of many of the other egg products were not established until the late 1960s. However, questions still remain about the pasteurization requirements of some of these products.

Desugaring

Desugaring egg products before drying was a major technical breakthrough for the egg-products industry when methods were developed to

FIG. 9.2.   Slotted trough for breaking eggs.
Courtesy of Sanovo (USA), Inc.

remove the reducing sugars. As a result of solving this problem, the shelf-life of dried products was improved.

Heat Treatment of Egg-White Solids

High-temperature storage of egg-white solids is an effective means of destroying *Salmonella*. *Salmonella*-negative products can be assured, and new products evolved, such as angel cake mixes.

Additives or Added Ingredients

The textural and functional properties of egg products can be altered by various additives. Gelation control has been previously mentioned. The functional properties of egg white solids can be improved by adding anionic surfactants, esters (such as triethyl citrate), and gums. Free-flowing products containing yolk can be made by adding sodium aluminum silicate. Scrambled-egg mixes containing several food ingredients have been produced commercially.

Salmonellae Control

The egg-products industry has played a major role in developing and standardizing the methods for the detection of salmonellae. The pre-enrichment, enrichment, and differential plating techniques developed for egg products have been adapted by other branches of the food industry.

Nutrient Composition

The nutrient composition of eggs and egg products has been thoroughly reviewed and re-evaluated (Posati and Orr 1976; Cotterill and Glauert 1979). Consequently, the data on these products are now very complete (see Table 7.2).

RECENT DEVELOPMENTS

In the last 5 years three major market developments have evolved as follows: (1) prepeeled hard-cooked eggs preserved in a solution of sodium benzoate and citric acid; (2) improved scrambled egg mixes, particularly those for cooking in a plastic bag; and (3) cooked-frozen-thaw-reheated egg products.

FUTURE DEVELOPMENTS

Although many problems with egg products have been resolved, there are others that still need attention. For example, the foaming properties of whole egg solids need to be improved. Likewise, products containing dried yolk solids are particularly susceptible to off-flavors and also do not reconstitute easily. The hydration and/or dispersibility of all egg solids is a problem.

Up until the present, the egg-products industry has supplied ingredients for the manufacture of other food products—for example, candy, baked goods, or prepared mixes. In the future, egg products for the foodservice and retail markets should become a reality. Each technological advancement increases the potential for new products. Stabilized egg white solids made possible the angel cake mix. *Salmonella*-negative egg products have made other products possible. If the nonpathogenic organisms that currently survive pasteurization can be controlled, liquid egg products for the retail trade will be possible. Ultra high-temperature pasteurization has a good potential for producing these sterile egg products.

SUMMARY

The egg-products industry involves much science and technology. Many chemical microbiological, and engineering problems have been resolved over the years so that safe products with improved functional qualities are now available. Egg products are very versatile and with the solution of additional problems, further products will evolve. In the past hundred years egg products have developed into about a one billion lb (454 million kg) industry in the United States alone.

REFERENCES

Cotterill, O. J. 1981. A Scientist Speaks about Egg Products. American Egg Board, 1460 Renaissance Drive, Park Ridge, IL 60068.

Cotterill, O. J., and Glauert, J. L. 1979. Nutrient values for shell, liquid/frozen and dehydrated eggs derived by linear regression analysis and conversion factors. Poult. Sci. *58*, 131–134.

Posati, L. P., and Orr, M. L. 1976. Composition of Foods: Dairy and egg products raw-processed-prepared. U.S., Dept. Agric., Agric. Handbook *8-1*. Agricultural Research Service, United States Department of Agriculture, Washington, DC.

Egg Breaking

Owen J. Cotterill

INTRODUCTION

Liquid egg products provide an outlet for surplus eggs, even though some shell eggs are produced commercially for this purpose. About 12–15% of the total egg supply in the United States is processed into egg products. The total quantity of liquid egg produced and its disposition is shown in Table 10.1. Almost 1 billion lb (454 million kg) of edible liquid products were produced in 1985 in the United States alone. In addition, about 68 million lb (31 million kg) of inedible liquid products and about 100 million lb (54 million kg) of eggshells (wet weight basis) were derived as by-products of this operation. Of the edible products, 58% was used for immediate consumption, 32% was frozen, and 10% was dried. Not only has there been a gradual increase in total liquid produced, but more important, there has been a gradual shift from frozen products to those for immediate consumption. Energy, availability of product, and need has caused this change. Inedible egg product accounts for about 7% of the liquid broken; much of this product is now pasteurized, dehydrated, and sold to manufacturers of pet foods.

REGULATIONS

On December 29, 1970, the Egg Products Inspection Act required the mandatory continuous inspection of all egg-product plants in the United States, unless exempted; it became effective July 1, 1971. Prior to this period, inspection was on a voluntary basis. Regulations governing the inspection of egg products are available through the Poultry Div-

193

EGG SCIENCE & TECHNOLOGY,
3rd Edition

TABLE 10.1. Total Liquid Egg Production and Disposition in the United States (in millions)

Period[a]	Total edible liquid		Inedible liquid		Nonegg ingredients		Liquid for immediate consumption or processing		Frozen product produced		Dried product produced	
	(lb)	(kg)	(lb)	(kg)	(lb)	(kg)	(lb)	(kg)	(lb)	(kg)	(lb)	(kg)
Jan. 1–Dec. 30												
1970	748	339	—	—	—	—	110	50	357	162	75	34
July 1–June 30												
1975	703	319	37	17	29	13	310	141	317	144	59	27
Oct. 1–Sept. 30												
1980	883	400	57	26	35	16	422	191	331	150	81	37
1981	921	419	59	27	36	16	450	205	342	155	82	37
1982	907	411	59	27	35	16	460	208	339	154	80	36
1983	897	404	59	27	45	20	494	222	304	137	86	39
1984	917	417	61	28	43	19	528	240	323	147	76	35
1985	999	454	68	31	44	20	578	262	324	147	96	44

[a] Year denotes end of period.

ision, Agricultural Marketing Service, United States Department of Agriculture, Washington, DC. 20250. It is recommended that a copy of these regulations be obtained to supplement this book, even though some main rulings in this Act are discussed in this chapter.

Plant Approval

Any person desiring to process and pack egg products under continuous inspection must receive approval of the plant and facilities. An initial survey is made, drawings and specifications must be furnished for review, and then a final survey is made to assure that the plant meets all requirements. Equipment installation and equipment should comply with the applicable E-3-A Sanitary Standards and Accepted Practices.

For international trade, another useful document is the "Recommended International Code for Hygienic Practice for Egg Products." In the United States, it can be obtained from UNIPUB, 205 42nd St., New York, NY 10017 (mailing address: P.O. Box 433, Murray Hill Station, New York, NY 10016). In addition to recommendations for good processing and transportation practices, this document describes the α-amylase test for adequate pasteurization of egg products.

Plant Inspector

After approval, a resident inspector is assigned to the plant. This person is responsible for observing the quality, type, and wholesomeness of raw materials and finished products; sanitation of premises, plant, and equipment; handling of ingredients, labeling, freezing, storing; and all general operations in the processing and production of egg products at the official plant.

SHELL EGGS FOR BREAKING

Usually the breaking plant is located close to the source of supply of shell eggs. Most eggs for breaking are derived from surpluses in major production areas. However, there is some contracting of production for breaking as well as shell-egg production by companies involved in breaking.

Shell eggs for breaking must be of edible interior quality, and the shell must be sound and free from adhering dirt and foreign materials, unless otherwise excepted. Detailed regulations have been established designating the types of shell eggs that cannot be used for breaking stock, including leakers, rots, eggs with developed embryos, stuck yolks, blood rings, and incubator rejects.

The nature or type of eggs used for breaking can affect yolk color, yield, and microbiological content. Yolk color is a function of feed as well as the rate and duration of lay. When the source of eggs is primarily small farm flocks, there is a strong seasonal variation in yolk color. Yield is affected by strain and age of bird, season, and age of egg. At one time most eggs were produced by layers confined under a floor system of management. Currently, the cage-layer system prevails. Cotterill and Geiger (1977) reported on yield trends that occurred between 1966 and 1975 (Fig. 10.1). Both egg size and yolk size decreased. The level of solids

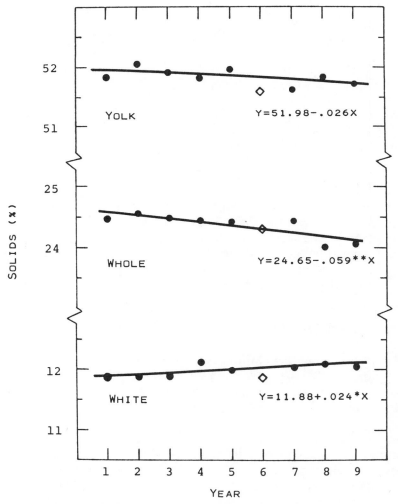

FIG. 10.1.   Percentage of solids in white, whole, and yolk.

in whole egg decreased from 24.7 to 24.1% (see Fig. 10.1). These workers also derived yield data based on the total liquid and solids per case and the amount produced per bird per year (see Table 10.2). Breaking procedures also affect yield. Very low bacterial counts exist in eggs that are produced in commercial cage operations and in eggs that are broken soon after laying. The number of bacteria in the raw liquid egg product is strictly a function of the care of the eggs before breaking and plant sanitation. Internal contamination of fresh unwashed, unbroken shell eggs rarely occurs.

EGG-BREAKING PLANT

Floor Plans

The floor plan shown in Fig. 10.2 was designed specifically for an egg-breaking plant. Note the flow of materials through the plant. Cross traffic is reduced to a minimum. Separate rest rooms and lunchrooms are available for employees handling the raw materials and the finished product. Positive ventilation is provided from the can draw-off room and holding-tank area, counter to the flow of product into these areas.

The plan shown in Fig. 10.3 is part of a large shell-egg grading and packing unit. A more detailed layout of the transfer, breaking, and draw-off rooms is shown in Fig. 10.4. This breaking plant is designed primarily to handle surplus eggs and those that are not suitable for the shell-egg operation. It was estimated that this plant (two machines) had a production rate of 5.25 cases per man-hour with a labor cost of 33¢ per case (1969 basis). Equipment costs (including operating costs) were 31¢

TABLE 10.2. Yield of Liquid and Solids on a per Case and per Bird per Year Basis

| | Liquid | | Solids | |
	(kg)	(lb)	(kg)	(lb)
Amount per case (30 doz)				
White	13.8	30.4	1.67	3.68
Yolk	6.1	13.5	3.12	6.87
Whole	19.9	43.9	4.79	10.55
Amount per bird per year				
White	8.4	18.5	1.02	2.24
Yolk	3.7	8.1	1.89	4.17
Whole	12.1	26.6	2.91	6.41

SOURCE: *Cotterill and Geiger (1977).*

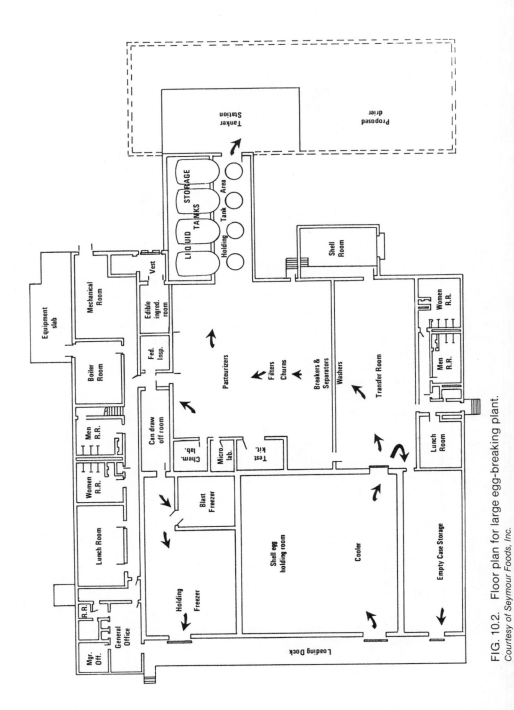

FIG. 10.2. Floor plan for large egg-breaking plant.
Courtesy of Seymour Foods, Inc.

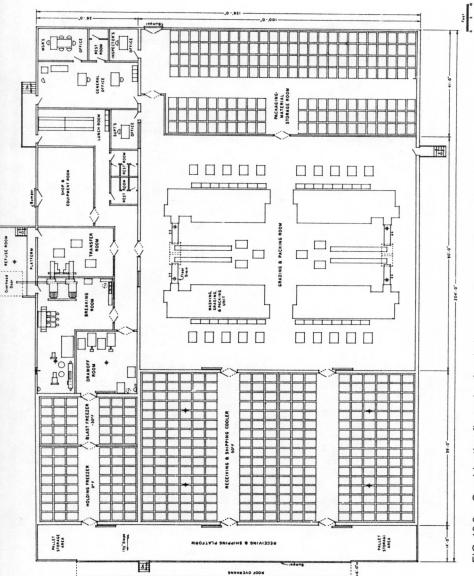

FIG. 10.3. Combination floor plan for shell-egg grading, packing, and breaking.
Courtesy of ARS—U.S. Department of Agriculture, from Harris (1969).

199

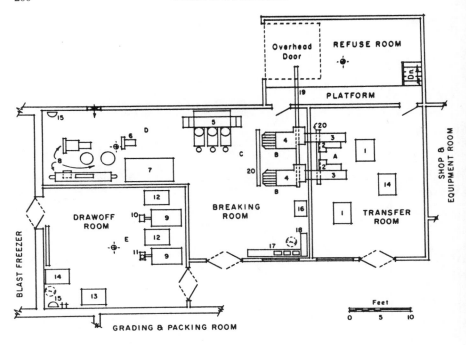

FIG. 10.4.   More detailed floor plan of breaking plant shown in Fig. 10.3.

per case for one shift and 28¢ for two shifts. However, a plant like the one shown in Fig. 10.2 should have lower operating costs.

Holding and Transfer Rooms

The shell-egg storage room should hold enough eggs to operate a plant for 5 to 10 days. If the eggs are held up to 1 week, the temperature should be 13°C; if over one week, 7°C storage is recommended. Maurer and Wisnieski (1981) found that more egg white adhered to the shells of small than large eggs. Also, refrigeration temperatures of 4° and 13°C did not affect the residual albumen in the shells. The storage room should be adjacent to the empty-case storage and transfer rooms. The transfer room is semi-isolated from the rest of the plant. After the eggs are transferred to the washers, undesirable eggs are separated by flash candling. Certain requirements must be met when washing and sanitizing shell eggs for breaking. The temperature of the water must be 32°C or higher and at least 11°C warmer than the eggs. The washing operation must be continuous. Eggs must not be allowed to stand or soak in water. Immersion-type washers are not permitted. Washed eggs must be spray-rinsed with a sanitizing agent of not less than 100 ppm nor more than 200 ppm of available chlorine or its equivalent. Shell eggs should not be washed in the breaking room and must be sufficiently dry at time of breaking to prevent contamination from free moisture on the outside of the shell.

Breaking Room

The regulations on breaking room facilities and operations provide for adequate lighting, ventilation, hand-washing facilities, and personnel hygiene, that are essential to any good food-processing operation. A separate draw-off room with filtered positive air ventilation must be provided for packaging liquid-egg products, except products packaged by automatic, closed packaging systems. Breakers must use a complete set of clean equipment when starting work and after lunch periods. Whenever an inedible egg is broken, the affected equipment must be replaced. This practice reduces bacterial counts and yolk contamination of egg white. Containers used for packaging should not pass through the breaking room or areas where shell eggs and cases are handled.

Tanks, vats, drums, or other containers used for holding liquid eggs must be of approved construction, fitted with covers and located in rooms maintained in a sanitary condition. The minimum cooling and temperature requirements for liquid-egg products are shown in Table 10.3.

Breaking Machines

The development of egg-breaking and separating machines was a major improvement in the efficient production of liquid-egg products. A successful and widely used unit was developed by Seymour Foods, Inc. of Topkea, Kansas, during the 1950s. By 1963, 200 of these machines had been leased to commercial egg processors. The first units had a

TABLE 10.3. Minimum Cooling and Temperature Requirements for Liquid Egg Products

Product	Unpasteurized product temperature within 2 hr from time of breaking			Temperature within 2 hr after pasteurization	Temperature within 3 hr after stabilization
	Liquid (other than salt product) to be held 8 hr or less	Liquid (other than salt product) to be held in excess of 8 hr	Liquid salt product		
Whites (not to be stabilized)	12.8°C or lower	7.2°C or lower		7.2°C or lower	
Whites (to be stabilized)	21.1°C or lower	12.8°C or lower		12.8°C or lower If to be held 8 hr or less, 7.2°C or lower. If to be held in excess of 8 hr, 4.4°C or lower[b]	[a] If to be held 8 hr or less, 7.2°C or lower. If to be held in excess of 8 hr, 4.4°C or lower
All other product (except product with 10% or more salt added)	7.2°C or lower	4.4°C or lower			
Liquid egg product with 10% or more salt added			If to be held 30 hr or less, 18.3°C or lower. If to be held in excess of 30 hr, 7.2°C or lower		

SOURCE: *Federal Register 36, No. 104, May 28, 1971.*
[a] Stabilized liquid whites shall be dried as soon as possible after removal of glucose. The storage of stabilized liquid whites shall be limited to that necessary to provide a continuous operation.
[b] The cooling process shall be continued to assure that any salt product to be held in excess of 24 hr is cooled and maintained at 7.2°C or lower.

capacity of about 15 to 20 cases per hour, and in 1963 were breaking about 40% of the eggs in the United States. The labor requirement was reduced by about one-fifth. Whole egg, egg yolk, and egg white could be produced simultaneously. Also, it was possible to segregate eggs with broken yolk membranes. The separator cups could be removed individually for cleaning and sanitization if they became contaminated by an undesirable egg. The three breaking and separating machines shown in Figs. 10.5, 10.6, and 10.7 account for most of the eggs broken at this time. They feature faster operation (over 50 cases/hr), easier cleaning and maintenance, better separation, and accommodation of a greater variation in egg sizes. Note that each separator cup has a separate cracking assembly and can be removed and cleaned individually. In-place washing and sanitizing is also provided for. Most plants have several of these machines operating simultaneously, as shown in Fig. 10.8. An individual cracker head from one of these machines is shown in Fig. 10.9.

Shell Room

The shells are conveyed from the breakers directly into a trailer or truck in another room. Sometimes these shells are centrifuged to recover any adhering egg white. This inedible material is then sold as technical albumen.

Inspector's Office, Laboratories, and Ingredient Room

The liquid products are pumped through a filter into mixing or holding tanks. If necessary, ingredients as salt and carbohydrates are added at this point. Note the separate ingredient storage room shown in Fig. 10.2. Also, the inspector's office and the quality control laboratories are conveniently located adjacent to the breaking room. This permits constant monitoring of the product and operations by the technical staff.

Draw-Off Room

Since some of the products of these primary processing plants are packaged and frozen in the plant, a separate draw-off room is mandatory. This helps reduce postcontamination of the pasteurized product. The packaging materials must not pass through the breaking room. Sometimes conveyors from another level of the plant are necessary to avoid cross traffic. In the plant shown in Fig. 10.2, these supplies enter the draw-off room from above. Egg products for freezing should be solidly frozen or reduced to a temperature of $-12°C$ or lower within 60 hr from time of breaking or pasteurization.

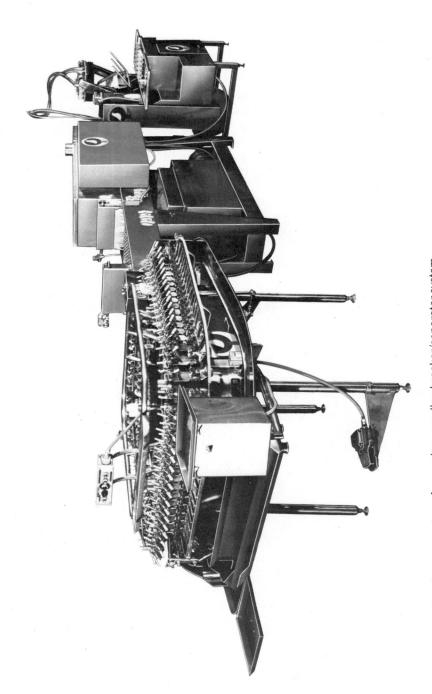

FIG. 10.5. Henningsen egg transfer, washer, candler, breaker/separator system.
Courtesy of Henningsen Foods, Inc.

FIG. 10.6.   Sanovo egg loading, washing and breaker/separator system.
Courtesy of Sanovo (USA), Inc.

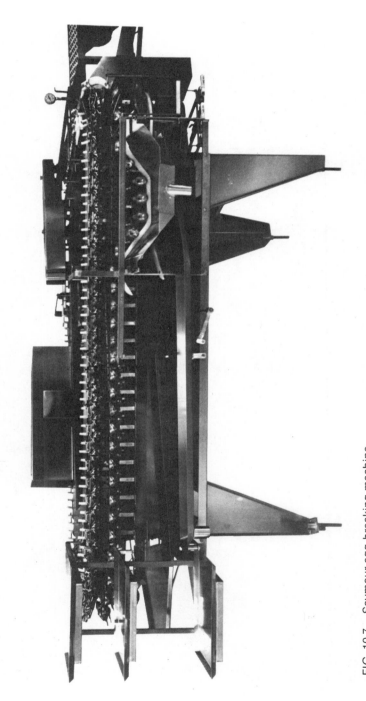

FIG. 10.7. Seymour egg-breaking machine.
Courtesy of Seymour Foods, Inc.

FIG. 10.8. A breaking room in which several machines are operating simultaneously.
Courtesy of Milton G. Waldbaum Co.

FIG. 10.9.   Cracker head and yolk-white separator for an egg-breaking machine.

Waste Water

The BOD (biological oxygen demand) of waste water from egg-breaking plants ranges from 1,000 to 22,000 mg per liter (Harris and Moats 1975). Egg solids are introduced into the waste water from broken eggs during washing and from the cleaning of equipment. These researchers were able to reduce the BOD by 76–97% in a process that involved acidification to pH 4.7 with H_2SO_4, heating to 75°C, and then removing the precipitated proteins by filtration or centrifugation.

BULK TANK SHIPMENT

Considerable amounts of egg products are shipped in liquid form in insulated bulk truck tankers to other plants for further processing (see Fig. 10.10). One company ships about 500 million eggs this way annually. Handling costs are 75% lower than shipping shell eggs. These tankers may be compartmentalized to handle egg white, yolk, and whole egg all at the same time. Sometimes the liquid is shipped over 1,000 miles at temperatures below 4°C. Tankers are also used to ship liquid to large bakeries or other outlets for immediate consumption.

FIG. 10.10.   Tanker used to transport liquid egg products.
Courtesy of Seymour Foods, Inc.

NATURAL EGG PRODUCTS

Five main egg products are produced by an egg-breaking plant as follows: (1) liquid egg white, (2) liquid egg yolk, (3) liquid whole egg, (4) egg shells, and (5) inedible egg products. Some of the properties of each of these products are discussed. In all cases, the products are derived from chicken eggs produced by commercial egg-laying stock in use in the 1980s.

Liquid Egg White

The solids content of liquid egg white is usually very close to 12%. The only way to change this is to vary the breaking stock and even then the potential variation would be less than 1%. Eggs that have lost considerable moisture (i.e., those with large air cells) or smaller pullet eggs with high interior quality will cause the solids level to be higher. To decrease the solids level, fresh eggs from older birds are used. Hot weather may also decrease the egg-white solids level.

The pH of egg white can vary from 7.6 to 9.3, but a 8.4–9.2 range is the most common. Most commercial liquid egg white will have a pH of 9.0 to 9.1. This increase is strictly a function of the amount of carbon dioxide lost from the egg. Its rate of loss depends on the temperature of the egg, the amount of carbon dioxide in the environment, and the degree of egg-shell sealing (oiling).

The bacterial quality of liquid egg white, which is usually very good, is strictly a function of the quality of the eggs being broken, the sanitation in the plant, and handling practices. It should be remembered that the interior of most shell eggs is sterile and that it is only during processing that bacteria are introduced. Also, small amounts of yolk will cause bacteria to grow much faster in egg white after processing.

Liquid Egg Yolk

The solids content of pure egg yolk from fresh eggs is about 51.9 ± 0.1%. It is very constant. Age of the egg, meaning the amount of water migrating from the white, is the only major factor affecting the initial egg-yolk solids. The solids level in egg yolk produced by egg-breaking machines can be affected by many things (the equipment, age of egg, size of egg, etc.). Usually this solids level is in the 46–48% range, depending on the amount of egg white adhering to the yolk. However, most "plain" egg yolk (no added nonegg ingredients) is standardized by the addition of egg white to a 43–44% solids level. Why this solids level? There is a great transitional change in the viscosity of egg yolk at the 45–48% solids level (Cunningham 1972), which it is easy to see when preparing egg yolk with different solids levels. The scientific basis for this is obscure and there is a need for a rheological study on the ratio of egg white to yolk. Then the question remains: How to equate these observations to practical applications such as pumping and fluid flow?

The pH of pure egg yolk is about 6.0 and changes very little during storage of the shell egg. Pure egg yolk contains very little carbon dioxide. Since commercial yolk contains egg white, its pH may be slightly higher than 6.0.

Liquid Whole Egg

This is a blend of egg white and yolk. In natural proportions, it may contain between 23 and 25% egg solids, yet, even without the addition of egg white, lower solids levels are possible. The standard, or basis, has changed over a period of many years. At one time, it was as high as 26%. If whole egg is formed by blending liquid egg white and yolk, the solids level must be standardized to 24.7% (USDA regulations). Generally, this requires the addition of more yolk than is present in natural proportions in the liquid from shell eggs. The 24.7% solids level is the basis for trading tanker trailer lots on the Egg Clearing House, Inc. (ECI), which has specifications for the solids levels in three liquid whole egg products:

Option 1. Standard Products. While 24.7% is the basis of trade, liquid with as low as 24.0% solids can be traded with an appropriate price adjustment. This product is unpasteurized.

Option 2. Product below 24.0% Solids. The solids level may range between 23.3 to 24.0%, and the appropriate price discount will apply. This product is unpasteurized.

Option 3. If the product is pasteurized, then it should be stated that it involves Option 3.

For further information, contact ECI, P.O. Box F, Durham, New Hampshire 03824 (telephone: 603-868-2899). Procedures have been established to protect the interests of both the buyer and seller. Trading zones are the basis for determining freight rates. The seller is responsible for delivering the product of appropriate specifications to a designated site. A standard trading unit is 45,000 lb (20,455 kg), which must be delivered within 72 hr after breaking at a temperature of 4°C or less.

The pH of liquid whole egg can vary from 7.0 to 7.6, depending on the condition of the egg (freshness, oiling, temperature of storage, etc.). Like egg white, the pH of whole egg is a function of the amount of carbon dioxide in the shell eggs at the time of breaking.

The bacterial quality of raw liquid whole egg is probably not as good as raw white or yolk. Yet, it is much better than many food products that have a high reputation for low total plate counts. If the initial total bacterial count is above 5,000/g, plant operators should suspect that processing conditions are less than optimal and they should take a careful look at plant sanitation, operating procedures, and the quality of product being broken. Pasteurization further reduces these counts to perhaps 500 to 1,000/g. If very fresh eggs are broken, counts of less than 100 in a pasteurized product can be expected.

Eggshells

Maurer and Wisnieski (1981) found that the amount of albumen adhering to the shell ranged from 2.55 to 2.93% on a total wet shell weight. Walton *et al.* (1973) reported that this waste material contained an average of 29.1% moisture as it came from the breakers and 16.2% water after centrifugation. (See Table 10.4 for other components in egg shells.) The amino acid composition of egg shells was also determined by Walton *et al.* (1973). The eggshell centrifuge (shown in Fig. 10.11) removes not only liquid albumen but increases the bulk density of the product (Fig. 10.12).

What to do with eggshells from breaking plants can be a major problem. Three common methods for their disposal are in current use: (1) as a fertilizer, (2) as a feedstuff, (3) as a waste product to be taken to municipal dumps. Methods 1 and 3 require immediate disposal, and at least four dehydrators have been installed for this purpose in the world. Vandepopuliere *et al.* (1978) described the installation of a triple-pass

TABLE 10.4.   Composition of Eggshell Waste from Egg-Breaking Plants (Dry Basis)

	With adhering albumen (%)	Centrifuged (%)	Washed (%)
Original moisture	29.1	16.2	
Protein	7.6	5.3	5.2
Lipid	0.24	0.30	0.05
Ash	91.1	94.2	95.4
CaCO$_3$	90.9	91.8	93.1
Calcium	36.4	36.7	37.3
Iron	0.0020	0.0022	0.0023
Potassium	0.097	0.072	0.060
Magnesium	0.398	0.400	0.407
Sodium	0.152	0.126	0.115
Sulfur	0.091	0.087	0.043
Phosphorus	0.116	0.104	0.117

SOURCE: *Walton et al. (1973).*

FIG. 10.11. Centrifuge used to continuously separate inedible ("technical") albumen from the shells.
Courtesy of Sanova (USA), Inc.

FIG. 10.12.   Appearance of shells before (right) and after centrifugation (left).

drum dryer in a breaking plant and reported that it cost about $16.00 to produce a metric ton of eggshell meal. The feedstuff value was estimated to be about $38.00/ton, if corn and soybean meal were priced at $92.00 and $210.00, respectively. The product had excellent storage and handling properties (exhaust temperature was 82°C). Its performance as a feedstuff was excellent. A single-pass drum dryer designed specifically for eggshells is shown in Fig. 10.13.

Inedible Egg

By law, at least in the United States, certain types of eggs are classed as inedible or unfit for human consumption. Most of this product is derived from leakers, eggs with blood and meat spots, and egg white from centrifuged eggshells. The solids content of inedible egg product may vary between that for egg white (12%) to whole egg (24.7%). There are no standards for this product except that it may be denatured, usually with a green dye.

The volume from large breaking operations is sufficient to justify bulk-tank collection. Since value at the site of production is nil, the plant must often pay for its disposal. Some inedible egg product is also dehydrated for use in pet foods or mink food.

FIG. 10.13.   A rotary dryer built specifically for drying eggshells.
Courtesy of Sanova (USA), Inc.

REFERENCES

Cotterill, O. J., and Geiger, G. S.   1977.   Egg product yield trends from shell eggs. Poult. Sci. *56*, 1027–1031.

Cotterill, O. J., Stephenson, A. B., and Funk, E. M.   1962.   Factors affecting the yield of egg products from shell eggs. Proc. World's Poult. Congr. *12*, 443–447.

Cunningham, F. E.   1972.   Viscosity and functional ability of diluted egg yolk. J. Milk Food Tech. *35*, 615–617.

Harris, C. E.   1969.   An egg grading and processing plant for high-volume production. U.S. Dep. Agric., Mark. Res. Rep. *837*, Feb.

Harris, C. E., and Moats, W. A.   1975.   Recovery of egg solids from wastewaters from egg-grading and breaking plants. Poult. Sci. *54*, 1518–1523.

Marion, W. W., Nordskog, A. W., Tolman, H. S., and Forsythe, R. H.   1964.   Egg composition as influenced by breeding, egg size, age and season. Poult. Sci. *43*, 255–264.

Maurer, A. J., and Wisniewski, G. D.   1981.   Effect of egg breakout temperature and other factors on residual albumen in the shell and on the yolk. Poult. Sci. *60*, 1254–1258.

Vandepopuliere, J. M., Walton, H. V., Jaynes, W., and Cotterill, O. J.   1978.   Elimination of pollutants by utilization of egg breaking plant shell-waste. U.S. Environ. Prot. Agency, Off. Res. Dev. [Rep.] EPA. *EPA-600/2-78-044*, Mar.

Walton, H. V., Cotterill, O. J., and Vandepopuliere, J. M.   1973.   Composition of shell waste from egg breaking plants. Poult. Sci. *52*, 1836–1841.

Freezing Egg Products

Owen J. Cotterill

PRODUCTS; PACKAGES; THERMAL PROPERTIES

Most frozen egg products are marketed as ingredients for use in other food products. About one-third of the total liquid production is frozen (see Table 10.1). Plain egg white, yolk, and whole egg, as well as a variety of blends of yolk and white that may include other food ingredients, are frozen. The annual U.S. production of frozen whole, white, and yolk products is shown in Table 11.1. The composition of some frozen products is shown in Table 7.2.

Specifications for some other products are shown in Chapter 15. Sodium chloride and sucrose are commonly added to frozen egg products (at levels of 10%) to control gelation of products containing egg yolk. Other additives include glycerin, syrups, gums, and sodium metaphosphate. Gums thicken whole egg. Phosphates make it possible to pasteurize at a lower temperature. Triethyl citrate improves the whipping properties of egg white (yolk contaminated or heat damaged). For further details, see the chapters on pasteurization, dehydration, and quality control.

Frozen eggs are packaged in a variety of containers (Fig. 11.1). Until recently, the 30-lb can predominated. However, it is difficult to open and is an inconvenient size for many uses. Cartons and plastic bags avoid the many disadvantages of cans. Also, 55-gal. drums with plastic bag liners are used for bulk shipments. Winter and Wrinkle (1949) reported on the freezing rates for several egg products in various containers at different temperatures.

Freezing causes major textural changes in some egg products, also, large reductions in bacterial counts occur. Although functional proper-

217

EGG SCIENCE & TECHNOLOGY,
3rd Edition

TABLE 11.1. Frozen Egg Production in the United States (in millions)

Period[a]	Whole[b]		White		Yolk[b]		Total	
	(lb)	(kg)	(lb)	(kg)	(lb)	(kg)	(lb)	(kg)
Jan. 1–Dec. 31								
1970	203	92	60	27	95	43	357	162
July 1–June 30								
1975	198	90	39	18	80	36	317	144
Oct. 1 through Sept. 30								
1980	209	95	44	20	79	36	331	150
1981	203	92	51	23	88	40	342	155
1982	210	95	55	25	80	36	339	154
1983	196	89	49	22	60	27	305	138
1984	220	100	48	22	61	28	323	147
1985	214	97	46	21	64	29	324	147

[a] Year denotes end of period.
[b] Includes plain and blends.

ties are only slightly affected. One group of workers (Zabik and Figa 1968; Zabik 1968, 1969; Wolfe and Zabik 1968; Zabik and Brown 1969; Zabik *et al.* 1969; Downs *et al.* 1970; Morgan *et al.* 1970; Funk and Zabik 1971) used frozen-egg products as a control to compare the functional performance of various dried products. Since this chapter features the changes that occur in egg products as a result of freezing and thawing, these papers are not reviewed herein.

The thermal properties of egg products vary widely, depending on the

FIG. 11.1. Frozen egg containers.
Courtesy of Missouri Agricultural Experiment Station.

moisture or solids content. The specific heats and latent heats of fusion can be approximated as follows:

$$\text{Specific heat} = \frac{\% \, H_2O + 0.5 \times \% \, \text{solids}}{100}$$

$$\text{Latent heat} = (\% \, H_2O/100)[144 - 0.5 \, (32 - \text{freezing point})]$$

This last formula assumes that all water in the product becomes ice when the product is frozen. Therefore, the latent heat values in Table 11.2, from Lineweaver et al. (1969) provide the maximum requirements for freezing.

PRODUCT CHANGES DUE TO FREEZING

Raw Egg White

Freezing causes only minor changes in raw egg white. Some thinning of thick white may occur. Husaini and Alm (1955) reported a decrease in "masked" sulfhydryl groups during 10 days of frozen storage. This may be related to the thinning. However, when milled or blended white is frozen, there is some thickening near the outside of the container, and a cloudy mass may form in the center. These changes may be related to the rate of freezing and migration of salts. Blending returns this egg white to its original consistency. We have not observed changes in electrophoretic and chromatographic patterns or functional property differences due to freezing. Clinger et al. (1951) stated that cakes made

TABLE 11.2. Heat Content of Egg Products

Egg product	Solids (%)	Freezing point (°F)	Specific heat[a] Above freezing	Below freezing	Latent heat of freezing[b] (Btu/lb)
Water	0	32	1.00	0.5	144
Whole eggs	25	31	0.88	0.5	108
Whites	12	31	0.94	0.5	127
Yolks	44	31	0.78	0.5	81
Sugared yolks	50	25	0.75	0.5	72
Salted yolks	50	1	0.75	0.5	64

[a] Btu/lb/°F or calories/gm/°C.
[b] To convert to calories per gram, multiply values by 0.555. These data, together with the specific heat values, can be used to calculate the Btu required to lower (or raise) the temperature of eggs including the freezing step. For example, the Btu that must be removed to lower the temperature of 30 lb of whole eggs from 40°F to 0° are obtained as follows: $30 \times 0.88 \times (40 - 31) + (30 \times 1.08) + (30 \times 0.5 \times 31) = 3{,}943$ Btu.

from frozen egg white had nearly as good cake volumes as those made from fresh egg white. Wootton *et al.* (1981), using differential scanning calorimetry, reported that the loss of enthalpy of denaturation was increased by slower freezing rates, higher thawing temperature, higher storage temperature, and longer storage times. Conalbumin suffered greater losses, and ovalbumin had smaller losses than egg white itself. Egg-white viscosity and foam instability was reduced by slow freezing rates, higher thawing temperatures, increased storage times, and lower storage temperatures. It is common practice to freeze large quantities of egg white for research purposes. This provides a uniform test material for repeated functional or other analytical studies.

Raw Yolk

Upon freezing and storing raw egg yolk below −6°C, the viscosity increases and gelation occurs. Eventually, all fluidity is lost. Hence, this gelled product cannot be processed like other liquids. It is difficult to mix with ingredients and has an undesirable appearance. Until recently, this gelation was considered totally irreversible. Palmer *et al.* (1970) observed that heating thawed yolk at 45° to 55°C for one hour partially reversed this gelation. Saari *et al.* (1964) reported that yolk plasma, after freezing and thawing, had a pasty consistency and that only 15% of the lipoproteins were insoluble in 10% NaCl.

Gelation is easily controlled. Sodium chloride and sucrose are most commonly used. Although 2% salt is as effective as 8% sucrose, a level of 10% of both ingredients is generally used. The only disadvantage of either of these ingredients is that their future use may be restricted to specific food products. Syrups, glycerin, phosphates, and other sugars can also be used. Syrups are primarily used in blends of whole egg, yolk, and smaller amounts of salts. Lopez *et al.* (1954) reported that treatment of yolk with proteolytic enzymes (papain, trypsin, or rhozyme) inhibited gelation. Only papain did not seriously affect its organoleptic properties. Apparently, these enzymes break down the components responsible for gelation. Also, Feeney *et al.* (1954) reported that gelation was reduced by incubation with crotoxin (lecithinase A) before or after freezing. Mechanical treatment, such as homogenization, colloid milling, or excessive mixing, also reduces the viscosity of frozen yolk (Thomas and Bailey 1933; Pearce and Lavers 1949; Lopez *et al.* 1954; Marion 1958).

Egg-yolk gelation is affected by rate and temperature of freezing and thawing as well as by storage time and temperature. The effects of these conditions are so interrelated that they are difficult to quantitate individually. The freezing point of egg yolk is about −1°C; yet, gelation does not occur until a temperature of −6°C is attained. Gelation of plain yolk takes place most rapidly at −18°C (Palmer *et al.* 1970). Fast freezing and thawing result in less gelation than slow freezing and thawing (Marion 1958). Apparently, smaller ice crystals are formed and less dehydration of the proteins occurs (Powrie *et al.* 1963). At lower temperatures, the rate of freezing seems more important than the temperature of freezing. Jaax and Travnick (1968) postulated that the heat of pasteurization may alter the susceptability of egg yolk to gelation. Also, sodium chloride lowered the pH. An explanation for the decreased gelation after rapid thawing was offered by Palmer *et al.* (1970), who indicated that if the temperature used for thawing is sufficiently high, an actual melting of the gel occurs. They also compared this to the melting of a gelatin gel.

The functional properties of plain egg yolk are little affected by freezing. Miller and Winter (1951) noted that mayonnaise made from frozen plain yolk was more stiff, but slightly less stable, than that made from unfrozen yolk. The addition of sodium chloride increased the emulsifying ability of defrosted yolk. Fructose decreased the emulsifying ability (Jaax and Travnick 1968). Davey *et al.* (1969) reported that freezing and thawing were not detrimental to the emulsion-stabilizing power of the lipovitellin, lipovitellenin, and livetin fractions of egg yolk. Likewise, Palmer *et al.* (1969A) reported that freezing and storage at 0° and −10°F for as long as 4 months caused no deterioration in the emulsifying properties of unpasteurized or pasteurized salted yolks. However, Palmer *et al.* (1969B) observed that freezing did adversely affect the emulsifying performance of acidified yolk. Also, Palmer *et al.* (1969C) reported that freezing and frozen storage damaged the performance of both unpasteurized and pasteurized sugared yolks in sponge cakes.

The mechanism of egg-yolk gelation is still not completely understood. Early workers (Lea and Hawke 1951; Urbain and Miller 1930) indicated that lecithin and/or a lecitho-protein complex were altered. Since only proteolytic enzymes are effective in inhibiting gelation (Lopez *et al.* 1954), it appears that a protein fraction is definitely involved. Feeney *et al.* (1954) suggested that crotoxin (lecithinase A) attacked the lipoprotein complexes directly without splitting the lipid from the protein. Powrie *et al.* (1963) proposed that water plays an important role in gelation. Ice-crystal formation would dehydrate the protein, increase salt concentration, and the breakdown of the water shell surrounding the molecules could promote rearrangement and aggregation of yolk lipoproteins. Meyer and Woodburn (1965) suggested that fructose, which reduces gelation, may prevent aggregation of the lipoproteins. They postulated that fructose may chelate the iron in egg yolk and limit cross-bonding of the protein structure. Reinke (1967) postulated that yolk gelation and the aggregated components observed during fractionation involved phosvitin and/or calcium as a bridging component. Mahadevan *et al.* (1969) attributed yolk gelation to interaction between protein molecules following disruption of lipid-protein bonds. Palmer *et al.* (1970) suggested that these interactions may involve hydrophobic bonding, hydrogen bonding, and electrostatic forces so that entangled strands are formed. They further supported the role of dehydration. The observations by Seideman and Cotterill (1969) that freezing and spray-drying affect the B fraction on an ion-exchange chromatographic pattern support the view that dehydration and molecular rearrangement are involved in gelation. Hasiak *et al.* (1972) concluded on the basis of scanning electron microscopy that the structure of frozen-thawed yolk was more open than unfrozen yolk. The open spaces indicated that large "droplets" of water were trapped in the structure. Also, transmission electron photomicrographs indicated that both the low-density fraction (LDF) and high-density fractions (HDF) were altered by freezing and that an aggregation of these fractions occurred.

Wakamatu *et al.* (1983) provided an excellent review of the gelation process and the role that salt plays in controlling this problem. His article makes the following points:

1. Low-density lipoprotein (LDL) is the primary egg-yolk component altered by freezing.
2. Freezing rate, storage temperature and time, as well as thawing rates, are important.
3. Addition of low molecular weight additives (salts, sugars, glycerin, etc.) reduces gelation.
4. Transformation of a definite amount of water from liquid to ice is necessary for gelation.
5. Concentration of salts in frozen yolk due to freezing may cause irreversible "precipitation" of lipoprotein.
6. LDL "aggregation" may be related to pH change in the unfrozen phase as well as salt concentration.
7. Granules of yolk are disrupted during freezing by a high concentration of soluble salts in the unfrozen phase.
8. The effect of salt concentration on the precipitation, gelation, or aggregation process is not clear.
9. The protein and phospholipid moieties of LDL may participate in the formation of a LDL-water-sodium chloride complex.

Raw Whole Egg

Liquid whole egg undergoes gelation upon freezing and thawing. The amount of gelation is less drastic than in yolk. Egg white, either by dilution of the yolk or by other means, decreases the apparent gelation in whole egg. Frozen and thawed whole egg has a lumpy or curdled appearance. A watery, light-orange liquid separates from the curdled aggregate material. The retarding effect of additives and mechanical treatment on gelation (Sugihara et al. 1966; Ijichi et al. 1970) is similar to that on yolk. They also noted that storing at −18°C for 1 month produced the most viscous product. McCready and Cotterill (1972) reported that centrifugation of liquid whole egg minimizes many of the adverse changes caused by pasteurization and freezing. The viscosity of the supernatant fraction (plasma) was about the same as that of unfrozen whole egg. After gentle mixing it had a good-appearing, creamy, uniform texture. While freezing does not gelatinize the plasma from whole egg, the plasma from egg yolk will gelatinize (Saari et al. 1964; Mahadevan et al. 1969). Surprisingly, the precipitate fraction, which presumably contains the yolk granules, from whole egg or from yolk does not gelatinize upon freezing. It would appear that the gelation mechanism in whole egg depends upon the interaction of components that are separated by centrifugation (McCready and Cotterill 1972). Parkinson (1968) observed that frozen whole egg increased the insoluble materials after dialysis against a glycine buffer and centrifugation.

Most functional properties of whole egg are not drastically affected by freezing (Miller and Winter 1950), although a few reports have indicated some changes take place. Pearce and Lavers (1949) observed that freezing reduced the baking quality of whole egg, but it improved after storage for 3 months and then decreased. There was no relation between viscosity and baking quality. Miller and Winter (1951) compared the performance of unfrozen and frozen whole egg in mayonnaise. Mayonnaise made from frozen egg was more stable than that made from unfrozen whole egg. However, freezing did not affect the performance of frozen whole egg in custards (Jordan et al. 1952A; Miller et al. 1959). Jordan et al. (1952B) found that untreated whole egg produced sponge cakes of good quality. Also, when data presented by McCready et al. (1971) and McCready and Cotterill (1972) are compared, it appears that freezing does not adversely affect the whipping properties or the volume of sponge cakes made from frozen whole egg.

Cooked Yolk

Bengtsson (1967) stated that cooked yolk is little affected by freezing or rate of freezing. Apparently, any changes that do occur are not serious enough to be a problem.

Cooked Egg White and Cooked Whole Egg

Freezing causes cooked egg white to become tough or rubbery, and water separates from the clumps or layers. The thick white of a hard-cooked fresh shell egg will separate into distinct layers. This damage is apparently caused by large ice-crystal formation during slow freezing. Davis *et al.* (1952) showed that factors that reduce ice-crystal size, such as supercooling or the addition of finely ground calcium carbonate, reduce this damage. Soluble additives, except possibly gelatin, did not alleviate this problem. Bengtsson (1967) reported that freezing in circulating gas temperatures of −80°C produced an acceptable product, but that freezing in an air blast at −35°C did not. However, dip-freezing in liquid Freon 12 at −30°C gave comparable freezing rates and quality to that done in nitrogen gas at −150°C. Hawley (1970) prevented the syneresis of cooked frozen egg white by adding 2 to 4% of a water-binding carbohydrate such as algin, carrageenan, agar, or starch. Subsequently, this product can be used in cooked frozen egg rolls (Fig. 11.2). A machine has been developed in Denmark that automatically makes egg rolls (Fig. 11.3). First, a cooked egg white cylinder is formed, and this cylinder then is filled with liquid yolk and cooked.

One variation of this product is to use a freeze-stable deviled-egg mixture in place of the yolk. Extra frozen mix is available to decorate slices for hor d'oeuvres.

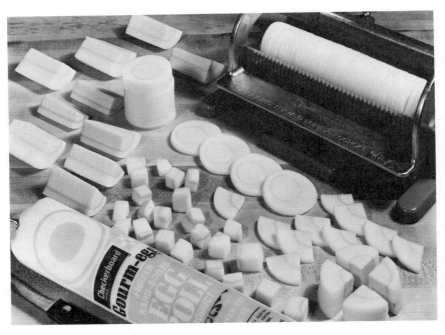

FIG. 11.2.   Cooked frozen egg roll.
Courtesy Ralston Purina Co.

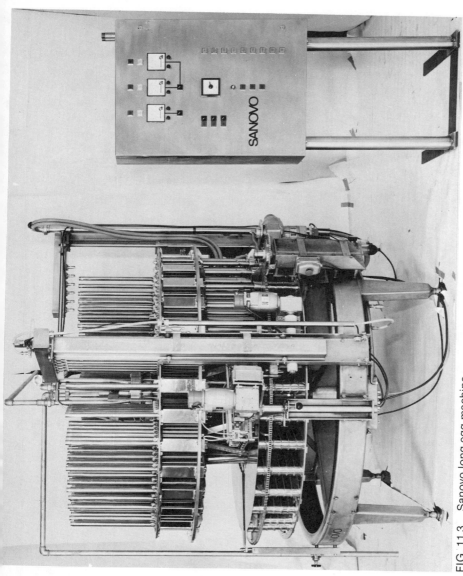

FIG. 11.3.   Sanovo long egg machine.
Courtesy Sanovo (USA), Inc.

Davis *et al.* (1952) observed that a yolk-white ratio of 40:60 to 80:20, when diluted with 20% water and adjusted to pH 6.0 to 7.0 before cooking, should be suitable in frozen foods. However, substituting whole egg for yolk decreased the stability of cooked frozen puddings or custards (Hanson *et al.* 1953). This would be expected, since whole egg contains white and when cooked is freeze-damaged. Jokay and Meyer (1961) stressed rapid freezing to avoid undesirable texture and color changes in freeze-dried cooked scrambled eggs. Cimino *et al.* (1967) reported that the addition of flour and methycellulose favorably affected the stability of baked unsweetened plain or cheese whole-egg soufflé. The stability of baked egg white soufflé was directly related to sugar concentration. Palmer *et al.* (1974) and Walker *et al.* (1976) developed, tested, and published production guides for several cook-freeze egg dishes and sauces (see Table 11.3 through 11.12) for U.S. Air Force Missile sites. Presumably, these products could also be produced for the institutional and retail markets.

Feiser and Cotterill (1982, 1983) reviewed the literature on cooked-frozen-thawed-reheated (CFTR) egg products. Food items in this category include scrambled eggs, quiches, souffles, meringues, omelets, and egg rolls, all of which have common problems. They analyzed the composition of the serum that was expressed from CFTR scrambled eggs. Various pH levels and salt concentrations were studied. The increase in pH due to CFTR is shown in Figure 11.4. Apparently, this change is due to the loss of carbon dioxide. Electrophoretically many of the proteins were still present in the serum. As expected, ovomacroglobulin and lipovitellin were the most heat sensitive and were retained in the coagulum. However, these proteins also appeared in the serum in samples containing more than 0.4% salt. The volume of serum and the amounts of total solids, protein, ash, lipids, iron, phosphorus, sodium, potassium, and chlorine in the serum were determined. While major changes were observed, it was not possible to adequately relate them to the freeze-damage problem.

O'Brien *et al.* (1982) reported that the addition of 0.1% xanthan gum, the use of moist heat in cooking, and freezing with liquid CO_2 or N_2 minimized moisture loss and shear force of omelets. Omelets that had been steamed for 5 min. and then cryogenically frozen had satisfactory organoleptic properties.

FROZEN SCRAMBLED-EGG MIXES

A wide range of ingredients has been added to whole egg to prepare commercial scrambled-egg mixes. The most common added components are nonfat dry milk (liquid or dried), whey, vegetable oil, water (if

TABLE 11.3. Production Guide for Creamed Egg and Beef, Chicken, or Turkey

Yield: 100 Portions **Each Portion: 200 g (7 oz)**

Ingredients	%	Pounds	Grams	Procedure
Ground beef, chicken or turkey meat, raw	23.32	11.03	5,003.2	1. Mix. Form a loaf covering the bottom of one or more baking pans
Egg, whole	2.59	1.22	553.4	
Salt	0.18	0.08	36.3	2. Cook by steaming for 45 min in a steam cooker at atmospheric pressure. Cool and refrigerate overnight
Onion flakes, dry	0.08	0.04	18.1	
Monosodium glutamate	0.03	0.01	4.5	
Pepper, black	0.03	0.01	4.5	
Thyme (for turkey)	0.01	0.01	4.5	3. Dice to 0.95 cm (0.375 in.)
Egg yolk	14.25	6.74	3,057.3	4. Stir until dry ingredients are dissolved
Egg white	6.22	2.94	1,333.6	
Water	5.18	2.45	1,111.3	5. Pour into one or more baking pans. Cook by steaming in a steam cooker at atmospheric pressure for 15 min. Cool and refrigerate
Salt	0.18	0.08	36.3	
Monosodium glutamate	0.03	0.01	4.5	
				6. Dice to 0.95 cm (0.375 in.)
Milk, whole	21.50	10.17	4,613.1	7. Blend dry ingredients with margarine in a steam-jacketed kettle or double boiler until uniform. Add liquids, continue stirring and scraping sides until thickened and smooth
Beef or chicken broth	20.72	9.81	4,449.8	
Flour, rice	2.33	1.10	499.0	
Margarine	1.68	0.80	362.9	
Flour, general purpose	1.16	0.55	249.5	
Salt	0.33	0.15	68.0	
Onion flakes, dry	0.08	0.04	18.1	
Monosodium glutamate	0.05	0.02	9.1	8. Mix white sauce, diced meat and diced egg together in a 2:1:1 ratio by weight
Pepper, black	0.03	0.02	9.1	
Thyme (for turkey)	0.02	0.01	4.5	
				9. Fill 200 g (7 oz) into 16 × 20 cm (6.5 × 8 in.) Scotch Pak pouches[a]
				10. Seal, mark, and freeze
				11. For reconstitution and serving, see Footnote b
Totals	100.0	47.29	21,450.6	

[a] Kapak Industries stock No. 201, 9809 Logan Ave. So., Bloomington, MN 55431.
[b] Reconstitution and serving:
1. Air convection oven: Punch 10–12 holes in one surface of pouch. Place in pan and heat in a 204°C oven for 35 min.
2. Microwave oven: Punch holes as above. Place pouch in a Pyrex baking dish. Heat for 1 min on, 1 min off. Repeat for three heating cycles.
3. Serve on two slices of toast.

dry ingredients are used), gums (CMC and xanthan are most common), organic acids or other chelators (citric acid, lactic acid, or phosphates), salt, and egg white. A summary of these mixes was published by Cotterill (1983) (see Table 11.13).

Most of the ingredients in the scrambled-egg mixes in use today can be traced to the art of good cooking, developed over many years or perhaps even centuries. The addition of milk not only provided added nutrients and served to extend the more costly egg component, but its fat helped form a soft coagulum. A small amount of lemon juice may have been added to improve color, thus controlling the greening problem

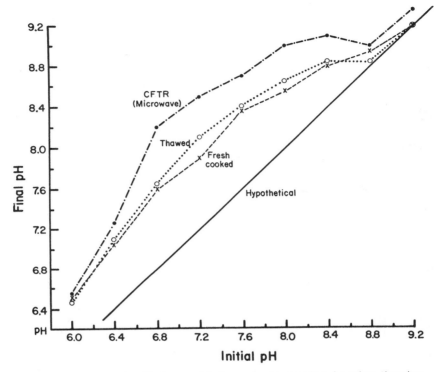

FIG. 11.4.   Increase in pH of scrambled egg due to cooking, freezing, thawing, and reheating (CFTR).
SOURCE: *Feiser and Cotterill (1982).*

prevented by lowering the pH. Butter was used to grease the skillet to reduce sticking, but it also may have improved flavor and texture. Salt added taste, but apparently had no other functional role in scrambled eggs. Flour, a forerunner of the gums that are found in some mixes today, could have been added to increase thickness and improve water-binding capacity.

Even scrambled eggs have their problems. When they are held at serving temperatures (at or above 145°F) for long periods of time, as in institutional feeding, a fluid may separate (from syneresis) and a green discoloration may appear (from an iron-sulfur compound). Although, the fluid separation can be controlled little is really understood about what happens to the egg-product system or structure. Scrambled-egg mixes have good steam-table stability, color, and texture. Nevertheless, the complaint is frequently heard that they lack flavor. This may be because less hydrogen sulfide is formed at the lower pH needed to avoid greening.

The problem of greening can be easily prevented. For example, ingredients can be added to lower the pH and chelate the iron so it will not react with sulfur and form the gray-green color. Gossett and Baker (1981) reported the optimum conditions needed to prevent this discoloration in liquid whole eggs (Table 11.14). The minimum amount of acid or

TABLE 11.4. Production Guide for Western or Denver Eggs

Yield: 100 Portions **Each Portion: 164.3 g (5.8 oz)**

Ingredients	%	Pounds	Grams	Procedure
Egg white/Methocel[a]	15.82	5.73	2,599.1	1. Heat 584.0 g (20.6 oz) of water to 90°C. Gradually add 120.8 g (4.3 oz) of Methocel and stir with a mechanical mixer using a flat paddle and 2,332.4 g (82.3 oz) of cold, blended egg white and chill to 10°C. Use 2,599.1 g (91.6 oz) of this mixture
Onion, dehydrated, minced	0.51	0.18	81.6	2. Rehydrate onions in 749.7 g (26.4 oz) of water
Peppers, green, finely chopped	4.56	1.65	748.4	3. Sauté onions and peppers in oil and butter
Oil, vegetable	1.98	0.72	326.6	
Butter	1.98	0.72	326.6	
Flour, rice	3.65	1.32	598.7	4. Add flour and water and cook briefly
Water	15.21	5.51	2,499.3	
Water chestnuts, canned, chopped	6.84	2.48	1,124.9	5. Blend egg white/Methocel, sautéed onions and peppers, flour and water, water chestnuts, soy sauce, black pepper, smoke salt, nonfat dry milk, and egg yolk
Soy sauce	1.82	0.66	299.4	
Monosodium glutamate	0.05	0.02	9.1	
Pepper, black	0.08	0.03	13.6	
Smoke salt	0.09	0.03	13.6	
Milk, nonfat, dry	1.52	0.55	249.5	
Egg yolk, blended	30.68	11.11	5,039.5	
Ham, chopped	15.21	5.50	2,499.3	6. Dice ham approximately 0.6 to 0.9 cm (0.25 to 0.375 in.) and weigh 35 g (1.2 oz) into each aluminum pan[b]
				7. Add 130 to 135 g (4.6 to 4.8 oz) of egg mixture. Stir to blend
				8. Cover with lid and aluminum foil
				9. Steam for 13 min [c]
				10. Cover, mark, and freeze
				11. For reconstitution, see Footnote d
Totals	100.0	36.22	16,429.2	

[a] Methocel MC, Dow Chemical Co., premium grade (food use), 15 cps.
[b] Ekco No. 7036 with foil/paper lid or equivalent.
[c] These eggs can be easily overcooked.
[d] Reconstitute by heating (lid on pan) in an air convection oven at 149°C for 35 to 40 min.

TABLE 11.5. Production Guide for Scrambled Eggs with Bacon

Yield: 100 Portions   **Each Portion: 164.3 g (5.8 oz)**

Ingredients	%	Pounds	Grams	Procedure
Eggs, whole	69.06	17.63	7,997.0	1. Mix eggs and milk in a mixer
Milk, whole	17.27	4.41	2,000.4	until smooth
Starch, freeze-thaw, stable	1.44	0.37	167.8	2. Blend dry ingredients well; add to egg-milk mixture with
Monosodium phos-phate monohydrate	0.43	0.11	50.0	continuous stirring
Salt	0.36	0.01	4.5	
Butter	1.08	0.27	122.5	3. Melt butter; add egg and continue mixing until dry ingredients are dissolved
Bacon bits[a]	10.36	2.64	1,197.5	4. Cook eggs as described in Footnote b 5. Pack 100 g (3.5 oz) of cooked egg[b] into each aluminum pan.[c] Sprinkle 12 g (0.4 oz) of bacon bits on top of eggs 6. Cover, mark, and freeze 7. For reconstitution, see Footnote d
Totals	100.00	25.44	11,539.7	

[a] Use real bacon bits.

[b] Cook eggs as follows
  1. Cut high-density polyethylene tubing 7.6 cm (3 in.) wide into 38-cm (15 in.) lengths. Seal one end with a metal clip.
  2. Fill tubings with egg mixture and clip the open end. The clip should be approximately 1 in. above the level of the egg mixture. Eliminate as much air as possible from the egg mixture to prevent buoyancy and bursting problems.
  3. Cook the eggs in 75°C water for 80 min.
  4. Cool cooked eggs to room temperature. Eggs may be kept at 2°C to 3°C for up to 30 days before further processing.
  5. Open tubings and put the egg content into mixer bowl equipped with a wire whip. Break the cooked eggs into the desired chunk size by control of speed and length of whipping.
  6. Package as given in Step 5 of procedure.

[c] Ekco No. 7036 with foil/paper lid or equivalent.

[d] Reconstitution:
  1. Air convection oven: Heat (lid on pan) at 204°C for 30 to 35 min.
  2. Cold air convection oven: Put pan (lid on) in cold oven. Set temperature to 232°C and heat 35 to 40 min.
  3. Microwave oven: Move eggs from aluminum pans into a nonmetal container with cover. Heat for 30 seconds. Turn off for 2.5 min. Heat for an additional 30 sec.

chelator varies between 0.17% for citric acid to 0.34% for monosodium phosphate. The pH ranged between 5.74 for lactic acid to 8.40 for $Na_2 +$ EDTA. Citric acid, the most common chelator in use today, lowered the pH to about 6.8 at a concentration of 0.17%. The use of monosodium phosphate in scrambled-egg mixes has been patented by Chin and Redfern (1968).

The increase in pH of scrambled egg due to cooking and the pH of the serum pressed from frozen, thawed, and reheated scrambled eggs is shown in Fig. 11.4 (Feiser and Cotterill 1982). With most acids, it is

TABLE 11.6. Production Guide for Puffy Omelet

Yield: 100 Portions
Each Portion: 150 g (5.3 oz)

Ingredients	%	Pounds	Grams	Procedure
Margarine	9.94	3.86	1,750.9	1. In a steam kettle, melt marga-
Milk, nonfat, dry	4.26	1.65	748.4	rine. Add dry ingredients
Methocel[a]	0.99	0.38	172.4	mixed together. Stir to blend.
Flour, rice	3.55	1.38	626.0	Add water gradually with
Flour, general purpose	3.55	1.38	626.0	continuous stirring and heat-
Salt	0.71	0.28	127.0	ing until mixture thickens and
Water	40.63	15.76	7,148.7	reaches a temperature of 80°C
Egg yolk, blended	13.92	5.40	2,449.4	2. Beat egg yolk at high speed for 200 sec (3.3 min), scraping bowl after 160 sec (2.7 min). Add hot white sauce and beat at medium speed for 100 sec (1.7 min). Scrape bowl after 50 sec
Egg white, blended	22.45	8.71	3,950.9	3. Beat egg white at high speed until stiff but not dry. Fold 3,120 cm^3 (13 cups) of egg white into yolk mixture, then fold in remaining whites
				4. Place 150 g (5.3 oz) into aluminum pans[b]. Add cover with a smaller pan with a 1-in. hole cut in the center
				5. Bake at 107°C for 1 hr
				6. Cover, mark, and freeze
				7. For reconstitution, see Footnote c
Totals	100.00	38.80	17,599.7	

[a] Methocel MC, Dow Chemical Co., premium grade (food use), 15 cps.
[b] Ekco No. 7036 with foil/paper lid or equivalent.
[c] Reconstitution: Remove lid and cover omelet loosely with aluminum foil. Heat at 149°C for 50 to 60 min. With sauce: Remove frozen sauce (Spanish, cheese, or chipped beef) from package and place on top of omelet. Heat 60 min at 149°C.

necessary to reduce the pH of the whole egg from 7.0–7.6 to about 6.6–6.8. Unless a strong chelator, like EDTA, is present, greening is likely if the pH of the cooked product exceeds 8.2.

The role of the other ingredients in scrambled-egg mixes is less understood. Nonfat dry milk, for example, like fluid milk, contains calcium, phosphates, and proteins that may react with egg proteins and alter the nature of the coagulum. Also, lactose from milk could increase the temperature of coagulation of the proteins. Gums may also react with the egg proteins, but their primary function is probably to improve water-holding capacity and thus reduce syneresis. Vegetable oils contain large amounts of polyunsaturated fatty acids which are essential in the diet. It should be noted that eggs contain almost the proper ratio of unsaturated-polyunsaturated fatty acids recommended by human nutritionists. Both butter and vegetable oils add softness to scrambled eggs.

TABLE 11.7. Production Guide for Potato Pancakes

Yield: 100 Portions   **Each Portion: 2 Pancakes**

Ingredients	%	Pounds	Grams	Procedure
Egg, whole	42.60	11.01	4,994.1	1. Mix milk, salt, and mono-
Milk, nonfat, dry	2.84	0.73	331.1	sodium glutamate with water
Water	2.84	0.73	331.1	and eggs until dissolved
Salt	0.43	0.11	50.0	
Monosodium glutamate	0.17	0.04	18.1	
Potatoes, shredded, hash browns	51.12	13.21	5,992.1	2. Mix portions of 30 g (1.1 oz) of potatoes and 25 g (0.9 oz) of egg mixture
				3. Drop onto greased skillet or griddle at 171°C. Form into 8.9-cm (3.5 in.) patties as they cook. Fry for 2 min on each side
				4. Pack 2 pancakes in each aluminum pan[a]
				5. Cover, mark, and freeze
				6. For reconstitution, see Footnote b
Totals	100.0	25.83	11,716.5	

[a] Ekco No. 7036 with foil/paper lid or equivalent.
[b] Reconstitution:
   1. Air convection oven: Heat 30 min at 205°C, covered with aluminum foil.
   2. Microwave oven: Remove patties from foil pans to glass dish. Heat for 30 sec, turn off for 30 sec. Repeat for 3 On-cycles.

TABLE 11.8. Production Guide for Spanish Sauce

Yield: 100 Portions   **Each Portion: 60 g (2.1 oz)**

Ingredients	%	Pounds	Ounces	Procedure
Onion flakes, dry	0.32	0.04	19.2	1. Soak dry vegetables in water for 10 min.
Pepper flakes, sweet, dry	0.32	0.04	19.2	
Parsley flakes, dry	0.03	0.004	1.8	
Water	4.78	0.63	286.8	
Margarine	4.78	0.63	286.8	2. Melt margarine and cook vegetables lightly for 5 min
Tomatoes, peeled, chopped, canned	89.24	11.72	5,354.4	3. Add tomatoes and spices Simmer until reduced to 2/3 volume
Salt	0.47	0.06	28.2	
Pepper, black	0.02	0.003	1.2	4. Fill 60 g (2.1 oz) into 4 x 6 inch pouches[a]
Oregano, ground	0.02	0.003	1.2	5. Seal, mark, and freeze
Garlic powder	0.02	0.003	1.2	6. For reconstitution, see Footnote b
Totals	100.00	13.133	6,000.0	

[a] Kapak Industries Stock No. 203; 9809 Logan Ave. So., Bloomington, MN 55431.
[b] Reconstitute by removing from pouch and placing on top of a fluffy omelet. Follow heating directions for omelet.

TABLE 11.9. Production Guide for Creamed Chipped Beef Sauce

Yield: 100 Portions   **Each Portion: 60 g (2.1 oz)**

Ingredients	%	Pounds	Grams	Procedure
Milk, nonfat, dry	6.75	0.89	405.5	1. Reconstitute milk
Water	64.82	8.50	3,889.2	
Margarine	8.10	1.06	486.0	2. Melt margarine
Flour, general purpose	8.10	1.06	121.8	3. Add flours and pepper. Stir
Flour, rice	2.03	0.27	121.8	until smooth
Pepper, black	0.07	0.01	4.2	4. Add milk gradually and stir over medium heat until mixture comes to a boil and thickens
Beef, chipped, dried	16.20	2.13	972.0	5. Shred beef. Add to white sauce and heat slowly for 5 min.
				6. Fill 60 g (2.1 oz) into 4 x 6 in. pouches[a]
				7. Seal, mark, and freeze
				8. For reconstitution, see Footnote b
Total	100.00	13.13	600.00	

[a] Kapak Industries Stock No. 203; 9809 Logan Ave., So., Bloomington, MN 55431.
[b] Reconstitute by removing from pouch and placing on top of fluffy omelet. Follow heating directions for omelet.

TABLE 11.10. Production Guide for Cheese Sauce

Yield: 100 Portions   **Each Portion: 60 g (2.1 oz)**

Ingredients	%	Pounds	Grams	Procedure
Cheese, processed American	66.64	10.01	4,540.5	1. Cut cheese into 0.5-in. cubes
Milk, whole	33.35	5.01	2,272.5	2. Combine ingredients in a double boiler or other suitable
Pepper, black	0.01	0.01	4.5	vessel. Heat and stir until cheese is melted and sauce is smooth
				3. Fill 60 g (2.1 oz) into 4 x 6 inch pouches[a]
				4. Seal, mark, and freeze
				5. For reconstitution, see Footnote b
Totals	100.00	15.03	6,817.5	

[a] Kapak Industries Stock No. 203; 9809 Logan Ave. So., Bloomington, MN 55431.
[b] Reconstitute by removing from pouch and placing on top of a fluffy omelet. Follow heating directions for omelet.

TABLE 11.11. Production Guide for French Toast

Yield: 100 Portions **Each Portion: 2 Slices**

Ingredients	%	Pounds	Grams	Procedure
Eggs, whole	50.33	12.40	5,624.6	1. Mix ingredients into a smooth batter
Milk, whole	48.55	11.95	5,420.5	2. Dip bread into batter for 30 to
Salt	1.12	0.27	122.5	45 secs per slice
Bread[a]				3. Fry slices in a thin layer of
Shortening or oil		2.03	925.0	shortening on a grill or electric skillet at 188°C for approximately 2.25 min per side, or until golden brown
				4. Cut cooked toast slices in half diagonally
				5. Pack 4-halves in each aluminum pan and cover with proper lid[b]
				6. Mark and freeze
				7. For reconstitution, see Footnote c
Totals	100.00	24.62	11,167.6	

[a] 200 slices
[b] Ekco No. 7036 with foil/paper lid or equivalent.
[c] Reconstitution:
1. Air convection oven: Place frozen slices on a sheet pan and heat in oven (204°C) for 11 min.
2. Microwave oven: In the presence of a beaker of water in the oven, place two frozen slices on Pyrex pan. Heat for 30 secs. Turn off for 30 secs. Repeat for four heat-cycles.

TABLE 11.12. Production Guide for Egg-and-Potato Patties with Bacon

Yield: 100 Portions **Each Portion: 2 Patties**

Ingredients	%	Pounds	Grams	Procedure
Egg, whole	40.57	11.01	4,994.1	1. Mix until milk, salt, and mono-sodium glutamate are dissolved
Milk, nonfat, dry	4.06	1.10	499.0	
Onion flakes, dry	1.62	0.44	199.6	
Salt	0.81	0.22	100.0	
Monosodium glutamate	0.16	0.04	18.1	
Pepper, black	0.03	0.01	4.5	
Potatoes, shredded, hash browns, thawed	48.69	13.21	5,992.1	2. Mix 30 g (1.1 oz) of potatoes and 25 g (0.9 oz) of egg mixture
Bacon bits	4.06	1.10	499.0	3. Add 2.5 g (0.09 oz) of bacon bits to each potato-egg portion
				4. Drop onto a buttered or oiled skillet or griddle at 171°C. Form into 9-cm (3.5-in.) patties as they cook. Fry for 2 min on each side
				5. Pack 2 patties in each aluminum pan[a]
				6. Cover, mark, and freeze
				7. For reconstitution, see Footnote b
Totals	100.00	27.13	12,306.1	

[a] Ekco No. 7036 with foil/paper lid or equivalent.
[b] Reconstitution:
1. Air convection oven: Heat for 30 min at 204°C, covered with aluminum foil.
2. Microwave oven: Remove patties from foil pan to glass dish. Heat for 30 sec. Turn off for 30 sec. Repeat for 3 heating cycles.
3. Serve with tomato catsup.

TABLE 11.13. Scrambled Egg Mixes

Source	Milk[a]	Oil	Water	Gum	Acids and/or chelator	Salt	Other
Chin and Redfern (1968)	—	—	—	—	0.1% NaH$_2$PO$_4$	—	—
Katz (1968)	—	—	—	—	—	—	1.2% pectin
Jones et al. (1963)	4-6% NFDM and/or whey	—	37%	0.05-0.7%	—	—	1.5% casein and soy 0.25-1.4% soy isolate
Iijichi et al. (1970)	25% LSM	—	—	—	—	0.6%	0.4% dextrose
Tuomy and Walker (1970)	30% CSM	4.8%	—	—	—	0.6%	—
Ziegler et al. (1971)	1.4% CSM	—	19.2%	0.1% CMC[b]	0.1% citric acid	—	2% potato flour
Chen (1981)	3.2% CSM	4.0%	12.1%	—	0.1% lactic acid	—	—

SOURCE: Cotterill (1983).

[a] NFDM, nonfat dry milk; LSM, liquid skim milk; CSM, concentrated skim milk.
[b] CMC, carboxymethyl cellulose.

235

TABLE 11.14. Optimum Conditions for Prevention of Gray-Green Discoloration in Cooked Liquid Whole Egg

Chelator	Initial pH	Concentration (%)
Acetic acid	6.42	0.19
Citric acid	6.83	0.17
Na_2EDTA	8.40	0.029
Lactic acid	5.74	0.27
Malic acid	6.53	0.22
NaH_2PO_4	7.08	0.34
Propionic acid	6.47	0.26
Succinic acid	6.15	0.27

SOURCE: Gossett and Baker (1981).

One of the early commercial mixes was prepared by Jones (1963). This mix was originally intended for drying and is similar to the U.S. Department of Agriculture (1981) Scrambled Egg Mix according to the specification of the U.S. government (Table 11.15). Also, this formulation can be used to prepare either a frozen product or a basic formula for further modification. Table 11.13 shows several formulations reported in the literature (including U.S. patents) and in industrial specification brochures. The most common added ingredient to liquid whole egg is skim milk—in dry, liquid, or concentrated form. The amount of water is varied to compensate for the level of solids in the skim milk. Oil, if included, is added at a level of 4–5%. Both xanthan and cellulose gums are used at a level of 0.1%. Citric and lactic acids are most commonly used to adjust the pH to about 6.6–7.0. While monosodium phosphate improves the color of the cooked product, as well as lowers the pH and chelates iron, it also adds sodium to the product, which may be a disadvantage. For this reason, there is a tendency to eliminate sodium

TABLE 11.15. Scrambled Egg Mixes (USDA Formulas)

Ingredient	Liquid egg product prepared from		Dried egg product (%)
	Liquid egg (%)	Dried egg (%)	
Liquid whole egg[a]	66.3	—	> 51
Dried whole egg	—	17.1	(51)
Nonfat dry milk	9.6	10.0	> 30
Vegetable oil	4.8	5.0	> 15
Salt	0.3	0.3	< 1
Water	19.0	67.6	< 3
	100	100	100

SOURCE: U.S. Department of Agriculture (1981).
[a] Liquid whole egg = 24.7% solids.

chloride in specifications of products for such institutions as hospitals. Two formulas contain monosodium glutamate as a flavor enhancer, but this additive, naturally, also contains sodium.

It should be noted in Table 11.13 that some commercial products are produced with about 20–21% total solids. Presumably, egg white is added to reduce the solids level. A dry pre-mix formula, developed at the University of Missouri-Columbia, is shown in Table 11.16. When used at a level of 3%, in a pre-mix containing 58% whole egg and 39% egg white, this scrambled-egg mix contained 21.1% total solids. The pH was 7.15 and 8.17 before cooking and after cooking, respectively.

EGG SUBSTITUTES

Several low-cholesterol "egg substitute" products are on the retail market. Most, or all, of these products are characterized by yolk replacement with such other ingredients as: vegetable oil (high in polyunsaturated fatty acids), nonfat dry milk, soy protein, gums, food coloring, artificial flavors, and vitamins and minerals (for nutritional fortification). The primary ingredient in these products is egg white, but they may also contain added egg-white solids or a small amount of yolk. Chen (1976) and Baker and Darfler (1977) have reported on various formulations, whose main use is as scrambled eggs. A dry-mix formulation was test marketed in the early 1960s and at least one such dry mix is on the market today. One other product is sold as a refrigerated liquid, whereas all the others are marketed in frozen form. The nutrient composition analyses of some of these products have been reported by Pratt (1975), Childs and Ostrander (1976), and Anonymous (1975). Leutzinger et al. (1977) found that scrambled eggs made from one "substitute" product had a texture and aroma comparable to that produced from whole egg, whereas another product was definitely poorer in all the organoleptic ratings.

TABLE 11.16. Dry Pre-mix to Formulate a Frozen Scrambled Egg Mix

Dry pre-mix	%	Scrambled egg mix	(%)
Nonfat dry milk	72.4	Whole egg	58
Salt	16.0	Egg white	39
Dried egg yolk	3.2	Premix	3
Puritein 29 whey[a]	3.2		100
Purity W starch[b]	2.6		
Citric acid	2.2		
NaH$_2$PO$_4$	0.3		
	99.9		

[a] Puritein 29 whey, Anderson Clayton Foods.
[b] Purity W starch, National Starch and Chemical Corp.

MICROBIAL CHANGES DUE TO FREEZING

Freezing reduces the number of bacteria in egg products. Winter and Wilkin (1947) reported reductions as high as 99%. Some increase in numbers occurs upon defrosting (Winter and Wrinkle 1949); the latter paper contains an excellent literature review. Wrinkle *et al.* (1950) found that most genera survive freezing, but only three genera survived both pasteurization and freezing. *Bacillus* represented 83.5% of the survivors and the remaining survivors were equally divided between *Alcaligenes* and *Proteus*. All these organisms produced spoilage when inoculated into sterile products. The relative reductions in total count during frozen storage were not product dependent (Winter *et al.* 1951). Baush and Goresline (1953) observed somewhat smaller reductions due to freezing than those reported above and stated that direct microscopic counts gave a more reliable estimate of total bacterial populations and sanitary history than total viable counts. Brooks (1954) and Brooks and Hale (1954) related the amount of mustiness to fluorescence in frozen egg. According to Vadehra *et al.* (1969), pasteurized unfrozen whole egg had a better shelf life than a similar frozen product. Incubation or holding temperature after thawing was more important than thawing temperature.

The approximate total number of bacteria (psychrophiles, thermophiles, and anaerobes) found in 44 pasteurized, frozen commercial egg products, as reported by Shafi *et al.* (1970), is given in Table 11.17. *Bacillus* was the predominate genus present in these products. It is generally agreed that freezing alone will not destroy all of any bacterial genus found in egg products (Wrinkle *et al.* 1950). Cotterill and Glauert (1972) recently confirmed this concept for *Salmonella oranienburg* for salted yolk during frozen storage. However, this organism can be destroyed by storage of salt yolk at 16°, 25°, and 36°C. Increasing the salt concentration made this organism more labile.

TABLE 11.17. Levels of Bacteria in Pasteurized Frozen Commercial Egg Products

Product	TPC[a]	Psychrophiles	Thermophiles	Anaerobes
Egg white	220[b]	2	25	18
Egg yolk	500	35	25	8
Whole egg	630	25	250	45
Blends	110	10	15	7

SOURCE: *Shafi et al. (1970).*
[a] Total plate counts.
[b] Average number per gram.

REFERENCES

Anon. 1975. Egg replacer provides caloric reduction, total functionality and convenience. Food Prod. Dev. 9(5), 12–14.

Baker, R. C., and Darfler, J. M. 1977. Functional and organoleptic evaluation of low cholesterol egg blends. Poult. Sci. 56, 181–186.

Baush, E. R., and Goresline, H. E. 1953. An analysis of commercial frozen egg products. U.S., Dep. Agric. Circ. 932, Oct.

Bengtsson, N. 1967. Ultrafast freezing of cooked egg white. Food Technol. 21(9), 95–97.

Brooks, J. 1954. The fluorescence of liquid egg. I. The relation between the fluorescence and the mustiness in frozen whole egg. Food Technol. 8, 400–405.

Brooks, J., and Hale, H. P. 1954. The fluorescence of liquid egg. II. The effect of specific bacteria. Food Technol. 8, 406–409.

Chen, T. C. 1976. Improvements of liquid egg products. Poult. Sci. 55, 2019–2020.

Chen, T. C. 1981. Observations on the weeping of double boiler scrambled egg products and its prevention. J. Food. Sci. 46, 310–311.

Childs, M. T., and Ostrander, J. 1976. Egg substitutes: Chemical and biologic evaluations. J. Am. Diet. Assoc. 68(3), 229–234.

Chin, R. G. L., and Redfern, S. 1968. Egg compositions containing soluble phosphorous compounds effective to impart fresh egg color. U.S. Pat. 3,383,221.

Cimino, S. L., Elliott, L. F., and Palmer, H. H. 1967. The stability of souffles and meringues subjected to frozen storage. Food Technol. 21(8), 97–100.

Clinger, C., Young, A., Prudent, I., and Winter, A. R. 1951. The influence of pasteurization, freezing, and storage on the functional properties of egg white. Food Technol. 5, 166–170.

Cotterill, O. J. 1983. Unscrambling frozen egg mixes. Poult. Tribune 89(6), 10.

Cotterill, O. J., and Glauert, J. 1972. Destruction of Salmonella oranienburg in egg yolk containing various concentrations of salt at low temperatures. Poult. Sci. 51, 1060–1061.

Davey, E. M., Zabik, M. E., and Dawson, L. E. 1969. Fresh and frozen egg yolk fractions; emulsion stabilizing power, viscosity and electrophoretic patterns. Poult. Sci. 48, 251–260.

Davis, J. G., Hanson, H. L., and Lineweaver, H. 1952. Characterization of the effect of freezing on cooked egg. Food Res. 17, 393–401.

Downs, D. M., Janek, D. A., and Zabik, M. E. 1970. Custard sauces made with four types of processed eggs. J. Am. Diet. Assoc. 57, 33–37.

Feeney, R. E., MacDonnell, J. R., and Fraenkel-Conrat, H. 1954. Effects of crotoxin (lecithinase A) on egg yolk and yolk constituents. Arch. Biochem. Biophys. 48, 130–140.

Feiser, G. E., and Cotterill, O. J. 1982. Composition of serum from cooked-frozen-thawed-reheated scrambled eggs at various pH levels. J. Food Sci. 47, 1333–1337.

Feiser, G. E., and Cotterill, O. J. 1983. Composition of serum and sensory evaluation of cooked-frozen-thawed scrambled eggs at various salt levels. J. Food Sci. 48, 794–797.

Funk, K., and Zabik, M. E. 1971. Comparison of frozen, foam-spray-dried, freeze-dried and spray-dried eggs. 8. Coagulation patterns of slurries prepared with milk and whole eggs, yolks or albumen. J. Food Sci. 36, 715–717.

Funk, K., and Zabik, M. E. 1969. Custards baked in quantity. J. Am. Diet. Assoc. 55, 572–577.

Gossett, P. W., and Baker, R. C. 1981. Prevention of the green-gray discoloration in cooked liquid whole eggs. J. Food Sci. 46, 328–331.

Hanson, H. L., Nishita, K. D., and Lineweaver, H. 1953. Preparation of stable frozen puddings. Food Technol. 7, 462–465.

Hasiak, R. J., Vadehra, D. V., Baker, R. C., and Loop, L. 1972. Effect of certain physical and chemical treatments on the microstructure of egg yolk. Food Sci. *37*, 913–917.

Hawley, R. L. 1970. Egg product. U.S. Pat. 3,510,315, May 5.

Husaini, S. A., and Alm, F. 1955. Denaturation of proteins of egg white and of fish and its relation to the liberation of sulfhydryl groups on frozen storage. Food Res. *20*, 264–272.

Ijichi, K., Palmer, H. H., and Lineweaver, H. 1970. Frozen eggs for scrambling. J. Food Sci. *35*, 695–698.

Jaax, S., and Travnick, D. 1968. The effect of pasteurization, selected additives and freezing rate on the gelation of frozen-defrosted egg yolk. Poult. Sci. *47*, 1013–1022.

Jokay, L., and Meyer, R. I. 1961. Process of preparing a dehydrated precooked egg product. U.S. Pat. 3,009,818, Nov. 21.

Jordan, R., Luginbil, B. N., Dawson, L. E., and Echterling, C. V. 1952A. The effect of selected pretreatments upon the culinary qualities of eggs frozen and stored in a home-type freezer. I. Plain cakes and baked custards. Food Res. *17*, 1–7.

Jones, E., Tracy, D. H. and Deffenbaugh, C. B. 1963. Egg products and processes for preparing same. U.S. Pat. 3,093,487.

Jordan, R., Dawson, L. E., and Echterling, C. J. 1952B. The effect of selected pretreatments upon the culinary qualities of eggs frozen and stored in a home-type freezer. II. Sponge cakes. Food Res. *17*, 93–99.

Katz, M. 1968. Process for making a freezable egg food product. U.S. Patent 3,408,207.

Lea, C. H., and Hawke, J. C. 1951. The relative rates of oxidation of lipovitellin and of its lipid constituent. Biochem. J. *50*, 67–73.

Leutzinger, R. L., Baldwin, R. E., and Cotterill, O. J. 1977. Sensory attributes of commercial egg substitute mixtures. Poult. Sci. *55*, 2057.

Lineweaver, H., Palmer, H. H., Putnam, G. W., Garibaldi, J. A., and Kaufman, V. F. 1969. Egg pasteurization manual. U.S. Agric. Res. Serv., *ARS 74–78*, Mar.

Lopez, A., Fellers, C. R., and Powrie, W. D. 1954. Enzyme inhibition of gelation on frozen egg yolk. J. Milk Food Technol. *18*(3), 77–80.

Mahadevan, S., Satyanarayana, T., and Kumar, S. A. 1969. Physicochemical studies on the gelation of hen's egg yolk. Separation of gelling protein components from yolk plasma. J. Agric. Food Chem. *17*, 767–770

Marion, W. W. 1958. An investigation of the gelation of frozen egg yolk after thawing. Ph.D. Thesis, Purdue Univ., Lafayette, IN.

McCready, S. T., and Cotterill, O. J. 1972. Centrifuged liquid whole egg. 3. Functional performance of frozen supernatant and precipitate fractions. Poult. Sci. *51*, 877–881.

McCready, S. T., Norris, M. E., Sebring, M., and Cotterill, O. J. 1971. Centrifuged liquid whole egg. 1. Effects of pasteurization on the composition and performance of the supernatant fraction. Poult. Sci. *50*, 1810–1817.

Meyer, D. D., and Woodburn, M. 1965. Gelation of frozen-defrosted egg yolk as affected by selected additives: Viscosity and electrophoretic findings. Poult. Sci. *44*, 437–446.

Miller, C., and Winter, A. R. 1950. The functional properties and bacterial content of pasteurized and frozen whole egg. Poult. Sci. *29*, 88–97.

Miller, C., and Winter, A. R. 1951. Pasteurized frozen whole egg and yolk for mayonnaise production. Food Res. *16*, 43–49.

Miller, G. A., Jones, E. M., and Aldrich, P. J. 1959. A comparison of the gelation properties and palatability of shell eggs, frozen whole eggs and whole egg solids in standard baked custards. Food Res. *24*, 584–594.

Morgan, K. J., Funk, K., and Zabik, M. E. 1970. Comparison of frozen, foam-spray-dried, freeze-dried eggs. 7. Soft meringues prepared with a carrageenan stabilizer. J. Food Sci. *35*, 699–701.

O'Brien, S. W., Baker, R. C., Hood, L. F., and Liboff, M. 1982. Water-holding capacity and textural acceptability of pre-cooked, frozen whole egg omelets. J. Food Sci. *47*, 412–417.

Palmer, H. H. *et al.* 1969A. Salted egg yolks. 1. Viscosity and performance of pasteurized and frozen samples. Food Technol. *23*, 1480–1485.

Palmer, H. H. *et al.* 1969B. Salted egg yolks. 2. Viscosity and performance of acidified, pasteurized and frozen samples. Food Technol. *23*, 1486–1488.

Palmer, H. H. *et al.* 1969C. Sugared egg yolks: Effects of pasteurization and freezing on performance and viscosity. Food Technol. *23*, 1581–1585.

Palmer, H. H., Ijichi, K., and Roff, H. 1970. Partial thermal reversal of gelation in thawed egg yolk products. J. Food Sci. *35*, 403–406.

Palmer, H. H., Tsai, L. S., Ijichi, K., and Hudson, C.A. 1974. Frozen egg products for air force missile sites. U.S. Agric. Res. Serv., West. Reg. [Rep.] ARS-W *ARD-W-*   .

Parkinson, T. L. 1968. Effect of pasteurization on the chemical composition of liquid whole egg. III. Effect of staling and freezing on the protein fraction patterns of raw and pasteurized whole egg. J. Sci. Food Agric. *19*, 590–596.

Pearce, J. A., and Lavers, C. G. 1949. Liquid and frozen egg. V. Viscosity, baking quality and other measurements on frozen egg products. Can. J. Res. Food *27*, 231–240.

Powrie, W. D., Little, L, and Lopez, A. 1963. Gelation of egg yolk. J. Food Sci. *28*, 38–46.

Pratt, D. E. 1975. Lipid analysis of a frozen egg substitute. J. Am. Diet. Assoc. *66*(1), 31–33.

Reinke, W. C. 1967. Investigations on egg yolk gelation. Ph.D. Thesis, Cornell Univ., Ithaca, NY.

Saari, A., Powrie, W. D., and Fennema, O. 1964. Influence of freezing egg yolk plasma on the properties of low-density lipoproteins. J. Food Sci. *29*, 762–765.

Seideman, W. E., and Cotterill, O. J. 1969. Ion-exchange chromatography of egg yolk. 2. Heating, freezing and spray-drying. Poult. Sci. *48*, 894–900.

Shafi, R., Cotterill, O. J., and Nichols, M. L. 1970. Microbial flora of commercially pasteurized egg products. Poult. Sci. *49*, 578–585.

Sugihara, T. F., Ijichi, K., and Kline, L. 1966. Heat pasteurization of liquid whole egg. Food Technol. *20*, 1076–1983.

Thomas, A. W., and Bailey, M. I. 1933. Gelation of frozen egg magma. Ind. Eng. Chem. *25*, 669–674.

Tuomy, J. M., and Walker, G. C. 1970. Effect of storage time, moisture level and headspace oxygen on the quality of dehydrated egg mix. Food Technol. *24*, 1287.

Urbain, O. M., and Miller, J. N. 1930. Relative merits of sucrose, dextrose and levulose as used in the preservation of eggs by freezing. Ind. Eng. Chem. *22*, 355–356.

U.S. Department of Agriculture 1981. Purchase of dried egg mix for distribution to eligible outlets. Announcement/Invitation PV-95. Food Safety and Quality Service, Poultry and Dairy Division. Washington, D.C. 20250.

Vadehra, D. V., Steele, F. R., Jr., and Baker, R. C. 1969. Shelf life and culinary properties of thawed frozen pasteurized whole eggs. J. Milk Food Technol. *32*, 362–364.

Wakamatu, T., Sato, Y., and Saito, Y. 1983. On sodium chloride action in the gelation process of low density lipoprotein (LDL) from hen egg yolk. J. Food Sci. *48*, 507–516.

Walker, G. C., Tuomy, J. M., and Kanter, C. G. 1976. Egg Products for Use in Cook-freeze System, Tech. Rep. 75-73-FEL. U.S. Army Natick Research Development Command (AD A009 749), Natick, MA.

Winter, A. R., and Wilkin, M. 1947. Holding, freezing, and storage of liquid egg products to control bacteria. Food Freezing *2*(4), 338–341.

Winter, A. R., and Wrinkle, C. 1949. Fast freezing at low temperatures protects frozen egg quality. U.S. Egg Poult. Mag. *55*(2), 20–23, 30–31.

Winter, A. R., Burkart, B., and Wrinkle, C. 1951. Analyses of frozen egg products. Poult.

Sci. *30*, 372–380.

Wolfe, N. J., and Zabik, M. E.   1968.   Comparison of frozen, foam-spray-dried, freeze-dried, and spray-dried eggs. 3. Baked custards prepared from eggs with added corn syrup solids. Food Technol. *22*, 1470–1476.

Wootton, M., Hong, N. T., and Thi, H. L. P.   1981.   A study on the denaturation of egg white proteins during freezing using differential scanning calorimetry. J. Food Sci. *46*, 1337–1338.

Wrinkle, C., Weiser, H. H., and Winter, A. R.   1950.   Bacterial flora of frozen egg products. Food Res. *15*, 91–98.

Zabik, M. E.   1968.   Comparison of frozen, foam-spray-dried, freeze-dried and spray-dried eggs. 2. Gels made with milk and albumen or yolk containing corn syrup solids. Food Technol. *22*, 1465–1469.

Zabik, M. E.   1969.   Comparison of frozen, foam-spray-dried, freeze-dried, and spray-dried eggs. 6. Emulsifying properties at three pH levels. Food Technol. *23*, 838–840.

Zabik, M. E., and Brown, S. L.   1969.   Comparison of frozen, foam-spray-dried, freeze-dried, and spray-dried eggs. 4. Foaming ability of whole eggs and yolks with corn syrup solids and albumen. Food Technol. *23*, 262–266.

Zabik, M. E., and Figa, J. E.   1968.   Comparison of frozen, foam-spray-dried, freeze-dried, and spray-dried eggs. 1. Gels prepared with milk and whole eggs containing corn syrup solids. Food Technol. *33*, 1169–1175.

Zabik, M. E., Anderson, C. M., Davey, E. M. and Wolfe, N. J.   1969.   Comparison of frozen, foam-spray-dried, and spray-dried eggs. 5. Sponge and chiffon cakes. Food Technol. *23*, 359–364.

Egg-Product Pasteurization

F. E. Cunningham

INTRODUCTION

Pasteurization of egg products in the United States became virtually mandatory on June 1, 1966. Regulations were passed for USDA-inspected plants (*Federal Register*, May 1, 1965; *Egg Products Inspectors Handbook*, August 1, 1965); new FDA standards of identity for egg products were adopted (*Federal Register*, March 19, 1966); and some state laws were enacted requiring pasteurization, such as the California Senate Bill No. 643. The purpose of these various measures was, of course, to ensure against the presence of any pathogenic bacteria in egg products.

Pasteurization of liquid whole egg and liquid yolk was first utilized by the egg-product industry in the 1930s. The first operations were small and used batch-type dairy pasteurizers, operated at temperatures up to 60°C. It was later found that plate-type high temperature–short-time (HTST) dairy pasteurizers could be used, provided that the operating temperatures were carefully controlled. In Europe, almost 30 years ago, liquid whole egg was successfully pasteurized in a continuous commercial installation. The unit consisted of a plate heat exchanger with sufficient holding tubes to maintain a temperature of 61°C for a minimum of 3 min. Although their main objective was not specifically the elimination of salmonellae, the time-temperature conditions used were in the range reported to effectively kill most of the salmonellae from contaminated eggs.

The USDA now requires that liquid whole egg be heated to at least

243

EGG SCIENCE & TECHNOLOGY,
3rd Edition

60°C and held for no less than 3.5 min for the average particle. British researchers claim that this is insufficient heat treatment to ensure a satisfactory product in the United Kingdom. Their data and recommendations resulted in the adoption of pasteurizing temperatures of 64°C for 2.5 min. An important advantage of this combination of time-temperature conditions is the fact that an enzymatic test for determining the adequacy of pasteurization is available. Unfortunately, this test does not work at lower time-temperature combinations and it is generally felt in the United States that the temperatures recommended by the British workers may be higher than necessary to accomplish satisfactory salmonellae reduction in whole egg. They claim that 2.5 min pasteurization at 64° to 65°C results in products which are generally acceptable for most bakery products, although some adjustment of the recipes may be desirable. Eggs pasteurized under commercial conditions in the United States show a small (5%) reduction in volume and a similar loss in other functional properties. Public health officials generally consider that eggs pasteurized under conditions described by the USDA do not pose any public health hazards.

The pasteurization of egg white has posed a considerably greater problem because of its great instability to heat in the range of effective pasteurization.

HEAT EFFECTS ON EGG-PRODUCT PERFORMANCE

Changes in egg products brought about by heat are of major concern to the processor and user alike. Products must not only be microbiologically sound but must also perform satisfactorily. Finished products are checked for functional damage using laboratory tests designed to detect small changes. When egg white is heated, the first noticeable change is in foaming power. The beating rate, meringue stability, and angel-cake volume are classic tests for egg white. Damage to whole egg is evaluated in similar tests, using layer cakes or sponge cakes instead of angel cakes and observing foaming and coagulating properties. The stability of mayonnaise and salad dressing is an indication of the emulsifying power of yolk and whole egg, and the stiffness of custards reflects coagulating ability.

Denaturation and performance impairment is a function of time and temperature. Changes in both the physical and functional properties of normal egg white due to heat have been described by many early researchers. (See the list of Selected References for further reading.) It has been reported consistently that heating egg white in the pasteurization range of 54° to 60°C damages foaming power. Most workers have agreed that egg white is impaired when heated for several minutes above 57°C. However, in the case of yolk-contaminated egg white, heating improves foaming properties. The extent of improvement varies with pH.

Egg white at pH 9 tends to increase in viscosity when heated to 56.7° to 57.2°C and coagulates rapidly at 60°C. The proteins in egg white vary greatly in their coagulation rates at various conditions of pH, in the presence of additives and in the presence of each other. The macromolecular changes in albumen ultrastructure were studied with an electron microscope following pasteurization with and without various additives used commercially to minimize heat denaturation. The changes were characterized by the formation of dense irregular clumps of approximately 0.1 μm diameter, which became more numerous and dense as heating time progressed. Experiments show that increase in whip time occurs when egg white is held for 2 or more minutes at 53°C (see Table 12.1). However, this treatment did not damage the volume or texture of angel cakes made from the whites. The use of acceptable whipping aids (triacetin and triethyl citrate) effectively reduces the whipping time.

Denaturation of whole egg, as indicated by a change in viscosity, has been reported to take place in the temperature range of 56° to 66°C. Above this range, fractional precipitation of proteins occurs and coagulation takes place rapidly above 73°C.

Studies on the effect of pasteurization of whole egg revealed that it exerted no harmful effect on custard-making properties until the temperature reached 71°C. Whole egg pasteurized in the temperature range of 60° to 68°C gave sponge cakes with volumes approximately 4% less than those made from control samples. Cakes made with whole egg pasteurized at 71°C had volumes 8% lower. Poorer textures were also noted in sponge cakes made from pasteurized whole egg. Pasteurized frozen whole egg has a different appearance when thawed from that of unpasteurized frozen whole egg; there is considerably more separation of a watery portion. When a temperature of 61°C with a holding time of

TABLE 12.1. Change in Whip Time on Heating Natural Egg White[a]

| | Whip time[b] | | |
Temperature (°C)	No additive (sec)	0.015% TEC[c]	0.03% TEC
Control	20		
53.3	45		
55.6	60	40	25
56.4	75		

[a] Flash-heated and held 2 min.
[b] Time required to produce maximum volume angel cake.
[c] Triethyl citrate (whipping aid).

3 min was used, other differences were noted. Pasteurization decreases the viscosity of liquid whole egg. Also, the combination of pasteurization and freezing reduces the viscosity of the product when defrosted. Also, pasteurization and freezing increased the beating time required for the preparation of a sponge cake, but improved the foam stability as measured by drainage.

When whole egg was pasteurized between 57°C and 66°C for up to 3 min, sponge cake volumes were actually improved by the heat process. Pasteurization at 61°C for 3 min did not damage the performance of whole egg in custards but did slightly impair the quality of sponge cakes. The performance and stability of both pasteurized whole egg and yolk were reviewed and it was concluded that whole egg commercially pasteurized in the United States performed satisfactorily in commercial baking tests. No significant damage to functional properties of frozen whole egg occurs until the egg is heated above 63°C for 3.5 min, or 74°C for 2 to 3 sec.

Liquid yolk was pasteurized using four time-temperature treatments as follows: 60°C for 4 min; 61°C for 2 min; 62°C for 1 min; 63°C for 0.5 min. The pasteurized yolks were tested for performance in doughnuts after spray-drying. The pasteurizing treatments caused no difference in the performance of the yolk solids. Changes in fat content were not great enough to be statistically significant, or to affect the eating quality of the finished doughnut.

Certain added substances help to stabilize liquid egg against heat denaturation. Carbohydrates such as sucrose, glucose, fructose, arabinose, mannitol, and xylose have been found to inhibit heat denaturation, as evidenced by preventing the formation of sulfhydryl groups. Carbohydrates also protect whole egg and yolk from coagulation by heat. Sugars have a marked influence on the stability of egg white to heat. For example, liquid egg white with 20% sucrose can be heated instantaneously to 65°C before a change in whipping power is noted.

The electrophoretic and chromatographic changes in egg-yolk proteins due to heat were reported by Dixon and Cotterill (1981). The temperatures studied ranged between those normally used for pasteurization to those causing coagulation. The products studied included plain, 10% salted, and 10% sugared yolk and whole egg. The USDA pasteurization temperatures of 61.0°C and 63.3°C for plain, 10% sugared, and 10% salted egg yolk, respectively, had little effect on their electrophoretograms. Only the γ-livetins in plain yolk were slightly affected. Higher temperatures progressively altered these proteins until they disappeared from the electrophoretograms from plain yolk at approximately 73°C, sugared yolk at approximately 76°C, and salted yolk at approximately 79°C. The lipovitellin fraction and phosvitins were similarly affected by increasing temperatures. Sugar and salt at 10%

levels had a protective effect against heat, allowing, respectively, higher temperatures of approximately 3° to 6°C before heat damage occurred. The chromatograms of pasteurized, plain, and sugared yolk showed almost no effects of such treatment. However, major changes were noted between unheated and pasteurized (63.3°C) salted yolk. Patterns for salted yolk, pasteurized at 63.3° and 68.0°C, were similar, which further indicated that this product can be pasteurized at 68.0°C, as recommended by Cotterill and Glauert (1969), without significant damage.

Woodward and Cotterill (1983) observed the electrophoretic and chromatographic changes in whole egg heated from 57° to 87°C for 3.5 min. Livetins and some globulins were the most heat sensitive, while conalbumin and ovalbumin were the most stable. Sugar and salt increased the heat stability of proteins by 1.6° and 7.9°C, respectively. Heat-sensitive proteins were most stabilized by sugar and salt. Heating whole egg and sugared whole egg changed the chromatograms substantially, while heating salted whole egg caused fewer changes.

A similar fractional, or stepwise, study on aggregation of egg-white proteins was reported by Matsuda et al. (1981), who used temperatures ranging from 54° to 90°C for various periods of time (1–180 min). Even with a heating time of 120 min, ovalbumin and globulins A1 and A2 failed to aggregate in egg white (pH 7 and 9) at 70°C, and ovotransferrin and ovomucoid also did not aggregate in egg white at 60°C (pH 9) and 76°C (pH 7 and 9), respectively. The ovoinhibitor was much more unstable than ovomucoid under heat treatment, and the time dependency of heat-induced aggregation of flavoprotein was greater than those of the other proteins in egg white.

Johnson and Zabik (1981) observed the gelation properties of egg proteins, both singly and in combination. Conalbumin was the least heat-stable protein with a denaturation temperature of 57.3°C. Globulins and ovalbumin ranked second with denaturation transition temperatures of 72.0° and 71.5°C, respectively. Lysozyme denatured at 81.5°C, whereas ovomucin and ovomucoid showed no coagulation abilities. Lysozyme produced the strongest gel followed by globulins, ovalbumin, and conalbumin.

DESTRUCTION OF BACTERIA IN EGG PRODUCTS

The reader is referred to Stumbo (1965) for a review of the methods for determining heat resistance of bacteria. However, a brief discussion follows of some of the techniques used by some workers on egg products. It appears to make little difference which *Salmonella* serotype (other than *S. senftenberg* 775 W) is used as a test organism. The two serotypes most commonly used are *S. typhimurium* Tm-1 and *S. oranienburg*. Many workers have used a 24-hr broth culture for an inoculum, although some workers prefer an older culture. Sterile eggs are easily prepared. The inoculum level of the test organism is usually about 10^6 to 10^7 cells per gram of product. The use of thermal death-time tubes and vials to determine heat resistance is also described. The laboratory tubular pasteurizer is an excellent tool for the preparation of heat-treated samples or to confirm data obtained from thermal death-time tubes.

The compositional differences of egg products account for the wide range of pasteurization conditions recommended. The pH of the product is most important. Salmonellae are most heat resistant at pH 5 to 6. This is one of the reasons for the greater heat resistance in yolk (pH 6.0) as compared to white (pH 9.1). The type of acid or additive is another important variable in heat resistance. The heat resistance of *S. typhimurium* was determined in whole egg in the presence of various antibiotics and acids. Heat resistance was found to be lower in the presence of acetic and lactic acids than in that of hydrochloric acid for the same pH. Acetic acid was used to lower the heat resistance of *S. typhimurium* in salted and sugared yolk. The difference in solids content (yolk 45 to 48% and white 12%), hence the presence of water, is another reason for differences in heat resistance. The fat of egg yolk has some protective effect and the biologically active proteins of white may aid in the pasteurization of the latter. The water activity is affected most by drying or by the addition of salt or sugar. Undoubtedly, this further accounts for the high heat resistance in salt yolk.

The main purpose of pasteurizing egg products is to create a wholesome product by eliminating pathogenic bacteria. Primary concern has been with salmonellae because this organism has been commonly associated with eggs and egg products. Fortunately, most types of salmonellae are relatively easy to kill. For example, it was reported that 14 salmonellae serotypes commonly associated with eggs were destroyed in liquid whole egg at 65°C for 0.3 min; at 64°C for 0.6 min; at 63°C for 0.8 min; at 62°C for 1.2 min; at 61°C for 2.0 min; at 60°C for 2.6 min; and at 59°C for 3.7 min. However, two types, *S. senftenberg* and *S. cerro*, required longer holding times at each temperature for their destruction. A similar study showed variation in heat resistance; *S. oranienburg, S. senftenberg*, and *S. paratyphi* B had the greatest thermal resistance. One particular strain of *S. senftenburg* 775 W has been found to be quite resistant to heat pasteurization.

Researchers have discussed the question of pasteurization adequacy as it applies to whole egg and also have reevaluated the effectiveness of the pasteurization process in pilot-plant studies. They concluded that the conventional treatment of 60° to 62°C for 3.5 to 4 min is more than adequate to destroy anticipated loads of salmonellae strains of ordinary heat resistance. They were able to reduce *S. typhimurium* added at a level of 3×10^6 per milliliter to an undetectable level when heating whole egg to 60°C for 2 min. As in previous studies, they found *S. senftenberg* 775 W more difficult to kill. However, they concluded that even this strain could be destroyed if present at low levels, 10 to 100 per milliliter.

Egg white, whole egg, and egg yolk all have different effects on the heat resistance of salmonellae, apparently because of the differences in pH, solids, and nature of constituents. The heat destruction of different serotypes of salmonellae at 60°C is reported to be 3 to 14 times greater in egg white (pH 9.0) than in whole egg (pH 7.6). The greatest difference was noted in *S. senftenberg* 775 W. This indicates that less severe pasteurization conditions can be used for egg white than for whole egg. On the other hand, it was found that heat destruction of salmonellae in egg yolk (pH 6.2) was half that of whole egg.

MICROORGANISMS SURVIVING PASTEURIZATION

Normally less than 1% of the bacteria in raw egg products survive pasteurization. Early workers observed that the principal genera found in pasteurized egg products are *Alcaligenes, Bacillus, Proteus, Escherichia, Flavobacterium*, and gram-positive cocci. However, the last three genera are not found after freezing. Between 1950 and 1970 little attention was devoted to the microorganisms surviving pasteurization. Then, in 1970, commercially pasteurized liquid egg products were reported to contain an average bacteria count of less than 500 cells/gram. Psychrophilic, thermophilio, and anaerobic counts were less than 25 cells/gram. The most frequently occurring organism were micrococci. No coagulase-positive *Staphylococcus* was found. In 1973, York and Dawson found that when commercial samples of pasteurized liquid whole eggs were stored at 2°C the total plate counts and odor were acceptable up to 12 days and for 5 days at 9°C. No *E. coli*, salmonellae, streptococci, or staphylococci organisms were found in the liquid egg. Several workers have shown that fecal streptococci can survive egg pasteurization (Barnes and Corry 1969; Shafi *et al.* 1970; Payne *et al.* 1979; Payne and Gooch 1980). Strock and Potter (1972) presented data indicating that poliovirus survived the HTST pasteurization of egg albumen treatments but did not survive pasteurization times in excess of 1 min at 138°F. Heat-peroxide pasteurization treatments were used; yet, this virus survived.

PASTEURIZATION METHODS

Pasteurization of Egg White

Heat Treatment of Raw Liquid Egg Whites. It has been consistently reported in the literature that heat treatment of raw liquid egg whites (i.e., not fermented, normal pH of about 9.0, and without additives) in the pasteurization range does damage the functional properties of the liquid whites. Most studies concluded that even momentary heating of egg white above 57°C resulted in a loss of foaming power. A similar loss resulted from heating whites to as low a temperature as 49°C for 1 hr or 43°C for 6 hr. The beating time for angel-cake preparation was approximately doubled by heating to 57°C for 4 min.

However, maximum heat treatment from the standpoint of mechanical problems will vary with the design of equipment, flow rate (or degree of turbulence), temperature differential between heating medium and egg white, temperature drop during the holding period, viscosity of the egg white, and probably other factors. For example, using a small-scale commercial plate heating-holding tube-plate cooling unit operated at a fixed flow rate, observations were made on the effect of processing temperature on viscosity and turbidity of egg white, buildup on the heating plates, and gelation. In this type of equipment, with no holding time, whites could be flash-heated without complication to 59°C, but at 60°C marked physical changes in the whites interfered with operation of the exchanger. There is evidence that flash-heating egg white to 62°C without turbidity or gelation and to hold egg white for 7.5 min at 58°C is possible without any change in viscosity. Other researchers found that, with a 2-min hold, pasteurization could be carried out at 56° to 57°C without mechanical difficulties or physical changes in the egg white, whereas treatment at (57° to 58°C) for 2 min resulted in viscosity and turbidity increases in the whites. In a later study by the same researchers, using the same commercial equipment, egg whites were pasteurized at 57°C for 4 min without encountering mechanical problems. Under these conditions, they observed no change in the appearance of the white and maintained that an ample operating margin above 57°C remained, although use of higher temperatures was considered inadvisable.

For a pasteurization process to be most efficient in destroying salmonellae in liquid egg white, either the pH should be observed and time-temperature adjustments made, or else the pH should be changed to the level specified by the process. The time-temperatures ratios for various pH levels are shown in Fig. 12.1.

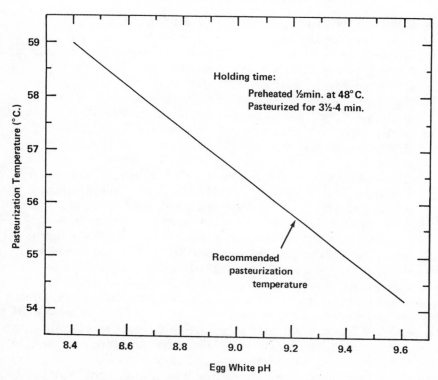

FIG. 12.1.   Recommended pasteurization temperatures for raw egg white at various pH levels.
Courtesy of Missouri Agr. Expt. Sta.

The Lactic Acid-Aluminum Sulfate (pH 7) Process. Use of this method permits pasteurization of liquid egg white in a manner identical to that used for whole egg in the United States (i.e., 60° to 62°C for 3.5 to 4 min). The maximum stability of most egg white proteins occurs at near neutral pH. A notable exception is conalbumin. This particular protein is very labile at pH 7, but can be stabilized by the addition of iron or aluminum salts. The metallic ions are capable of combining with conalbumin to form a complex that is more heat stable than the free protein.

Stabilization of egg white, according to this process, is accomplished by the addition of an aluminum sulfate solution to egg white prior to heat treatment. The solution consists of 1 oz of aluminum sulfate $[Al_2(SO_4)_3 \cdot 18H_2O]$ dissolved in 1 lb of 25% lactic acid. About 6.5 lb of this solution is required per 1,000 lb of liquid egg white. This addition should be made slowly, with rapid stirring of the whites, to avoid protein coagulation by local high concentrations of acid and aluminum. The pH of the whites at pasteurization should be in the range of 6.6 to 7.0.

If desired, a suitable whipping aid may be incorporated in the stabilizing solution. A stabilizer solution containing 7% of a whipping aid added to egg white at a rate of 7 lb/1,000 lb will give a whipping aid concentration of about 0.05% in the finished product.

The process is protected by a government patent and is available for license without charge. More than 45 commercial firms in the United States were issued licenses to use this process.

Heat-plus-Hydrogen Peroxide Process.   Hydrogen peroxide is a well-known bactericidal agent and was suggested some time ago as having possible application in destroying microbiological contamination in liquid egg whites. However, since the hydrogen peroxide treatment often resulted in heavy foam formation due to the evolution of oxygen, the process was not considered commercially practical. Developments in recent years, however, have resulted in the process becoming an acceptable method of pasteurization for whites.

Armour and Company was issued a patent for a process combining heat treatment and hydrogen peroxide. Liquid egg white, at normal pH, is heated to 52° to 53°C and held at this temperature for 1.5 min; according to the findings, this largely inactivates the indigenous catalase activity, thereby eliminating the problem of excessive foam formation. At this point in the process, sufficient hydrogen peroxide (as a 10% solution) is added to give a concentration of 0.075 to 0.10% in the egg whites, the peroxide being allowed to react for 2 min at the elevated temperature. The white is then cooled and catalase is added to destroy the residual peroxide. Successful commercial application of this process has been reported.

Other modifications of this procedure have been developed to utilize the bactericidal effect of peroxide. One such process incorporates sufficient peroxide with liquid whites to give a concentration of 0.0875%. The egg white is then heated to 52° to 54°C and held for not less than 3.5 min, thereby exposing the whites to peroxide action for a full 3.5-min period. Catalase activity in the whites is not reduced before addition of peroxide, but sufficient peroxide is used to compensate for any of the material destroyed by catalase during the first part of the holding period. Following heat and peroxide treatment, the egg white is cooled and catalase is added to remove any residual peroxide.

The heat-plus-hydrogen peroxide process has been used by many commercial firms and is widely accepted in the egg-processing industry.

Heat-plus-Vacuum Process.   Tests conducted by the Ballas Egg Company in Zanesville, Ohio, have resulted in another USDA-approved method for pasteurizing egg white. This system utilizes a typical high temperature-short time plate pasteurizer equipped with a vacuum chamber in which 17 to 20 in. of vacuum are applied to liquid egg white before heat treatment (see Fig. 12.2). The egg white is then heat-treated at 57°C for 3.5 min. One important objective of the vacuum is to remove air from the albumen and permit lower temperatures to achieve the same microbiological results. It is reported that inclusion of the vacuum chamber reduces the problem of coagulation of egg on the plates during heating.

Heat Treatment of Dried Egg White.   The storage of dried egg white at elevated temperatures has been shown to be an effective means of freeing egg-white solids of pathogens. The effect of moisture content and temperature of storage of dried albumen on the viability of S. *senften-*

FIG. 12.2.   HTST plate pasteurizer equipped with vacuum chamber.
Courtesy Ballas Egg Products Corp., Zanesville, OH.

berg, S. oranienburg, and *S. pullorum* and on solubility, specific gravity of meringue, and angel-cake making properties of the reconstituted product was studied. Results showed that albumens can be stored at elevated temperatures (50°, 60°, or 70°C) to eliminate salmonellae without significant impairment of functional properties.

In 1961 Bergquist was issued a patent for the pasteurization treatment of liquid egg white at relatively low temperatures, followed by a heat treatment of the egg white solids, for the production of egg white solids of low bacteria count. In 1964 Smith obtained a patent for a procedure involving storage for 8 hr at 71°C and for 1 hr at 99°C. The reduction of viable bacteria in egg-white solids was the principal claim of this patent.

The high-temperature storage (53.3°C for 5 days) of albumen powder, heated at 54°C for 3 min before spray-drying, does not affect angel-cake volume adversely.

The effect of high-temperature storage on spray-dried egg white, with and without yolk contamination, adjusted to various pH levels before drying, has been studied. Albumen is not damaged by storage at 54°C for as long as 60 days. There is no evidence of sulfhydryl activity of spray-dried egg white subjected to the elevated storage temperatures; but degradative changes such as decreased lysozyme mobility, iron-conalbumin complexing, and solubility, as well as increased evolution of ammonia during storage, were enhanced by high pH values of the albumen solids. The F 60°C values for *S. oranienburg* in egg white solids containing 2.60, 6.18, and 9.63% moisture were 4.4, 3.4, and 2.4 days, respectively. The z value for the thermal destruction curves was 32°C.

The earlier studies led to a general adoption by industry of some form of dry heat treatment in order to ensure a salmonellae-negative product. Following such treatment, albumen solids must be sampled for bacteriological examination.

Pasteurization of Whole Egg Products

In the United States, present pasteurization requirements for whole egg are described in Section 55.101 paragraph (6) of the Regulations Governing Grading and Inspection of Egg Products (see *Federal Register*, December 23, 1969). The temperatures and holding times listed in Table 12.2 are minimum.

These time-temperature conditions for pasteurizing whole egg have not been considered satisfactory by other countries. Researchers used temperatures of 61° to 65°C and holding periods up to 5 min in a variety of pilot-plant and commercial operations. On the basis of these studies, pasteurizing temperatures of 64° to 65°C for 2.5 min were recommended. These findings were confirmed in extensive plant trials. As a result, all liquid whole egg in the United Kingdom must be pasteurized at 64°C for 2.5 min.

Whole egg pasteurization requirements for other countries are 66° to 68°C for 3 min in Poland; 63°C for 2.5 min in China; 62°C for 2.5 min in Australia; and 65° to 69°C for 90 to 180 sec in Denmark. However, in the United States, the need for such high temperatures has not been demonstrated.

Studies in the United States have shown that heat treatment with or without homogenization had no significant effect on the viscosity of fresh whole egg. Changes in the viscosity or "body" of the product are primarily due to mechanical action and freezing. When tested in a true sponge cake, the performance of frozen whole egg decreases at pasteurization temperatures above 63°C. The performance at temperatures below 63°C is maintained, because homogenization was included in the operation.

Pasteurization of Yolk Products

Salmonellae are more heat resistant in yolk than in whole egg. This increased resistance is due to the lower pH and higher solids content of egg yolk. Hence, yolk must be pasteurized at a higher temperature than whole egg or egg white. Yolk is less sensitive to heat and thus has higher practical temperature limits for operating conventional plate-type pasteurizing units.

Additives such as sugar or salt increase the thermal resistance of microorganisms in egg products. Salt increases thermal resistance much more than sugar, but resistance is decreased by the addition of acetic acid. Thermal resistance is also affected by time of exposure to salt. Research has shown that S. oranienburg can be destroyed in 10% salted yolk by a terminal heat treatment in the package. Ijichi et al. (1973) observed that in pasteurized salted yolk held at 25°C (80°F) or less for 10 days neither spoilage nor staphylococci organisms grew.

In 1974, a "hot-pack" process was developed by Cotterill et al. for the pasteurization of egg products containing salt. The process is as follows: (1) Formulate product to desired specification (they experimented with 10% salted whole egg and 45% solids yolk). (2) Preheat (important) to "hot-room" temperature (52°C). (3) Package in a container that can be heat treated and is resistant to salt corrosion. (4) Place in "hot-room" (2 days for salted whole egg and 3 days for yolk). (5) Remove the product from "hot-room" and cool. Compared to pasteurization with heat-exchange plates and holding tubes, the advantages are the following: (1) Inexpensive and less complicated equipment are needed. (2) Product wastage is virtually eliminated. (3) Less specialized technical skills are needed. (4) There is good assurance of a salmonellae-negative product.

(5) An essentially sterile product results. (6) It is a terminal pasteurization process so recontamination is unlikely. (7) Cooling time after pasteurization is not critical. (8) Product can be stored at higher temperatures than currently being rcommmended. (9) Product can be processed "in transit."

The emulsifying properties of egg yolk products are little affected by heat treatments. Pasteurization at 62° to 64°C had no effect on performance of salted unfrozen yolks in mayonnaise and cream puffs. However, salted yolks acidified from pH 6.2 to 5.0 and then pasteurized at 60°C suffered a loss in emulsifying capacity. Freezing and storage further increased this damage. Sugared yolks, pasteurized at temperatures from 60° to 64°C performed satisfactorily in sponge cakes, orange chiffon cakes, and yellow layer cakes. Cotterill *et al.* (1976) found that even higher temperatures (up to 72°C) did not harm the emulsifying properties of 10% salted yolk. They used a phase-inversion titration technique to observe the emulsifying ability of both the oil-in-water and water-in-oil systems. Storage of product may be more important than temperature of pasteurization.

Thermal Destruction Curves

These curves reflect the relative heat resistance of microorganisms. They are a plot of log D or F values vs. temperature on a linear scale. D is the time required to destroy 90% of the viable cells in a product at a specified temperature. F is a parallel expression for the time required to kill all test organisms inoculated into a product. The temperature change required for these curves to traverse one log cycle is referred to as z. Large z values indicate greater heat resistance.

Thermal destruction curves for a wide range of egg products are illustrated in Fig. 12.3. The solid portion of the curves are based on observed F values and represent equivalent pasteurization conditions. Water activity has a major influence on z. Note that all lines tend to intersect within a given area. The best point of intersection (X_{PI}, Y_{PI}) was near 29,000 min and 43°C. Hypothetical curves were then constructed (Fig. 12.4). This represents a new concept for the construction of thermal destruction curves. Pasteurization conditions of other egg products can be estimated by drawing lines through this point of intersection and points intermediate to similar products. However, actual pasteurization conditions should eventually be confirmed by experimentation for each product.

Effective February 1, 1969, specific temperatures and holding times were required by the USDA for the pasteurization of egg products (Table 12.2). Except for products containing salt, there is a fairly good agreement between research workers, regulatory agencies, and industry.

PASTEURIZATION EQUIPMENT

Batch Pasteurizers

Although batch pasteurization, per se, is not a USDA-approved process, individual use of this technique has been approved. In the USDA Regulations Governing the Grading and Inspection of Egg Products (7 CFR) Part 55.101, paragraph C states: "Other Acceptable Methods: Other methods of pasteurization may be approved by the national supervisor upon receipt of satisfactory evidence that such methods will result in a salmonellae negative product."

One of the very first uses of pasteurization to reduce bacterial counts in whole egg was by Henningsen Brothers in 1938 using the batch method. Gibbons et al. in 1946 used batch pasteurization to heat whole egg to 60°C for at least 30 min. They suggested this as a minimum process.

Good kills of Salmonella in liquid whole egg at 57°C for 10 min following a come-up time of 30 min. have been reported. In experiments with batch pasteurizing of liquid whole egg, heating to 57°C and holding for 6 min after a 30-min come-up time reduced S. typhimurium by a factor of at least 10^6. Furthermore, no significant damage occurred, as judged by sponge-cake volume.

A typical batch-pasteurization system requires a processing vat (Fig. 12.5), a source of heat, and a supply of refrigerated water for cooling. Dairy-type, pressure-wall process tanks are available with 25- to 400-gal. capacity equipped with agitator, space heater, and temperature sensors. It is important that the head space in a batch pasteurizer be heated and that foaming of the egg product be held to a minimum.

The cutaway schematic (Fig. 12.6) shows an automated cooler-pasteurizer-cooler now in use by some commercial processors. The diagram reveals the clean-in-place spray ball, agitator, and adjustable baffle.

For batch pasteurization, whole egg would need to be heated to 56°C for 35 min or at 57°C for 15 min, to receive a treatment equivalent to 60°C for 3.5 min.

High-Temperature–Short-Time Equipment

This process is based upon the HTST pasteurizing system that has long been in general use for the processing of milk. A stainless steel plate-heat exchanger (Fig. 12.7) is highly efficient as a heating, cooling, and regeneration unit. The heart of the unit consists of a stainless steel-clad frame in which stainless steel-gasketed plates are clamped, allowing liquid egg and the heating or cooling water to flow between

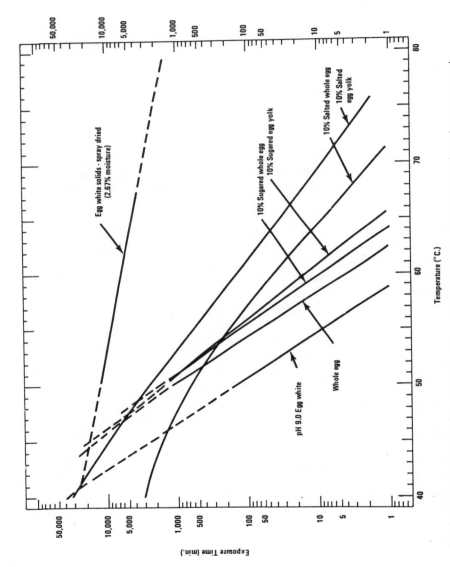

FIG. 12.3.   Observed thermal destruction curves for several egg products plotted on one graph.
From Cotterill and Glauert (1973).

258

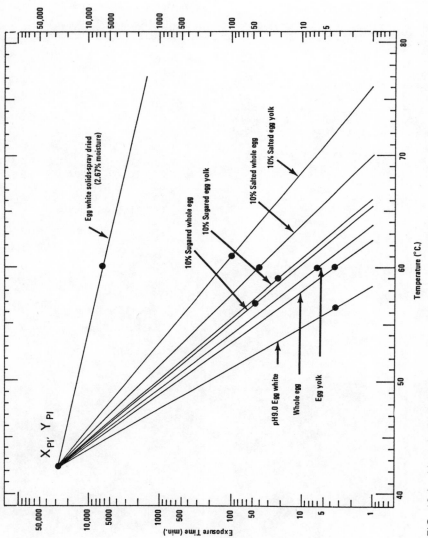

FIG. 12.4. Hypothetical thermal destruction curves for several egg products with z adjusted so all curves pass through one point of intersection.
From Cotterill and Glauert (1973).

259

TABLE 12.2. Pasteurization Requirements

Liquid egg product	Minimum temp. reqts (°C)	Minimum holding time	
		Fastest particle (min)	Average particle (min)
Albumen (without use of chemicals)	57	1.75	3.5
	56	3.1	6.2
Whole egg	60	1.75	3.5
Whole egg blends (less than 2% added	61	1.75	3.5
nonegg ingredients)	60	3.1	6.2
Fortified whole egg and blends (24 to 38% egg	62	1.75	3.5
solids, 2 to 12% added nonegg ingredients)	61	3.1	6.2
Salt whole egg (with 2% or more salt added)	63	1.75	3.5
	62	3.1	6.2
Sugar whole egg (2 to 12% sugar added)	61	1.75	3.5
	60	3.1	6.2
Plain yolk	61	1.75	3.5
	60	3.1	6.2
Sugar yolk (2% or more sugar added)	63	1.75	3.5
	62	3.1	6.2
Salt yolk (2 to 12% salt added)	63	1.75	3.5
	62	3.1	6.2

FIG. 12.5.   Process vat for batch pasteurization.
Courtesy Girton Manufacturing Co.

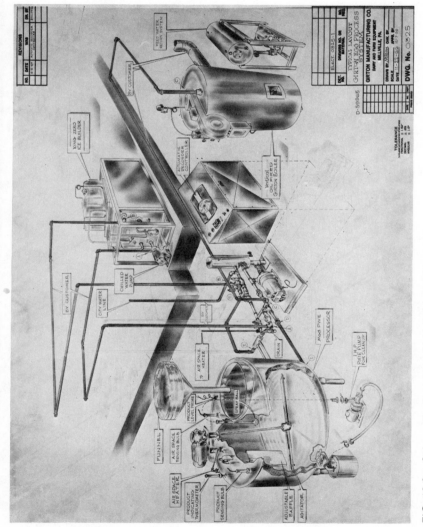

FIG. 12.6. Schematic diagram of batch pasteurizing unit.
Courtesy of Girton Manufacturing Co.

FIG. 12.7. Liquid egg passes through a clarifier (*right*), then to a holding tank and into a plate heat exchanger (*center*).
Courtesy of De Laval Separator Co.

alternate plates. An excellent discussion of what an HTST system consists of, the operational functions, the components, the required controls, the capital expenditures, and the economics of the operation has been written. A plate-heat exchanger, with holding tubes, vacuum chamber, and accessory equipment, is illustrated in Fig. 12.8.

Another type of heat exchanger is the so-called triple-tube. In actual construction the tube combination (shown in Fig. 12.9) is a 2½-in. outer tube surrounding a 2-in. intermediate tube which surrounds a 1½-in. inner tube. The outer space contains the heating or cooling medium which is transferred from one tube to another by a crossover called an H fitting, gasketed by Teflon seals and bushings. The liquid to be pasteurized flows between the intermediate tube and the inner tube. An

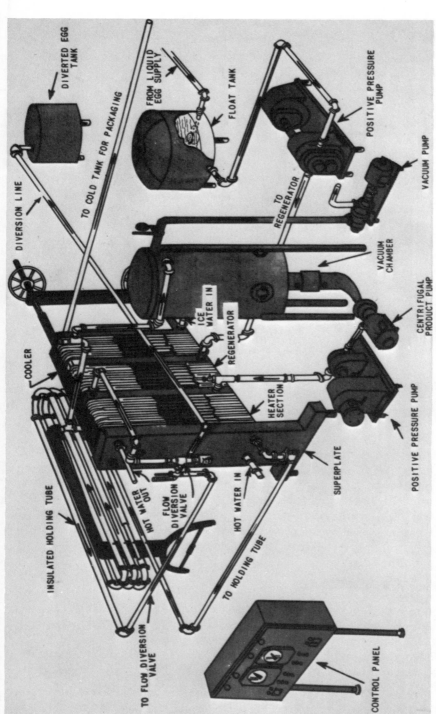

FIG. 12.8. Plate heat exchanger, holding tubes, vacuum chamber, and accessory equipment.

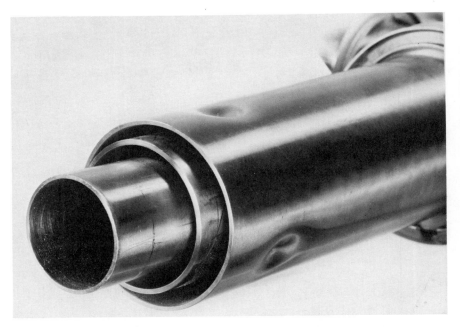

FIG. 12.9.   Triple-tube heat exchanger.
Courtesy of De Laval Separator Co.

H fitting also serves here as a crossover. Within the inner tube the heating or cooling medium is transferred from one tube to another by a U bend. All flows within the heat exchanger are countercurrent (as illustrated in Fig. 12.10). One advantage of the triple-tube heat exchanger is that the entire unit can be suspended from the ceiling, thereby conserving plant floor space. There are several triple-tube operations being used by egg processors; all systems are 100% clean-in-place and all are suspended from the ceiling.

Laboratory Pasteurizer

A tubular laboratory-scale pasteurizer (designed by O. J. Cotterill and L. M. McBee, University of Missouri-Columbia) was described by Cotterill (1968). This unit performs the three basic functions of a HTST pasteurizer: (1) preheating, (2) holding, and (3) cooling. The following are the component parts (shown in Fig. 12.11):

    1. *Pump.* Positive displacement, variable speed to control flow rate, stainless steel (Zenith Products Company, West Newton 65, MA:

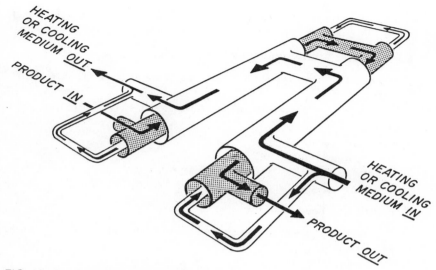

FIG. 12.10. Schematic of triple-tube heat exchanger.
Courtesy of De Laval Separator Co.

Zero-Max Variable Speed Transmission, Type 1 ZM with Pump No. 3).
2. *Heat sources.* Water baths with good temperature controls.
3. *Heat exchangers.* Stainless steel tubing, OD = 6.35 mm, ID = 4.30 mm. Length: preheating = 344 cm, holding = 1290 cm, cooling = 230 cm.
4. *Fittings.* Stainless steel. Locate a pressure gauge and rubber plug

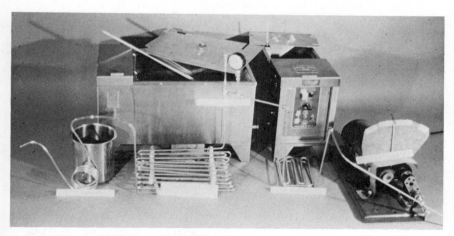

FIG. 12.11. Tubular laboratory pasteurizer.

in tees between preheating and holding water baths. Air or a dye can be injected through plug to determine flow time.

The operating conditions for the pasteurizer are:

1. *Preheating time.* About 0.5 min with a water temperature of 48° to 50°C.
2. *Holding time.* 3.75 min, water temperature varied depending on product.
3. *Cooling.* Ice water.
4. Flow rate. About 44 ml/min.
5. *Cleaning.* Alkaline and acid cleaners. Also, all parts coming in contact with the product can be sterilized in an autoclave.

This unit has been used to:

1. Determine pasteurization times and temperature to kill pathogens in various products.
2. Prepare heat-treated samples for future research.
3. Demonstrate pasteurization efficiency to classes.

Flow Problems and Holding Times. The physical properties of egg products vary widely. When considering flow problems and holding times, two major concerns are viscosity and specific heat. In regards to pasteurization, viscosity is most important, because it influences the flow characteristics in the pasteurizer, particularly the holding tubes. If the product is very viscous, laminar flow may occur unless the flow rate is increased. Hence, some product may move through the unit faster than expected. A turbulent flow pattern is desired, which will ensure thorough heat penetration to all components and a uniform holding time. Kaufman *et al.* (1968A) observed the flow characteristics in holding tubes of commercial egg pasteurizers and concluded that most plants were operating above the required holding temperatures and times. Their data on various products should be consulted regarding pasteurizer design and operation. Cunningham (1972) reported that the viscosity of 43% solids yolk is only about one-fourth that of pure yolk; hence, the lower solids products are much easier to pasteurize. Pitsilis *et al.* (1975) reported on the apparent viscosity of egg white at various temperatures and pH levels. Several methods have been used to determine flow characteristics. A dye may be injected into egg white to observe absorption as the product passes through the unit. A "cold shot" and fluorocarbon tracer and a slug flow of water and carboxymethylcellulose solution have been used to measure minimum holding times. Flow characteristics of the product also determine the position or location of the temperature sensing element that activates the flow diversion valve of the pasteurizer.

TESTS FOR DETERMINING THE ADEQUACY OF PASTEURIZATION

In the pasteurization of food products, there is a need for a simple test to check the efficiency of the pasteurization process. The enzymes phosphatase and α-amylase, as well as the catalase activity, have been studied as possible indicators of underpasteurization of egg products. Phosphatase activity remains in whole egg heated to 60°C for 20 min, and in that heated to 70°C for 5 min. Thus the phosphatase test so useful for milk appears to be unsuitable for eggs. The critical heat treatment for destruction of the activity of α-amylase in whole eggs is 64°C for 2.5 min. Thus, amylase, used by British workers to indicate treatment of whole egg at 64°C, cannot be applied to pasteurization of egg products in the U.S., because the minimum treatment of 60°C for 3.5 min will not inactivate the enzyme. The catalase activity of whites and whole egg is substantially reduced by heat treatments as low as 54°C. The catalatic activity of egg white can be used to indicate efficiency of pasteurization by the Cunningham-Lineweaver method. The smaller reductions of activity at lower temperatures and the inclusion of hydrogen peroxide in the egg white would preclude the use of catalatic activity to test the efficiency of other egg-white pasteurization processes. It is recommended to run assays of activity before and after pasteurization to compensate for variations in the catalatic activity found in egg white.

An automated enzyme activity test was devised in a USDA laboratory (Hansen 1971) to test egg-white pasteurization. This test is based on the assay of p-nitrophenol released by the enzymes in egg white from p-nitrophenyl-N-acetyl-β-glucosaminide before and after pasteurization. A 60% inactivation of β-N-acetyglucosamidase activity upon heating appears to be a suitable indication that egg white has been held at an adequate time-temperature combination to meet USDA requirements. Also, since one of the minor protein components of egg white, designated as "line 18," on starch-gel electrophoretic patterns is absent in whites and whole eggs heated to 61°C for 3.5 to 4.0 min, but not 60°C can be used as a test heat treatment.

It is unfortunate that, at present, suitable tests for determining the adequacy of pasteurization of egg white, whole egg, and yolk (as pasteurization is practiced in the United States) have not been found, although research work is still in progress.

OTHER PASTEURIZING METHODS

Because of the limited heat tolerance of liquid egg white, investigators have tried various approaches to eliminate *Salmonella*, such as the use of hydrogen peroxide. Coupled with the application of heat, acidification of liquid egg white to pH 7 has been suggested for increased stability; also recommended is the addition of carbohydrates, sodium citrate, or sodium hexametaphosphate, and soaps of fatty acids to diluted whites.

Colicins, which are produced by certain strains of *Escherichia coli* during fermentation, are reported to destroy *Salmonella* in egg white. There are claims that egg white containing 10^5 cells/ml of *S. senftenberg* ATCC8400 could be sterilized in 12 hr at 10°C by 0.1% β-propiolactone. However, there has been some question as to the safe use of this chemical in foods.

High-energy radiation has been used experimentally to pasteurize frozen, dried, and liquid egg products. Ultraviolet light has also been used to destroy *Salmonella* in liquid egg white. Radiation-treated egg products are satisfactory in many cases, but flavor changes can be expected. The flavors are volatile and therefore are not objectionable in products that are subjected to considerable heat. Off-flavors produced by pasteurizing doses of radiation are also volatilized when liquid egg white is spray-dried. An advantage of high-energy radiation, particularly gamma rays, is that it can be applied to frozen products, thereby eliminating the costs of thawing, heat pasteurization, and refreezing.

Other chemicals that have been used to pasteurize egg products include ethylene oxide and propylene oxide. Gaseous sterilization of egg powders has been successful experimentally in destroying *Salmonella* without damaging the products, but it is unlikely that the use of gaseous sterilants will be permitted.

The increased susceptibility of *Salmonella* to heat destruction at higher pH values has been the basis for proposals involving the addition of ammonium hydroxide to increase the pH of egg white to 10.5 or above. This treatment allows effective pasteurization at rather low temperatures, usually below 54°C, and is generally recommended for liquid egg white destined to be spray-dried, at which time the ammonia is lost by volatilization.

In most cases, the various methods cited above have resulted in good kills of bacteria, but none has been considered entirely satisfactory and none has been generally adopted by industry.

SELECTED REFERENCES

Anellis, A., Lubas, J., and Rayman, M. M.   1954.   Heat resistance in liquid eggs of some strains of the genus Salmonella. Food Res. *19*, 377–395.

Ayres, J. C. and Slosberg, H. M.   1949.   Destruction of *Salmonella* in egg albumen. Food Technol. *3*, 180–183.

Baldwin, R. E., Cotterill, O. J., Thompson, N. N., and Myers, M. 1967. High temperature storage of spray-dried egg white. I. Whipping time and quality of angel cake. Poult. Sci. *46*, 1421–1430.

Ball, H. R., and Cotterill, O. J. 1971. Egg white catalase. 1. Catalatic reaction. Poult. Sci. *50*, 435–448.

Banwart, G. J. 1955. Microbiological and functional changes in dried egg albumen stored at elevated temperatures. Ph.D. Thesis, Iowa State Coll., Ames.

Banwart, G. J., and Ayres, J. C. 1956. The effect of high temperature storage on the content of Salmonella and on the functional properties of dried egg white. Food Technol. *10*, 68–73.

Barnes, E. M., and Corry, J. E. L. 1969. Microbial flora of raw and pasteurized egg albumen. J. Appl. Bacteriol. *32*, 193–205.

Bergquist, D. H. 1961. Method of producing egg albumen solids with low bacteria count. U.S. Pat. 2,982,663.

Bernard, C., Slosberg, H. M., Lowe, B., and Stewart, G. F. 1948. Factors influencing performance of egg white in angle cakes. Poult. Sci. *27*, 653–654.

Brant, A. W., Patterson, G. W., and Walters, R. E. 1968. Batch pasteurization of liquid whole egg. 1. Bacteriological and functional property evaluation. Poult. Sci. *47*, 878–885.

Brown, S. L., and Zabik, M. E. 1967. Effect of heat treatments on the physical and functional properties of liquid and spray-dried albumen. Food Technol. *21*, 87–92.

Bruch, C. W., and Koesterer, M. G. 1962. Destruction of enteric bacteria in liquid egg with beta-propiolactone. Appl. Microbiol. *10*, 123–128.

Clinger, C. A., Young, A., Prudent, I., and Winter, A. R. 1951. The influence of pasteurization, freezing and storage on the functional properties of egg white. Food Technol. *5*, 166–170.

Cotterill, O. J. 1967. Pasteurization requirements to destroy *Salmonella* in egg products containing sugar and salt. Poult. Sci. *46*, 1248.

Cotterill, O. J. 1968. Equivalent pasteurization temperatures to kill salmonellae in liquid egg white at various pH levels. Poult. Sci. *47*, 352–365.

Cotterill, O. J., and Glauert, J. 1969. Thermal resistance of salmonellae in egg yolk products containing sugar or salt. Poult. Sci. *48*, 1156–1166.

Cotterill, O. J., and Glauert, J. 1971. Thermal resistance of salmonellae in egg yolk containing 10% sugar or salt after storage at various temperatures. Poult. Sci. *50*, 109–115.

Cotterill, O. J., Cunningham, F. E., and Funk, E. M. 1963. Effect of chemical additives on yolk-contaminated liquid egg white. Poult. Sci. *42*, 1049–1057.

Cotterill, O. J., Seideman, W. E., and Funk, E. M. 1965. Improving yolk-contaminated egg white by heat treatments. Poult. Sci. *44*, 228–235.

Cotterill, O. J., Baldwin, R. E., and Myers, M. 1967. High temperature storage of spray-dried egg white. 2. Electrophoretic mobility, conalbumin-iron complexing, sulfhydryl activity, and evolution of volatile bases. Poult. Sci. *46*, 1431–1447.

Cotterill, O. J., Glauert, J., and Krause, G. F. 1973. Thermal destruction curves for *Salmonella oranienburg* in egg products. Poult. Sci. *52*, 568–577.

Cotterill, O. J., Glauert, J., Steinhoff, S. E., and Baldwin, R. E. 1974. Hot-pack pasteurization of salted egg products. Poult. Sci. *53*, 636–645.

Cotterill, O. J., Glauert, J., and Bassett, H. J. 1976. Emulsifying properties of salted yolk after pasteurization. Poult. Sci. *55*, 544–548.

Cunningham, F. E. 1966. Possible tests for determining the adequacy of pasteurization of liquid egg white. Poult. Sci. *45*, 1079.

Cunningham, F. E. 1972. Viscosity and functional ability of diluted egg yolk. J. Milk Food Technol. *35*, 615–617.

Cunningham, F. E., and Lineweaver, H. 1965. Stabilization of egg white proteins to pasteurizing temperatures above 60°C. Food Technol. 19, 1442-1447.

Cunningham, F. E., and Lineweaver, H. 1967. Inactivation of lysozyme by native ovalbumin. Poult. Sci. 46, 1471-1477.

Cunningham, F. E., Lineweaver, H., Ijichi, K., and Garibaldi, J. A. 1964. Pasteurization of liquid egg white above 140° F. Poult. Sci. 43, 1311.

Cunningham, F. E., Garibaldi, J. A., Ijichi, K., and Lineweaver, H. 1965. Pasteurization of liquid egg white. World's Poult. Sci. J. 21, 365-369.

Dickerson, R. W., Jr., and Read, R. B., Jr. 1972. A flow diversion valve control for pasteurization of egg products. J. Food Sci. 37, 175-176.

Dixon, D. K., and Cotterill, O. J. 1981. Electrophoretic and chromatographic changes in egg-yolk proteins due to heat. J. Food Sci. 46, 981.

Garibaldi, J. A. 1968. Acetic acid as a means of lowering the heat resistance of Salmonella in yolk products. Food Technol. 22, 1031-1033.

Gibbons, N. E., Fulton, C. O., and Reid, M. 1946. Dried whole egg powder. 21. Pasteurization of liquid egg and its effect on quality of the powder. Can. J. Res. 24, 327-337.

Hansen, L. P. 1971. Automation study of beta-N-acetylglucosamidase activity as an indicator of egg white pasteurization. J. Food Sci. 36, 600-603.

Hanson, H., Lowe, B., and Stewart, G. F. 1947. Pasteurization of liquid egg products. 5. The effect on performance in custards and sponge cakes. Poult. Sci. 26, 277-283.

Heller, C. L. et al. 1962. The pasteurization of liquid whole egg and the evaluation of the baking properties of frozen whole egg. J. Hyg. 60, 135-143.

Henderson, A. E., and Robinson, D. S. 1969. Effect of heat pasteurization on some egg white enzymes. J. Sci. Food. Agric. 20, 755-760.

Ijichi, K., Garibaldi, J. A., Kaufman, V. F., Hudson, C. A., and Lineweaver, H. 1973. Microbiology of a modified procedure for cooling pasteurized salt yolk. J. Food Sci. 38, 1241-1243.

Johnson, T. M., and Zabik, M. E. 1981. Gelation properties of albumen proteins, singly and in combination. Poult. Sci. 60, 2071-2083.

Kaufman, V. F. 1972. Model of slug flow holder for commercial egg pasteurization. J. Milk Food Technol. 35, 725-729.

Kaufman, V. F., Ijichi, K., and Putnam, G. W. 1968A. Flow characteristics in holding tubes of commercial egg pasteurizers. J. Milk Food Technol. 31, 269-273.

Kaufman, V. F., Putnam, G. W., and Ijichi, K. 1968B. Test methods for measuring minimum holding times in continuous egg pasteurizers. J. Milk Food Technol. 31, 310-314.

Kline, L., Sugihara, T. F., and Ijichi, K. 1966. Further studies on heat pasteurization of raw liquid egg white. Food Technol. 20, 98-100.

Lategan, P. M., and Vaughn, R. H. 1964. The influence of chemical additives on the heat resistance of Salmonella typhimurium in liquid whole egg. Food Sci. 29, 339-344.

Lineweaver, H. 1969. Egg pasteurization Manual, ARS 74-48, U.S. Pep. Agric., Albany, CA.

Lineweaver, H. et al. 1965. Heat pasteurization of natural and modified liquid egg white. 1965-1.

Matsuda, T., Watanabe, K., and Sato, Y. 1981. Heat-induced aggregation of egg white proteins as studied by vertical flat-sheet polyacrylamide gel electrophoresis. J. Food Sci. 46, 1829-1834.

McBee, L. E., and Cotterill, O. J. 1971. High temperature storage of spray-dried egg white. 3. Thermal resistance of Salmonella oranienburg. Poult. Sci. 50, 452-458.

Meyer, R., and Potter, N. N. 1974. Changes in ultrastructure of egg albumen pasteurized with additives used to minimize heat denaturation. Poult. Sci. 53, 761-765.

Mickelson, M. N., and Flippin, R. S. 1960. Use of Salmonellae antagonists in fermenting egg white. 2. Microbiological methods for the elimination of *Salmonellae* from egg white. Appl. Microbiol. *8*, 371-377.

Miller, C., and Winter, A. R. 1950. The functional properties and bacterial content of pasteurized and frozen whole egg. Poult. Sci. *29*, 88-97.

Nichols, M. L. 1966. Egg product pasteurization efficiency. M.S. Thesis, Univ. of Missouri, Columbia.

Osborne, W. W., Straka, R. P., and Lineweaver, H. 1954. Heat resistance of strains of Salmonella in liquid whole egg, egg yolk, and egg white. Food Res. *19*, 451-463.

Palmer, H. H., Ijichi, K., Cimino, S. L., and Roff, H. 1969A. Salted egg yolks. I. Viscosity and performance of pasteurized and frozen samples. Food Technol. *23*, 148-153.

Palmer, H. H., Ijichi, K., Cimino, S. L., and Roff, H. 1969B. Salted egg yolks. II. Viscosity and performance of acidified, pasteurized and frozen samples. Food Technol. *23*, 154-156.

Palmer, H. H., Ijichi, K., Roff, H., and Redfern, S. 1969C. Sugared egg yolks. Effects of pasteurization and freezing on performance and viscosity. Food Technol. *23*, 85-89.

Patterson, G. W., Brant, A. W., and Walters, R. E. 1966. Batch pasteurization of liquid whole egg. Poult. Sci. *45*, 1113.

Paul, P., Varozza, A., Stewart, G. F., and Bergquist, B. H. 1957. Effects of several pasteurization schedules on performance of egg yolk solids in doughnuts. Food Technol. *11*, 508-509.

Payne, J., and Gooch, J. E. T. 1980. Survival of faecal streptococci in raw and pasteurized egg products. Br. Poult. Sci. *21*, 61-70.

Payne, J., Gooch, J. E. T., and Barns, E. M. 1979. Heat-resistant bacteria in pasteurized whole egg. J. Appl. Bacteriol. *46*, 601-613.

Pitsilis, J. G., Walton, H. V., and Cotterill, O. J. 1975. The apparent viscosity of egg white at various temperatures and pH levels. Trans. ASAE *18*, 347-349.

Rogers, A. B., Sebring, M., and Kline, R. W. 1966. Hydrogen peroxide pasteurization process for egg white. West. Exp. Stn. Collaborators Conf. 68-72.

Romanoff, A. L., and Romanoff, A. J. 1944. A study of preservation of eggs by flash heat treatment. Food. Res. *9*, 358-366.

Sebring, M., Rogers, A. B., Heck, J. G., and Pankey, G. 1965. Low temperature pasteurization of egg products. 1. Effect of heat and hydrogen peroxide on functional and bacteriological properties of liquid whole egg. Poult. Sci. *44*, 1414.

Sebring, M., Heck, J. G., Rogers, A. B., and Pankey, G. R. 1966. Low temperature pasteurization of egg products. 2. Effect of heat and hydrogen peroxide on functional and bacterial properties of liquid whole egg. Poult. Sci. *45*, 1124.

Seideman, W. E., Cotterill, O. J., and Funk, E. M. 1963. Factors affecting heat coagulation of egg white. Poult. Sci. *42*, 406-417.

Shafi, R., Cotterill, O. J., and Nichols, M. L. 1970. Microbial flora of commercially pasteurized egg products. Poult. Sci. *49*, 578-585.

Shrimpton, D. H., Monsey, J. B., Hobbs, B. C., and Smith, M. E. 1962. A laboratory determination of the destruction of alpha-amylase and salmonellae in whole egg by heat pasteurization. J. Hyg. *60*, 153-162.

Smith, C. F. 1964. Treatment of egg albumen. U.S. Pat. 3,161,527.

Solowey, M., Rosenstadt, A., Spalding, E. H., and Chemerda, C. 1948. Occurrence of multiple Salmonella types in spray-dried whole egg. Poult. Sci. *27*, 12-16.

Stewart, G. F. 1949. Heat treating liquid egg—practical quality control. U.S. Egg Poult. Mag. *55*, 10-13.

Strock, N. R., and Potter, N. N. 1972. Survival of poliovirus and echovirus during simulated commercial egg pasteurization treatments. J. Milk Food Sci. *35*, 247-251.

Stumbo, C. R. 1965. Thermobacteriology in Processed Foods. Academic Press, New York.

Sugihara, T. F., Ijichi, K., and Kline, L. 1966. Heat pasteurization of liquid whole egg. Food Technol. 20, 100-107.

Van Olphen, H. A. 1953. Pasteurizing egg products. Dutch Pat. 71,791.

Varadarajulu, P., and Cunningham, F. E. 1972. A study of selected characteristics of hen's egg yolk. 2. Influence of processing procedures, pasteurization and yolk fractionation. Poult. Sci. 51, 941-945.

Walters, R. E., Brant, A. W., and Patterson, G. W. 1968. Batch pasteurization of liquid whole egg. 2. Equipment design and operation. Poult. Sci. 47, 885-891.

Wilkin, M., and Winter, A. R. 1947. Pasteurization of egg yolk and white. Poult. Sci. 26, 136-142.

Winter, A. R. 1952. Production of pasteurized frozen egg products. Food Technol. 6, 414-415.

Winter, A. R., Greco, P. A., and Stewart, G. F. 1946A. Pasteurization of liquid egg products. I. Bacteria reduction in liquid whole egg and improvement in keeping quality. Food Res. 11, 229-245.

Winter, A. R., Stewart, G. F., McFarlane, V. H., and Solowey, M. 1946B. Pasteurization of liquid egg products. 3. Destruction of Salmonella in liquid whole egg. Am. J. Public Health 36, 451-459.

Woodward, S. A., and Cotterill, O. J. 1983. Electrophoresis and chromatography of heat treated plain, sugared and salted whole egg. J. Food Sci. 48, 501-506.

Wrinkle, C., Weiser, H. H., and Winter, A. R. 1950. Bacterial flora of frozen egg products. Food Res. 15, 91-98.

York, L. R., and Dawson, L. E. 1973. Shelf-life of pasteurized liquid whole egg. Poult. Sci. 52, 1657-1658.

Desugarization
of Egg Products

W. M. Hill
M. Sebring

INTRODUCTION

Dehydrated egg prepared from native liquid egg is subject to undesirable changes during storage; these include loss of solubility, decreased functionality, and the formation of off-color and objectionable flavor.

The United States dried-egg industry began an era of expansion in the early 1930s. During this period, imports from China gradually decreased as a result of the Chinese Civil War, high import duties on frozen and dried egg, and low shell-egg prices in the United States. Because of a previous dependency upon the Chinese products and the success of the Chinese in keeping their methods secret, the domestic producers of dried albumen lacked a fundamental knowledge of their product. They were unable to produce dried albumen comparable to the Chinese product in storage stability and functionality.

American scientists and dried-egg producers knew that the Chinese process involved fermentation of the albumen prior to drying. However, it was generally believed that fermentation was necessary only for "thinning" the thick albumen to render it suitable for drying and to improve the solubility and whipping qualities of the powder. Not surprisingly, the uncontrolled fermentation of liquid egg by organisms that just happened to be present was considered objectionable by some individuals. However, attempts to produce acceptable dehydrated egg products without a fermentation process were not successful.

World War II created a strong demand for dried-egg products, particu-

EGG SCIENCE & TECHNOLOGY,
3rd Edition

larly powdered whole egg, for use by the armed forces. The need for an egg product that remained palatable under severe conditions of storage and transportation stimulated research on the physical, chemical, and biological changes associated with the dehydration of egg products. The results of numerous studies elucidated the role of glucose in the storage stability of dried egg products.

REACTIVITY OF GLUCOSE IN DRIED EGGS

Much research has been conducted on the interaction of glucose with other egg components which result in deterioration in the quality of dried-egg products. Two major reactions have been defined—the glucose-protein (Maillard) reaction and the glucose-cephalin reaction.

Interaction of Glucose and Proteins

The condensation of sugars with amino acids was first studied by Maillard, thus the reaction bears his name. The initial bonding between the glucosidic hydroxyl groups of sugars and the amine groups of peptides and proteins is followed by other changes which result in the formation of brownish-colored products. Because of these end products, the overall process is often called the "browning" reaction.

Unfermented egg white powder darkens when heated to high temperatures for a substantial length of time. While conducting studies on the "thinning" of egg white by added trypsin, researchers noted that, although the enzyme liquefied the albumen, the subsequent dried powder developed off-colors during storage. It was suggested that this color change was due to a combination of the egg sugar with amino nitrogen. As further evidence of the importance of glucose in dried albumen, fermentation of egg white converts the glucose present to acid, and this process ensures that solubility and color stability is maintained while the product is in storage. The reaction between glucose and the amino groups of proteins is responsible, at least in part, for both discoloration in dried eggs and fluorescence in salt extracts. Glucose reacts with several of the egg proteins.

A sequence of reactions was postulated to explain the various deteriorative changes occurring in dried egg albumen. It was suggested that the initial reaction between glucose and the egg proteins is followed by additional reactions, one resulting in fluorescence and color alterations and another in insolubility. The former reaction was believed to proceed at a faster rate than the latter. Additional evidence substantiated the theory that the Maillard reaction was responsible for the changes in dried albumen.

Interaction of Glucose and Cephalin

Various researchers have presented evidence that some of the deteriorative changes occurring in dried whole egg and yolk are independent of the glucose-protein reaction. An ether-soluble brown substance was extracted from darkened whole egg powder. Chemical analysis of the substance indicated its derivation from a phospholipid, specifically cephalin. The results of further testing suggested that the discoloration of the dried whole egg was due, at least in part, to a reaction between a cephalin amino group and aldehydes. Major changes resulting in loss of palatability take place in the fatty constituents of the egg. Ether extracts were more reliable than salt-water extracts for the determination of fluorescence values as indices of palatability.

Glucose is the reactive aldehyde involved in the cephalin amine-aldehyde reaction. The changes that occurred in the phospholipid fraction of stored whole egg powder were essentially eliminated by the removal of glucose from the liquid before drying. In a study on the relative influence of the glucose-protein and glucose-cephalin reactions in whole egg powder, it was shown that loss in baking quality is associated chiefly with the glucose-protein reaction, while the glucose-cephalin reaction is involved in off-flavor development.

METHODS OF DESUGARIZATION

The factors that influence the quality and storage stability of dried eggs have been studied extensively. Moisture level, storage temperature, particle size, acidity, carbohydrate additives, and gas packing have been considered individually and in combination as to their influence on the stability of dried eggs. While each of these factors can be utilized in some beneficial manner, none has proven as successful in producing a stable egg product as the removal of glucose from the liquid prior to drying.

Spontaneous Microbial Fermentation

The fermentation of liquid eggs by contaminating microorganisms was practiced by the egg-drying industry until about the mid-1940s. The natural fermentation of albumen at 23.9° to 29.4°C can be described as follows:

The whites usually remain quiescent for several days (the exact time depending on the amount of bacterial contamination), during which time there is some increase in the amount of thin, watery white at the expense of the thick, jellylike white. Gradually an acid fermentation with a mild, characteristic odor sets in and the thick white begins to gather at the surface. The mucin fibers in this thick white contract under the influence of the acid produced. Usually after 6 to 7 days these fibers are practically free of thin white and collect as a layer of very stringy and cloudy material on the surface of the fermenting albumen. A considerable amount of the carbon dioxide present in the egg white is liberated as the acid increases, but the evolution usually stops about the sixth or seventh day of fermentation. The body of the thin white is at first very stringy, but after the fermentation is complete, it becomes watery and drips freely from the end of a glass rod. After this period the egg white has reached a stage when it is usually considered to be ready for drying. If the albumen is allowed to continue fermenting, there is a loss of acid (increase in pH), and usually very objectionable odors begin to develop.

Early bacteriological tests on samples of commercially fermenting egg white revealed a predominance of *Aerobacter aerogenes* or *Escherichia freundii* (these organisms are now classified in the genera *Enterobacter and Citrobacter*, respectively) with few other contaminants. Egg white fermented by either of these organisms yielded a bright, crystalline, granular product when pan dried. Fermentation by proteolytic bacteria such as *Proteus, Serratia,* or *Pseudomonas* resulted in an inferior product.

While natural fermentation did make possible the production of egg powders with adequate storage stability, it often caused problems in the freshly dried material. Not the least of these problems was the potential health hazard resulting from the growth of pathogenic bacteria, such as *Salmonella* in the fermenting liquid. The dehydration process, as commonly practiced in the production of powder of 4 to 6% moisture content, was not highly destructive to *Salmonella* organisms.

Controlled Bacterial Fermentation

A patent issued in 1931 described a process whereby liquid egg white was inoculated with an acid-producing organism such as "lactic acid bacillus." The following advantages were claimed for this process over spontaneous fermentation: (1) reduction of the time required from 48 to 60 hr or more to 24 hr or less, (2) production of a more uniform powder, (3) less hazard from pathogens, and (4) reduced possibility of the production of putrefactive odors.

An acceptable dried albumen was obtained from liquid that had been fermented by pure cultures of coliform organisms.

Stabilization of egg white was achieved by inoculating the liquid with a culture of *Streptococcus*, although some flavor changes resulted from the fermentation process. Also, desugarization of liquid egg yolk by

Pseudomonas was not highly beneficial to storage stability and resulted in the formation of an off-flavor.

Species of *Streptococcus* and *Lactobacillus* were utilized to desugar whole egg. It was found that these organisms could eliminate practically all the sugar from whole egg within 24 hr, but that a similar amount of sugar in inoculated egg white was decreased very little. Evidently, *Streptococcus lactis* does not successfully remove glucose from egg white unless large numbers of the organism are utilized or yeast extract is added to the cells as a growth promoter prior to inoculation. *Enterobacter aerogenes* did multiply in egg white and simultaneously converted the egg sugar to acid. The latter has been reidentified as *Klebsiella pneumoniae* and can be obtained from the Culture Collection, Northern Regional Research Center, United States Department of Agriculture, Peoria, Illinois 61604. All requests should state that the bacterial culture will be used to remove glucose from egg white.

More recently, Galluzzo *et al.* (1974) were able to simultaneously desugar whole egg and impart flavors and/or aromas characteristic of dairy products by fermentation with *Streptococcus diacetilactis*. In order to obtain the best growth and diacetyl production, it was necessary to heat the whole egg for 20 min at 65°C and adjust the pH to 5.5 with citric acid.

A 1% by weight of resting cells of *Streptococcus lactis* was used to desugar egg white. The glucose level was reduced from 0.320% to 0.006% in 1.5 hr at 37°C. The high numbers of streptococci (5×10^9 per milliliter of egg white) and short fermentation time prevented multiplication of gram-negative organisms. Angel food cakes prepared from pan-dried albumen fermented by concentrated streptococci were of good quality.

Much of the work conducted in the development of cultures for the bacterial fermentation of egg albumen has been carried out by the commercial segment of the egg industry. The results of these studies, and the cultures evolved, are generally regarded as trade secrets.

Yeast Fermentation

The use of pure cultures of yeast to remove glucose from liquid egg was introduced in the mid-1940s. Albumen and whole egg were desugared with *Saccharomyces apiculatus*. The glucose level of the albumen was reduced from 0.5 to 0.05% after 3-hr incubation at 37°C. However, the high level of yeast cells (1%) added to the liquid white resulted in a powder that had an objectionable yeasty flavor. In another study, a relatively small inoculum of *Saccharomyces cerevisiae* was used to remove the glucose from egg white. The glucose conversion was enhanced greatly by the presence of 0.1% yeast extract. The acid produced was not sufficient to cause precipitation of the mucin.

The development of yeast fermentation by *Saccharomyces cerevisiae* is credited to studies carried on in the laboratories of Armour and Company, Chicago, Illinois. Fermentation of whole egg with 0.2 to 0.4% by weight of moist baker's yeast at 22° to 23°C depleted the sugar within 2 to 4 hr. The final product was substantially free of yeast flavor. Later, the flavor of yeast-fermented whole egg was improved by centrifuging the sugar-free liquid to remove the yeast cells.

Strains of *Saccharomyces* produce a more palatable whole egg powder than strains of *Torulopsis*. Acidification of the whole egg melange to a pH below 6.0 increased the rate of fermentation, the optimum temperature of incubation being 30°C. A spray-dried whole egg with no significant yeast flavor was produced from liquid fermented with 0.07 to 0.15% dry weight of yeast in 2 to 3 hr.

Yeast fermentation is a practical means of removing the glucose from whole egg, although an occasional development of mustiness in yeast-fermented whole egg powder during storage is noted. Studies show that consumers have a strong preference for angel cakes made with frozen egg whites over angel cakes prepared with yeast-fermented, pan-dried albumen. However, the angel cakes made with the yeast-fermented product did not contain vanilla flavoring. When this ingredient was incorporated into the formulation, the differences between the two samples of cakes were not significant.

The advantages of yeast fermentation can be summarized as follows:

1. It causes little change in acidity and may thus eliminate the need for neutralization.
2. The loss of mucin is minimized.
3. The growth of contaminating organisms is reduced because of the short incubation time required by the process.
4. Moist yeast is readily available in a convenient form.
5. Fermentation without artificial means of aeration favors conversion of the sugar rather than the multiplication of yeast cells.
6. The process does not cause development of objectionable odors or flavors nor does it introduce undesirable by-products.

Enzyme Fermentation

Glucose oxidase was found in cultures of *Aspergillus niger* and *Penicillium glaucum* by Müller in 1928. He showed that in the presence of molecular oxygen the enzyme catalyzed the oxidation of glucose to gluconic acid. Franke and Lorenz in 1937 further characterized the reaction by finding that hydrogen peroxide was also a product of the enzyme activity. By using radioisotopes, Bentley and Neuberger (1949) showed that the oxygen atoms of the hydrogen peroxide produced by the reaction were derived from molecular oxygen, and that the enzyme catalyzes the transfer of hydrogen from glucose to oxygen. Thus glucose oxidase is, in actuality, an oxidoreductase.

An enzyme system comprised of glucose oxidase and catalase, an enzyme which catalyzes the decomposition of hydrogen peroxide to water and oxygen, was developed for use in the deoxygenation of foods and beverages. The activity of the glucose oxidase-catalase system in equation form is:

$$C_6H_{12}O_6 + O_2 + H_2O \xrightarrow[\text{oxidase}]{\text{glucose}} C_6H_{12}O_7 + H_2O_2$$

$$2H_2O_2 \xrightarrow{\text{catalase}} 2H_2O + O_2$$

Net reaction: $\underset{\text{glucose}}{2C_6H_{12}O_6} + O_2 \xrightarrow[\text{system}]{\text{enzyme}} \underset{\substack{\text{gluconic}\\ \text{acid}}}{2C_6H_{12}O_7}$

From the net reaction it is apparent that the enzyme system can be used to stabilize foods by (1) the removal of oxygen in the presence of excess glucose, and (2) the removal of glucose in the presence of readily available oxygen. For the latter usage, it is convenient to supply excess oxygen by the addition of hydrogen peroxide to the liquid being desugared.

The first report of a comprehensive study on the enzymatic desugaring of an egg product was that of Baldwin et al. in 1953, who used a commercial preparation of glucose oxidase-catalase to remove the glucose from large volumes of albumen. They found the process easy to control, reproducible, and capable of yielding a uniform powder with no objectionable odors traceable to the desugaring process. A few months later, other researchers reported that the powder made from enzyme-treated albumen had no off-odors or flavors and produced acceptable angel cakes. Empirical relationships were then developed among glucose level, time, enzyme level, and hydrogen peroxide demand in order to utilize the enzyme method for desugaring albumen more efficiently.

Enzyme treatment of egg yolk was demonstrated to improve the storage stability of the dry solids at 35°C, as determined in doughnut mixes. The untreated control solids yielded doughnuts of lower fat content and decreased palatability. Similar, but less dramatic, results have been obtained with yolk solids held at 20°C.

CURRENT INDUSTRY GLUCOSE-REMOVAL PRACTICES

A detailed description of glucose-removal practices currently used in the egg industry is difficult because of their confidential nature. A few generalizations may be made, however, concerning the overall procedures employed in desugarization.

Egg White

The removal of glucose from egg white is done almost entirely by the controlled bacterial fermentation process. While both the glucose oxidase enzyme and yeast methods may be used, the bacterial culture method has both functional and financial advantages. The use of bacterial fermentation results in a high whipping egg white solids product with good flavor, good solubility, and high whipping qualities at a nominal cost in terms of labor and materials. However, bacterial fermentation of yolk-containing egg is considered an unsatisfactory method, because it yields products with off-flavors and odors. In some companies the bacteria culture, or "bug," is specific, while in others it is merely characterized by its ability to convert the glucose and does not belong to a specific genus or species.

It is common practice for companies using the bacterial fermentation process to initiate growth of the culture in a small batch of albumen. The albumen is first pasteurized and then acidified to pH 7.0 to 7.5 with food-grade citric or lactic acid. The liquid is then inoculated with the proper culture and held at 30°–33°C. Extreme caution is taken to prevent contamination of the albumen by foreign bacteria, yeasts, or molds. Additional transfers of the actively fermenting culture to pasteurized acidified egg white can be made to increase the quantity of culture. The fermentation is stopped short of complete sugar depletion by reducing the temperature while the bacteria are in a high state of metabolic activity. The culture is then frozen in small quantities for use as inocula in the fermentation of large batches of egg white. This stepwise preparation of inocula gives the time needed for bacteriological analyses of the culture to ensure the absence of undesirable organisms.

Often, in large commercial operations, a portion of the fermenting batch of egg white is transferred to a second tank just prior to glucose depletion and serves as an inoculum for the second batch. With this technique, a single healthy culture may be carried for considerable time. This procedure eliminates the need for starting large fermentation batches from the "mother" or original pure culture. The level of transferred actively fermenting culture is usually 10 to 15% of the total batch weight. A properly designed bacterial monitoring system for the fermented egg whites, as well as careful attention to the pH, temperature and sugar-free status of the fermenting liquid, will yield a consistent, high-quality finished product.

Since egg whites are normally unpasteurized prior to bacterial fermentation, the microbial profile of the raw egg white is of utmost importance in the production of a final product of good quality. Large numbers of undesirable bacteria in the unfermented raw whites are capable of utilizing glucose and will compete with the culture bacteria during fermentation and give rise to a multitude of undesirable qualities in the finished product. In addition, fermentation rates will become inconsistent and, more often than not, undesirable bacterial contamination will be difficult to eliminate using standard time and temperature relationships during dry pasteurization.

Whole Egg and Yolk

The glucose oxidase-catalase enzyme system is used almost exclusively on whole egg and other yolk-containing egg products. This procedure may be carried out at an elevated temperature of 30° to 33°C, or at a low temperature of 10°C. The latter temperature requires a longer fermentation time but the low temperature is a deterrent to the growth of undesirable microorganisms during the processing period. The enzyme method may be employed in any practical size batch of whole egg or yolk. Adjustment of pH (required in egg white) is not necessary in yolk, which is naturally close to the optimum glucose oxidase reaction pH of 6.0. However, pH adjustment with citric or lactic acid may be required in some whole egg. Since the level of enzyme added is determined by the rate of reaction desired, temperature of egg, strength of enzyme purchased, and the amount of native glucose to be removed (yolk has a higher glucose content than whole egg), it is impossible to specify the exact amount of enzyme required for the process. Exact inoculum levels are best obtained from suppliers or determined by personal experience.

When pH, temperature, and enzyme are properly adjusted, hydrogen peroxide is metered into the product at a level determined by the amount of enzyme and type of egg product. Considerable caution is required in adding hydrogen peroxide to the egg because of foaming from evolved oxygen. Normally, the rate of peroxide addition is reduced during the latter part of the fermentation as the amount of glucose in the egg is decreased.

The use of yeast in glucose removal, while applicable to all liquid egg, is employed primarily in the desugaring of yolk-containing products. This method is probably the simplest of all to perform. Commercially, common baker's yeast is employed at a level of 3.4 lb per 1,000 lb of egg. Both the amount of yeast and the temperature are controlling factors in the rate of glucose removal. A temperature of 30°C is considered optimum for this process. If carefully controlled, the yeast fermentation process results in no multiplication of yeast cells during fermentation (resting fermentation) with little or no yeast flavor in the finished product. The yeast procedure removes not only the glucose present but other trace reducing sugars which are not removed by the glucose oxidase method (an interesting sidelight, but of little practical importance).

The fermentation process is deemed complete when glucose is not detected by the Somogyi sugar reagent or Clinistix®. A quick test that is particularly suited for albumen involves heating 1 to 2 ml of fermented liquid on a glass Petri dish with an infrared lamp for 10 to 15 min. The absence of browning shows that insufficient glucose is present to give a reaction.

In summary, the successful removal of glucose from egg products (in particular egg whites) requires thorough microbial control, good quality raw material, control of fermentation conditions (temperature and pH), and the appropriate healthy culture recommended for the particular procedure utilized.

REFERENCES

Ayres, J. C. 1958. Methods for depleting glucose from egg albumen before drying. Food Technol. *12*, 186–189.

Ayres, J. C., and Stewart, G. F. 1947. Removal of sugar from raw egg white by yeast before drying. Food Technol. *1*, 519–526.

Baker, D. L. (to B. L. Sarett) 1952. Enzymatically deoxygenated product and process. U.S. Pat. Reissue No. 23, 523 (original Pat. No. 2,482,724), July 22.

Baldwin, R. R., Campbell, H. A., Thiessen, R., Jr., and Lorant, G. J. 1953. The use of glucose oxidase in the processing of foods with special emphasis on the desugaring of egg white. Food Technol. *7*, 275–282.

Balls, A. K., and Swenson, T. L. 1936. Dried egg white. J. Food Sci. *1*, 319–325.

Bentley, R., and Neuberger, A. 1949. The mechanism of the action of notatin. Biochem. J. *45*, 584–590.

Boggs, M. M., Dutton II, J., Edwards, B. G., and Fevold, H. L. 1946. Dehydrated egg powders. Relation of lipid and salt-water fluorescence values to palatability. Ind. Eng. Chem. *38*, 1082–1084.

Carlin, A. F., and Ayres, J. C. 1951. Storage studies on yeast-fermented dried egg white. Food Technol. *5*, 172–175.

Carlin, A. F., and Ayres, J. C. 1953. Effect of the removal of glucose by enzyme treatment on the shipping properties of dried albumen. Food Technol. *7*, 268–270.

Carlin, A. F., Ayres, J. C., and Homeyer, P. G. 1954. Consumer evaluation of the flavor of angel cakes prepared from yeast-fermented and enzyme-treated dried albumen. Food Technol. *8*, 580–583.

Edwards, B. G., and Dutton, H. J. 1945. Role of phospholipides and aldehydes in discoloration. Ind. Eng. Chem. *37*, 1121–1122.

Fevold, H. L., Edwards, B. G., Dimick, A. L., and Boggs, M. M. 1946. Dehydrated egg powders. Sources of off-flavors developed during storage. Ind. Eng. Chem. *38*, 1079–1082.

Flippin, R. S., and Mickelson, M. N. 1960. Use of salmonellae antagonists in fermenting egg white. I. Microbial antagonists of salmonellae. Appl. Microbiol. *8*, 366–370.

Franke, W., and Lorenz, F. 1937. Cited by Bentley and Neuberger (1949).

Galluzzo, S. J., Cotterill, O. J., and Marshall, R. T. 1974. Fermentation of whole egg by heterofermentative streptococci. Poult. Sci. *53*, 1575–1584.

Hanson, H. L., and Kline, L. 1954. Consumer-type appraisal of whole egg powders stabilized by glucose removal (yeast fermentation) and by acidification. Food Technol. *8*, 372–376.

Hawthorne, J. R. 1950. Dried albumen. Removal of sugar by yeast before drying. J. Sci. Food Agric. *1*, 199–201.

Hawthorne, J. R., and Brooks, J. 1944. Dried Egg. VIII. Removal of the sugar of egg pulp before drying. A method of improving the storage life of spray-dried whole egg. J. Soc. Chem. Ind. *63*, 232–234.

Josh, G., Harriman, L. A., and Hopkins, E. W. (Armour and Company) 1949. Drying eggs. U.S. Pat. 2,460,986, Feb. 8.

Kaplan, A. M., Solowey, M., Osborne, W. W., and Tubiask, H. 1950. Resting cell fermentation of egg white by streptococci. Food Technol. *4*, 474–477.

Kline, L., and Sonoda, T. T. 1951. Role of glucose in the storage deterioration of whole egg powder. I. Removal of glucose from whole egg melange by yeast fermentation before drying. Food Technol. *5*, 90–94.

Kline, L., Gegg, J. E., and Sonoda, T. T. 1951A. Role of glucose in the storage deterioration of whole egg powder. II. A browning reaction involving glucose and cephalin in dried whole eggs. Food Technol. *5*, 181–187.

Kline, L. *et al.* 1951B. Role of glucose in the storage deterioration of whole egg powder. III. Effect of glucose removal before drying on organoleptic, baking, and chemical changes. Food Technol. *5*, 323–331.

Kline, L., Sonoda, T. T., Hanson, H. L., and Mitchell, J. H. 1953. Relative chemical, functional and organoleptic stabilities of acidified and glucose-free whole egg powders. Food Technol. *7*, 456–462.

Kline, L., Sonoda, T. T., and Hanson, H. L. 1954. Comparisons of the quality and stability of whole egg powders desugared by the yeast and enzyme methods. Food Technol. *8*, 343–349.

Kline, R. W., and Stewart, G. F. 1948. Glucose-protein reaction in dried egg albumen. Ind. Eng. Chem. *40*, 919–922.

Lightbody, H. D., and Fevold, H. L. 1948. Biochemical factors influencing the shelf life of dried whole eggs and means for their control. Adv. Food Res. *1*, 149–202.

Mickelson, M. N., and Flippin, R. S. 1960. Use of salmonellae antagonists in fermenting egg white. II. Microbiological methods for the elimination of salmonellae from egg white. Appl. Microbiol. *8*, 371–377.

Müller, D. 1928. Cited by Bentley and Newberger (1949).

Paul, P., Symonds, H., Varozza, A., and Stewart, G. F. 1957. Effect of glucose removal on storage stability of egg yolk solids. Food Technol. *11*, 494–498.

Pearce, J. A., Brooks, J., and Tessier, H. 1946. Dried whole egg powder. XXII. Some factors affecting the production and initial quality of dried sugar-and-egg mixtures. Can. J. Res. *24*, 420–429.

Scott, D. 1953. Food stabilization. Glucose conversion in preparation of albumen solids by glucose oxidase-catalase system. J. Agric. Food Chem. *1*, 727–730.

Solowey, M., Spaulding, E. H., and Goresline, H. E. 1946. An investigation of a source and mode of entry of *Salmonella* organisms in spray-dried whole egg powder. J. Food Sci. *11*, 380–390.

Stewart, G. F., and Kline, R. W. 1948. Factors influencing rate of deterioration in dried egg albumen. Ind. Eng. Chem. *40*, 916–919.

Stuart, L. S., and Goresline, H. E. 1942A. Bacteriological studies on the "natural" fermentation process of preparing egg white for drying. J. Bacteriol. *44*, 541–549.

Stuart, L. S., and Goresline, H. E. 1942B. Studies of bacteria from fermenting egg white and the production of pure culture fermentation. J. Bacteriol. *44*, 625–632.

Egg Dehydration

D. H. Bergquist

The removal of water from foods to a low enough level stops the growth of microorganisms and slows chemical reaction rates. Thus, dehydration can be used for preservation of food. It is nature's way of keeping certain foods, such as grains, and man learned centuries ago about the value of drying fruits, meats, and other foods.

Dehydration is a successful way of preserving eggs, and the egg-drying industry has developed over several years. Research has played a major role in solving problems which involve chemical, functional, and microbiological properties of dried egg products. Considerable improvements have been made in the production of dried eggs and they are now used extensively in bakery foods, bakery mixes, mayonnaise and salad dressings, confections, ice cream, pastas, and many convenience foods.

Dried-egg products have the following advantages:

1. They can be stored at low cost under dry storage or refrigeration and require less space than shell or liquid eggs.
2. Transportation costs for them are less than frozen or liquid eggs.
3. They are easy to handle in a sanitary manner.
4. They are not susceptible to bacterial growth when held in storage.
5. They permit a precise control over the amount of water used in a formulation.
6. They have good uniformity.
7. They have made possible the development of many new convenience foods.

In addition to research, progress in the production of dried egg

EGG SCIENCE & TECHNOLOGY,
3rd Edition

products has been due to several factors (Bergquist 1979). For example, users of eggs have demanded dried-egg products of high quality and with good functional properties. Dried-egg products are used by many segments of the food industry and many of these customers now have their own specifications. Dried-egg products must now meet strict chemical, physical, functional, and microbiological specifications that include moisture, fat, protein, ash, glucose, reconstituted viscosity, whipping ability, functional performance in the food in which it is used, and microbiological standards such as total plate counts of coliforms, yeasts, and molds, and *Salmonella*.

Government regulations have also had considerable influence on the development of the dried egg industry. Mandatory inspection of egg-drying plants by the USDA became a requirement in 1971. Under this law, regulations include minimum sanitary standards for construction of plant facilities, construction of equipment, requirements for processing of egg products, minimum pasteurization conditions, and sampling and testing procedures for *Salmonella* (Anon. 1975). The USDA requires that equipment meet the E-3A Standards and Practices that were developed through the cooperation of the International Association of Milk, Food, and Environmental Sanitarians, the Dairy and Food Industry Supply Association, the Poultry and Egg Institute of America, and equipment manufacturers.

Early records of egg dehydration in this country date back to about 1880, when Charles LaMont was issued a U.S. patent. At about the same time, W. O. Stoddard started an egg-drying operation in St. Louis, Missouri, and commercial production of dried egg had its beginning. In the early 1900s, China had a large supply of eggs at a very low cost. Germans, Americans, and others developed processes to produce dried-egg products, and they set up plants in China. The pan drying of egg white was developed using an egg white that had been naturally fermented. The stability of this product was excellent, and it was not known until many years later that this product was stable because the natural glucose had been removed in the fermentation (Kline and Stewart 1948). Whole egg and yolk products were dried in China by a belt-drying method, as well as by pan drying.

Prior to 1930, the low cost of egg products from China gave little incentive to dry eggs elsewhere. Tariffs on eggs coming into this country from China were increased during the 1920s. Drying of whole egg and yolk products in the United States began about 1930, and has expanded since that time. A great surge came during World War II, when there was a tremendous need for dried eggs by the military and also by the Lend-Lease Program. In 1944, there were 135 spray dryers producing dried whole egg. Total production in that year was over 300 million pounds (Koudele and Heinsohn 1960).

Pan drying of white also began in the United States in the early 1930s.

Although it was thought at first that this was the only way to dry whites, spray drying was tried in the late 1930s and was successful. However, full-scale production of spray-dried whites did not start until the late 1940s, when the cake-mix industry was demanding a better product for this purpose.

TYPES OF DRIED-EGG PRODUCTS

There are a number of different types of dried-egg products that have been developed through the years and that are now being produced for many different uses (Toney and Bergquist 1983). Some of the more important of these products are described in the following sections.

Dried Egg White

The most important types of dried egg-white products are:

Spray-dried egg white, whipping type
Spray-dried egg white, nonwhipping type
Pan-dried egg white
Instant-dissolving egg white

Most dried egg-white products are spray dried and are either the whipping or nonwhipping type, depending on the functional properties required. For example, there is a demand for an excellent whipping dried egg white for use in angel food cake and angel food cake mixes. These products usually have a whipping aid, such as sodium lauryl sulfate, added. On the other hand, there are many uses where whipping ability is not essential but where heat coagulation is important, such as in layer-type cakes (Miyahara and Bergquist 1961A).

Pan-dried egg white may be available in three forms: (1) flakes, (2) granules, and (3) powders. These products are commonly used for making aerated confections (Bergquist 1961B). A convenient procedure for using flake albumen is to allow it to soak in water overnight for reconstitution.

For some uses, spray-dried egg whites are sometimes difficult to solubilize, because the particles have a tendency to clump together and form balls when mixed with water. Instant egg white, which is an agglomerated product, overcomes this problem, dispersing and dissolving rapidly when added to water (Forsythe and Bergquist 1973; Bergquist *et al.* 1978).

Almost all dried egg-white products have had the natural glucose removed before drying and are shelf stable under almost any storage conditions.

Dried Plain Whole Egg and Yolk

The products in this category include:

Standard whole egg
Stabilized (glucose-free) whole egg
Standard egg yolk
Stabilized (glucose-free) egg yolk
Free-flowing whole egg
Free-flowing egg yolk

These products are all produced by spray drying and have no other added ingredients that might affect their function. If extended shelf life is needed at room temperature for a particular use, glucose is removed from the liquid or converted to an acid before drying to make the "stabilized" products. Flowability can also be greatly improved by adding a free-flowing agent. Either sodium silicoaluminate or silicon dioxide is permitted to be added at levels of less than 2% and 1%, respectively, with the total of moisture and free-flowing agent of less than 5.0%. Plain whole egg and yolk products do not have good whipping ability by themselves, but exhibit excellent binding, emulsifying, and heat-coagulating properties that make them desirable in many food products as for example layer cakes, pound cakes, doughnuts, cookies, mayonnaise and salad dressing, and egg noodles (Miyahara and Bergquist 1961B).

Dried Blends of Whole Egg and Yolk with Carbohydrates

The products in this category include:

Whole egg plus sucrose
Whole egg plus corn syrup
Whole egg plus yolk plus corn syrup
Egg yolk plus corn syrup

These products possess certain functional properties and handling characteristics which make them useful in baked goods. They can be used in any bakery item calling for whole egg or yolk, including those in which foaming is essential (Miyahara and Bergquist 1961C).

Specialty Dried Egg Products

The products in this category include:

Scrambled-egg mix

Imitation whole egg from egg whites plus ingredients that substitute for egg yolk

Eggs with other functional ingredients

Scrambled-egg mix is a product which has been in U.S. government purchase programs since 1967. (From 1957 to 1964 the U.S. government purchased stabilized whole egg from which the glucose had been removed by either the yeast fermentation or the enzyme desugaring method.) The scrambled egg mix specified by USDA contains 51% whole egg, 30% skim milk, 15% vegetable oil (containing an antioxidant), 1.5% salt, and 2.5% moisture on a finished dried basis. This product exhibits good storage stability.

Imitation whole egg products usually have an egg-white base and contain vegetable oil and skim milk as substitutes for the yolk portion of the egg. One product contains a small amount of whole egg. These products are primarily for low-cholesterol diets.

Eggs can be mixed and co-dried with many other food ingredients—for example, soy protein, shortenings, and emulsifiers. Fat can be extracted by different means from whole egg and yolk to produce dried, defatted egg products (Melnick 1967; Kandatsu and Yamaguchi 1973; Levin 1975; Larsen and Froning 1981). Although these products have certain desirable nutritional and functional characteristics, none has been produced and marketed commercially on a regular basis. By-products from fat extraction could include certain lipid fractions, such as egg lecithin.

Work continues in the development of new types of dried-egg products. One of these new products is egg in the form of a dried, puffed snack item (Froning *et al.* 1981). The production of dried egg products from 1940 is shown in Table 14.1. U.S. government purchases have greatly influenced these production figures. The U.S. military has purchased egg products, and the USDA has, at times, procured egg products in order to support prices of eggs and has used them for supplying products for school lunches and programs for the needy. During World War II, the period 1941–1946, military procurement and USDA purchases for lend-lease requirements boosted dried-egg production to the highest levels ever.

DRYING CHARACTERISTICS OF EGG PRODUCTS

The properties of eggs are very delicate, and thus the final quality of the dried product can be affected by several factors, including the quality of the shell eggs used, handling methods, sanitation practices, condi-

TABLE 14.1. Dried Egg Production in the United States (in millions)

	Whole[a]		White		Yolk[a]		Total	
	(lb)	(kg)	(lb)	(kg)	(lb)	(kg)	(lb)	(kg)
Jan. 1–Dec. 31								
1940	0.4	0.2	1.9	0.9	5.2	2.4	7.5	3.4
1945	97.0	44.0	1.7	0.8	7.2	3.2	105.9	48.0
1950	87.0	39.5	3.6	1.6	2.8	1.3	93.4	42.4
1955	3.2	1.5	10.6	4.8	9.3	4.2	23.1	10.5
1960	6.9	3.1	8.0	3.6	10.5	4.8	46.1	20.9
1965	22.1	10.0	13.6	6.2	13.4	6.1	49.1	22.3
1970	48.0	21.8	15.0	6.8	12.0	5.4	75.0	34.0
July 1–June 30								
1975	27.8	12.7	16.7	7.6	14.7	6.7	59.2	26.9
Oct. 1–Sept. 30								
1980	37.4	17.0	22.4	10.1	21.3	9.7	81.1	36.8
1981	34.3	15.5	25.7	11.7	22.2	10.1	32.2	37.3
1982	36.6	16.7	22.3	10.1	21.3	9.7	80.2	36.4
1983	48.9	22.2	20.3	9.2	20.4	9.2	81.9	37.2
1984	36.3	16.5	21.2	9.6	19.2	8.7	75.7	34.4
1985	50.2	22.8	25.0	11.4	21.5	9.8	96.8	44

a Includes plain and blends.

tions during processing, pasteurization procedures, drying, and conditions under which the dried-egg products are held in storage (Bergquist 1980B).

Eggs can be divided into two basic categories when considering their drying characteristics: (1) egg-white products, and (2) whole egg and yolk products. Egg-white products are virtually fat free, while whole egg and yolk products contain the highly emulsified lipids that are closely associated with the proteins and other components of the yolk. This makes quite a difference in the drying characteristics and the effect of drying on the properties of the dried-egg products.

Almost all dried-egg products are uncooked; in fact, in processing and drying eggs, every effort is made to preserve the uncooked state. For eggs to be useful, the native characteristics of the raw egg must be preserved. These characteristics include ability to coagulate with heat, production of stable foams when whipped, emulsifying power, color, and flavor.

In the drying of eggs, moisture is removed from the liquid by evaporation until only the solid portion, with a small quantity of moisture, remains. The rate of evaporation from the liquid depends on the following factors: temperature of the liquid, condition of the surroundings, and surface area of the liquid. Heat must be added to the liquid for evaporation of water to continue. Heat transfer to the liquid can take place by conduction, convection, and radiation. Energy absorbed by the liquid is balanced by the energy consumed in evaporation of water from the liquid.

There is no difference in the rate of evaporation from a quiescent egg albumen solution and from pure water. This applies to all the liquid egg products at the initial stages of drying, and is known as the constant drying rate period. The rate of evaporation balances the rate of heat transfer, and the temperature of the saturated surface is constant. The constant drying rate period continues until the entire evaporating surface can no longer be kept saturated by moisture movement from within the material. Then the evaporation rate continuously decreases until the final moisture content is reached. This is known as the falling drying rate period.

The moisture-vapor pressure relationships of the egg products being dried are important because they determine the drying conditions necessary to achieve the desired moisture levels. Tables 14.2, 14.3, and 14.4 show the equilibrium moisture contents of dried whole egg, yolk, and whites at various relative humidities and temperatures (Gane 1943). The difference in the equilibrium moisture content of egg yolk solids and egg white solids is due to the presence of inert fat in the yolk. By calculating the equilibrium moisture content of yolk on a fat-free basis, results are similar to egg white over a wide range of conditions. Dena-

TABLE 14.2. Equilibrium Moisture Content of Whole-Egg Solids as Related to Humidity and Temperature

Relative humidity (%)	Equilibrium moisture content (%) at					
	10°C	21°C	32°C	43°C	60°C	77°C
10	2.7	2.6	2.4	2.0	1.8	1.4
20	3.9	3.7	3.4	3.2	2.6	2.0
30	5.1	4.8	4.4	4.0	3.4	2.8
40	6.4	6.0	5.6	5.2	4.2	3.4
50	7.4	7.0	6.6	6.2	5.4	4.6
60	9.0	8.6	8.2	7.8	7.0	5.6
70	10.7	10.5	10.2	9.6	8.6	6.8

turation of the protein in both whites and yolks has little effect on the moisture-vapor pressure relationship.

The vapor pressure of dehydrated whole eggs in a low moisture range, 0.5 to 5.5%, was the subject of early studies (Makower 1945). The ratio of the heat of sorption to the heat of condensation of water vapor at the same temperature was 2.09 at a moisture content of 0.5%. This ratio was 1.2, with a moisture of 5.5%. These results indicate the energy necessary to vaporize water. For example, the heat of vaporization for eggs containing 0.5% water is more than twice the heat of vaporization of pure water.

In studies on the vapor pressure relationship of dried whole egg containing sucrose and corn syrup solids, it was found that carbon dioxide contributed significantly to the total pressure, as measured by a manometric method (Davis and Kline 1965). Although the water vapor pressure of a solution is lower than the vapor pressure of pure water, the vapor pressure of liquid egg whites, whole eggs, and yolks is substantially the same as that of pure water. This is because vapor-pressure lowering is about proportional to the number of molecules of solute in relationship to those of the solvent. Since the molecules in eggs are large, this ratio is very low until the final stages of drying.

TABLE 14.3. Equilibrium Moisture Content of Egg-Yolk Solids as Related to Relative Humidity and Temperature

Relative humidity (%)	Equilibrium moisture content (%) at					
	10°C	21°C	32°C	43°C	60°C	77°C
10	1.6	1.5	1.4	1.3	1.1	0.8
20	2.5	2.4	2.2	2.0	1.7	1.3
30	3.1	2.9	2.7	2.4	2.1	1.7
40	3.7	3.5	3.3	3.0	2.5	2.0
50	4.3	4.1	3.9	3.6	3.1	2.7
60	5.6	5.0	4.7	4.5	4.1	3.3
70	6.9	6.7	6.6	6.3	5.6	4.4

TABLE 14.4. Equilibrium Moisture Content of Egg-White Solids as Related to Relative Humidity and Temperature

Relative humidity (%)	Equilibrium moisture content (%) at					
	10°C	21°C	32°C	43°C	60°C	77°C
10	5.6	5.4	5.0	4.1	3.7	2.9
20	6.8	6.5	6.0	5.6	4.6	3.5
30	8.4	8.0	7.3	6.6	5.6	4.6
40	10.5	9.9	9.2	8.6	6.9	5.6
50	11.8	11.1	10.6	9.9	8.6	7.4
60	14.6	13.0	12.2	11.8	10.6	8.5
70	18.0	17.6	17.2	16.5	14.4	11.4

The shape of the surface from which drying takes place has an effect on drying characteristics. In the case of egg products, there are two types of surface: (1) flat, as in the case of pan drying, and (2) spherical, as in the case of spray drying. The drying characteristics of egg white from a flat surface in pan is illustrated in Fig. 14.1 (Bergquist 1964). The drying characteristics of whole egg are similar. It is noted that there is a constant rate drying period until a certain water content is reached; the drying rate then falls off abruptly due to the formation of a skin on the surface. The thicker the liquid in the pan, the higher the water content when the falling rate period occurs. The evaporation rate from the skin is relatively slow, and as more water evaporates, the skin becomes

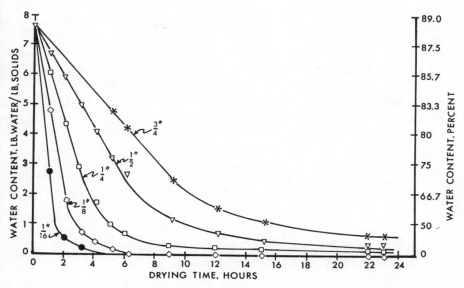

FIG. 14.1. Drying-rate curves for egg white poured at different depths on pans. Air velocity over pans—5 ft/sec at 54°C air temperature.

thicker and the drying rate becomes less and less. As the skin of egg white loses water, it becomes "crystalline" or solid in nature and eventually breaks into granular material, and finally into a powder.

In spray drying, evaporation rate from the sprayed particles is very rapid because of the large surface area formed. For example, a typical mean particle diameter in spray drying is 24 μm. This would give a surface area of 2,500 cm² per cubic centimeter of material. With this much surface, drying takes place within a fraction of a second.

The formation of new surfaces influences the drying rate. When new surfaces are formed, proteins spread and denature on the surface in a monomolecular layer. The rate of evaporation from egg albumen has been observed to be greater during surface formation than from free water similarly exposed (Bull 1938).

Rate of heat transfer to the egg and rate of water transfer from the egg during drying are affected by certain properties of the egg material, i.e., specific heat, viscosity, density, and surface tension. The specific heat of the egg product is at a maximum before drying begins and decreases as water is removed. The specific heat for all egg products can be approximated by using the following formula:

$$\text{Specific heat} = \frac{\% \text{ water} + (0.5 \times \% \text{ solids})}{100}$$

The viscosity of egg products varies over a wide range (Anon. 1969). Liquid egg white has a viscosity of about 12 centipoises (cP) at 5°C and 5 cP at 50°C, whole egg 20 cP at 5°C and 7 cP at 60°C, yolk 260 cP at 5°C and 45 cP at 60°C. The viscosity of each individual liquid egg product depends on various factors such as temperature, solid contents, prior physical treatment, prior heat treatment, and added ingredients. The reconstituted viscosity of a dried-egg product is also affected by drying and storage conditions, as will be discussed later.

The density of liquid egg white, whole egg, and yolk is approximately the same, 1.035 g/ml. Of course, the addition of sugar, corn syrup, or salt to the liquid egg products increases density. For example, the 33% blend products with corn syrup are 1.05 g/ml, 10% sugared whole egg or yolk is 1.07 g/ml, and 10% salted yolk or whole egg is 1.10 g/ml (Anon. 1969).

EFFECT OF DRYING AND SUBSEQUENT STORAGE ON PROPERTIES OF EGG PRODUCTS

The objective when drying egg products is to obtain a finished product whose properties, when reconstituted, are close to those of the raw, uncooked liquid egg. The properties of dried eggs can be affected by the following treatments: (1) drying per se; (2) physical forces, such as occur in pumping, homogenizing, and atomization; (3) heating of the liquid; and (4) heating of the dried material.

Functional Properties

Dried-egg products are useful because of the qualities they impart to the various food products in which they are used. Their performance is related to the following functional properties: whipping (aerating or leavening power), coagulation (binding and thickening power), emulsifying, flavor, nutrition, and color. How processing and drying of egg products affect these functional properties is discussed below.

Whipping. The whipping properties of dried-egg products are important for angel food cake, sponge cake, meringues, and confections. Eggs have the ability to incorporate air and form a relatively stable foam when beaten with a wire whip or other mechanical device. Egg white has whipping properties different from whole egg and yolk products and is affected in a different way by drying. Some of the effects of drying on egg white will be discussed below.

The proteins in egg white hold the foam structure produced by the tremendous surfaces formed during the process of whipping. Unfolding and spreading of the protein occurs at the surfaces within the foam, that is, the liquid/air interface. The unfolded protein surface denatures to form a structure which is relatively stable.

Egg white usually has a loss of whipping properties when dried without pretreatment. This loss apparently results from a combination of the drying process itself and any physical treatment therein. Heat could also have an effect if a high enough temperature is reached in the product during drying. The effect of heat depends on the water content of the product. When dried, a glucose-free egg white can withstand relatively high temperatures without damage. In fact, heating of dried egg white is used as a method to reduce bacteria count and to lessen the possibility of *Salmonella* contamination.

The small amounts of yolk that inadvertently contaminate egg white during egg breaking and separating can adversely affect the whipping properties of egg white (Bergquist and Wells 1956). For this reason, processors usually specify a maximum egg yolk content for egg whites—typically less than 0.03% yolk on a liquid basis.

Heating egg white contaminated with yolk at 54°C for 15 min and at 38°C for 24 hr improves the whipping properties of the liquid, but the benefits are not apparent after spray drying (Cotterill *et al.* 1965). One-tenth as much yolk is required to reduce foaming properties of spray-dried egg whites as is needed to affect liquid egg white (Cotterill and Funk 1963). Pancreatic lipase improves the foaming of egg white contaminated with yolk (Imai 1976). This is a good review and evaluation of the effects of bacterial fermentation and lipase treatment on the whipping properties of egg white.

In spray drying, eggs are subjected to certain physical forces that can affect their functional properties. Pumping and atomization are examples of treatments which can be relatively severe. Pumping subjects a liquid to shear forces, while atomization subjects it to both shear forces and to new surface formation.

How shear forces affect egg white can be demonstrated by what happens during homogenization. Homogenization with conventional equipment can have an adverse effect on the whipping properties of egg white (Slosberg 1946; MacDonnell et al. 1950; Forsythe and Bergquist 1951). The degree of change is related to homogenization pressure. The damage is actually attributed to shear forces, which mechanically disrupt the physical structure of the proteins. On the other hand, pressure under 5,000 psi causes no change in the viscosity and functional property of egg white; pressure above 5,000 psi can cause coagulation of egg white.

Surface formation during atomization will result in surface denaturation and may cause some change in the functional properties of egg white (Bergquist and Stewart 1952A). Large new surfaces are formed, and egg white protein spreads as a monomolecular layer over these surfaces and becomes irreversibly denatured. The amount of area depends on particle size. This is illustrated in Fig. 14.2. Although the area is great, the actual percentage of protein that is denatured under normal atomization for spray drying is relatively small. No adverse effects on whipping properties are noted until the amount of new surface formed is above 4,000 cm^2 per cubic centimeter (Bergquist and Stewart 1952B). This is equivalent to a mean particle diameter of 15 μm.

Many different conditions and ingredients have an affect on egg white when spray dried. Chemical additives, such as acids and whipping aids, are discussed later. Studies on the spray drying of egg whites over a pH range of 4.0 to 10.0 showed the highest angel cake performance to be at pH 8.5. At this pH, there was also an indication of less protein damage and greater heat coagulability (Hill et al. 1965).

Heat can also have an adverse effect on the whipping properties of egg white. The conditions during drying capable of denaturing the egg protein by heat are preheating the liquid, heat generated by friction from physical handling or shear forces, and heat transferred to the product during drying.

During spray drying, the temperature of the atomized particles ap-

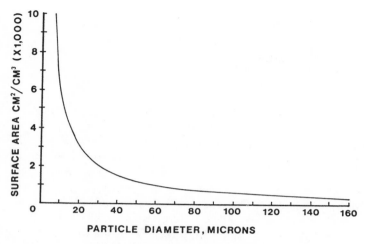

FIG. 14.2.   Surface area as determined by particle diameter.

proaches the wet-bulb temperature of the drying air. This temperature would be less than 49°C, even when atmospheric conditions are relatively humid. Thus, under most conditions, the temperature would not be expected to go high enough to cause damage to egg white. When the product becomes dry, the temperature rises. However, at this point, the dried egg white is quite stable. Preheating of the liquid or heat from friction due to pumping and atomization would be more likely causes for heat denaturation during spray drying. On the other hand, during pan drying, where the rate of evaporation is slow by comparison, it is possible for too rapid heat transfer at too high temperature to result in heat denaturation.

Heating egg white above 57°C can cause loss of whipping properties (Slosberg *et al.* 1948). Loss of whippability has been observed when heating egg white to only 49°C for 1 hr or even at 43°C for 6 hr. Egg white pasteurized at 57°C for 4 min produces poor angel cakes. Increased whip time and reduced angel cake volume results when heating egg white at 54°C for 3 min, but a recovery of volume occurs when the liquid is spray dried (Clinger *et al.* 1951). However, egg white can be pasteurized at its natural pH (9.0) below 57°C and spray dried with little effect on its whipping and functional properties, as measured by angel cake tests (Brown and Zabik 1967).

A beating time increase, but no effect on angel cake volume, results from heating egg white liquid 2 min at 53°, 56°, and 57°C (Kline *et al.* 1965B, 1966). Triethyl citrate and triacetin, but not sodium lauryl sulfate, restore beating time. Egg white is stabilized when its pH is reduced to neutral and a metal salt, such as aluminum sulfate, is added. The egg white proteins, except for conalbumin, are more stable at this pH and the metal ion complexes with conalbumin to make it more heat stable (Cunningham and Lineweaver 1965). (This is discussed in Chapter 12.) Damage to the whipping properties of egg white during pasteurization is caused by heat denaturation of the ovomucin-lysozyme complex (Garibaldi *et al.* 1968). Ovomucin helps stabilize egg white foam, and the increase in beating time of egg white when pasteurized is apparently due to loss of foam stability because of damage to ovomucin.

As noted before, egg white is very stable in the dry state, particularly with the natural glucose removed. This stability can be demonstrated at moisture levels below 20%. Thus, whipping properties are not affected unless excessively high temperatures are reached. When dried egg whites were heated to 82°C for 2 hr, no adverse effect on whipping and angel cake-making properties were found (Smith 1964).

The effect of drying on the whipping properties of yolk-containing products will now be considered although yolk is a complex balance of proteins, lipids, and other constituents it is its fat constituents that adversely affect whippability. Thus, the loss in whipping properties during drying is apparently due to the breakdown of finely emulsified fat globules into coalesced free fat. When the drying of whole egg or yolk is observed under the microscope, one can actually see the finely emulsified fat globules breaking down and coalescing as water evaporates from the liquid (Joslin and Proctor 1954). Even mild conditions of drying, such as in freeze drying, have been found to cause changes in the functional properties of whole egg and yolk (Rolfes *et al.* 1955). Carbohydrates are added to whole egg and yolk to prevent this loss in foaming properties, as will be discussed in a later section. The foaming ability of plain whole-egg solids can be improved by increasing the temperature at which the reconstituted material is whipped.

Whole egg and yolk liquids are much more resistant to damage by physical treatment than egg white. The physical forces normally encountered in processing and drying have little effect on their functional properties. For example, homogenization at pressures as high as 3,000 psig will cause little or no change in whipping ability of whole egg.

As with egg white, the excessive temperatures developed during drying can have an adverse effect on whipping and other properties of whole egg and yolk products. Denaturation of liquid whole egg is a function of time and temperature and has been observed to occur in the temperature range of 56° to 66°C by measuring changes in viscosity (Payawal *et al.* 1946). Just above this range fractional precipitation of protein takes place, and above 73°C coagulation is almost instantaneous. Yolk denatures in the range 63° to 70°C. No change occurs in the foaming properties of whole egg liquid pasteurized under standard conditions, but a change has been noted if whole egg is frozen. Homogenization after pasteurization before freezing restores foaming ability (Sugihara *et al.* 1966).

Dried whole egg and yolk products are more susceptible to heat damage than dried egg white. The amount of heat the dried-egg product absorbs during the final stages of drying is important and depends on the method of drying, the dryer design and conditions of its operation, and how rapidly the product is cooked after drying. Also, the conditions under which the dried product is stored will also affect is whipping performance.

Emulsifying. Egg yolk, whole egg, and egg white are all good emulsifiers for food products. Egg yolk is rated as four times as effective as egg white as an emulsifier, and whole egg is intermediate. The very excellent emulsifying properties of egg yolk are attributed to its lecithoproteins. However, the ether-insoluble portion of whole egg is the most important emulsifying substance. In fact, drying actually enhances the emulsifying properties of whole egg (Chapin 1951).

The emulsion stability of mayonnaise made with dried whole egg and dried egg yolk was found to decrease when these egg products were stored at 35°C for 3 months (Lieu *et al.* 1978). On the other hand, little change was noted in emulsion stability when these products were stored at 23.9°C or less for up to 6 months. Heating of liquid eggs under certain conditions can result in product changes. Liquid salted egg yolk has been pasteurized at relatively high temperatures with short holding time without affecting its emulsifying properties. However, storage at 52°C up to 8 days does have an adverse effect (Cotterill *et al.* 1976).

Coagulation. The ability of egg products to coagulate and bind pieces of food together or to thicken foods such as cakes, custards, omelets, and puddings is one of their most important functional properties. Egg proteins denature and coagulate over a wide range of temperatures. Under proper drying and storage conditions, dried-egg products retain their heat-coagulating properties quite well. However, if drying conditions are too severe or if storage conditions are adverse, whole egg and yolk products can lose their heat-coagulating properties as well as their solubility.

The effect of natural glucose on egg products is covered in Chapter 13. Generally, if the egg product contains its natural glucose, it will be more adversely affected by drying and subsequent heat treatment or storage than if the glucose had been removed (Kline and Sonoda 1951; Kline *et al.* 1954). However, moisture level, pH, and the inclusion of other ingredients will also influence the final result. For example, the browning reaction will be retarded at a low moisture level (below 2.0%), or at a low pH level (below pH 6.0). The browning reaction is also inhibited by co-drying eggs with carbohydrates (Brooks and Hawthorne 1943).

Flavor. Eggs not only have a distinct natural flavor, but they also contribute to the flavor of products in which they are used. Flavor of eggs can be changed by adverse drying conditions and also by storage. One of the most important factors causing poor storage stability of dried-egg products is the presence of natural glucose (Stewart and Kline 1941; Bate-Smith and Hawthorne 1945).

Almost all egg white products have the free glucose removed before drying. Thus, there is little, if any, change in flavor during dehydration. Also, dried egg white is very stable because of this and can be stored indefinitely at room temperature. Any off-flavors noted in dried egg white are probably caused by the methods used for processing. For example, some types of fermentations used to remove glucose may cause the development of off-flavors.

In the case of dried whole egg and yolk products, the reaction of glucose can cause the development of off-flavors and off-odors during drying, but more often during storage. Lightbody and Fevold (1948) did a literature review dealing with the shelf life of dried whole egg. Much work on this subject was carried out between 1942 and 1948, because of the need during this period for an improved whole-egg product to meet the requirements of the armed forces, the Lend-Lease Program, and later, the government price-support program. Almost all the work reported up to this period dealt with whole-egg solids containing their natural amount of glucose, the stability of which is considerably affected by time, temperature of storage, and moisture content. Changes in flavor and odor during storage are quite noticeable. The flavor stability of whole-egg powder can be improved to a certain extent by acidifying the liquid to pH 5.5 before drying. This inhibits the browning reaction involving the glucose and protein, but does not completely prevent it. Lowering of pH was adopted by both the Quartermaster Corps and the USDA in the early specification for a whole-egg product with improved stability. However, these specifications were later superseded by requirements for the removal of glucose from whole-egg products.

Although the stability of yolk and whole-egg products can be improved considerably by removal of glucose, there are still certain oxidation changes that can occur during storage. Thus, storage can be further improved by packing the product in an inert atmosphere. Nitrogen and carbon dioxide, or a mixture of the two, have been used for this purpose. Flavor scores of dried whole egg correlate better with UV absorption of volatiles than with any other chemical test (Privett et al. 1964).

The flavor stability of whole egg and yolk products can also be improved or reduced by the addition of carbohydrates, such as sucrose and different corn syrups. The relationship between type and concentration of carbohydrate, degree of protection afforded to functional properties, and rate of off-flavor development will be discussed in the section "Role of Chemical Additives."

Nutrition. Drying of eggs under normal conditions causes little, if any, loss of the nutritional properties of the eggs. Vitamin A, vitamin B, thiamine, riboflavin, pantothenic acid, and nicotinic acid have been determined in dried whole egg and found to be essentially the same as that in the fresh egg product (Klose *et al.* 1943; Everson and Souder 1957). Nutrient compositions of commercially spray-dried whole egg, yolk, and whites were recently studied and found to be quite comparable to their liquid counterparts (Cotterill *et al.* 1978). The protein value of dried eggs remains essentially unchanged. Adverse drying conditions or poor storage conditions can however, damage nutritional properties. However, this would be extreme, and any egg product with suitable flavor will probably have all its nutritional properties.

Color. Under normal drying and storage conditions, the color of whole egg and yolk remains unchanged. The pigments in eggs are very sensitive to oxidation, and excessive exposure to heat in a dryer will tend to bleach them. For example, in a spray dryer the fine particles will remain in the drying air of the system longer than the larger particles, which separate out rather easily. In some systems, the fines will be carried to the last collector and may even be lost out the stack. These fines will have lost much color. With proper design of the spray dryer, however, particles are separated out rapidly from the drying air and little, if any, change in color is noted.

Physical Properties

Whole egg and yolk products are more susceptible to changes in spray drying than is egg white, even though the proteins of egg white coagulate at a lower temperature. The temperature which the dried particles achieve within the drying system and the length of time the product is held in the highest temperature zone is important to the quality of dried whole egg and yolk products. High-temperature conditions at localized points within a drying system may also have a marked effect on these products. One manifestation of excessive heat in drying of plain whole egg and plain yolk is an increase in their reconstituted viscosity. Change in viscosity of dried plain yolk is much greater than in dried plain whole egg.

This change in viscosity is noted in both native and desugared whole egg and yolk products subjected to high temperatures during drying. Egg yolk spray dried with an outlet temperature of 60°C has a reconstituted viscosity of 3,000 cP and a soluble nitrogen content of 95.4% (Kline *et al.* 1963B). On the other hand, yolk dried with an outlet temperature of 107°C has a reconstituted viscosity of 80,000 cP and a soluble nitrogen content of 39.6%. Functional properties, as measured by performance in cake-type doughnuts, have also been found to be affected.

An increase in viscosity in whole egg and yolk products can also be observed during storage. Viscosity in the reconstituted product increases quite rapidly at temperatures above 38°C. Changes in viscosity and doughnut-making properties take place as a result of storage of dried egg yolk at elevated temperatures. Increased viscosity is accompanied by reduced protein solubility.

The density of egg products is not affected by dehydration. When a dried-egg product is reconstituted to its natural solids, it has about the same density as the liquid from which it was dried. On the other hand, the bulk density of the dried-egg product can vary considerably, depending upon the methods and conditions of drying. Pan-dried egg products are much higher in bulk density than their spray-dried counterparts. Freeze-dried egg products are lowest in bulk density. The bulk density of the spray-dried product depends upon the type of spray nozzle, temperature of the liquid being sprayed, particle size, particle-size distribution, drying temperatures, and any damage suffered by the dried product in the powder-handling equipment. Bulk density of a spray-dried product can generally be increased by milling the material and decreased by injecting gas into the liquid before drying (Kline *et al.* 1963A).

The flow and water-sorption properties of egg powders have also been studied (Passy and Mannheim 1982). The method of drying, fat content, and percentage of free-flowing agent, such as sodium silicoaluminate, can all have an effect on these properties.

Chemical Properties

The changes in functional properties noted above are undoubtedly caused by changes that occur in the chemical properties of the various components of the egg. Some of these changes have already been discussed.

Egg white is about 10% proteins. Thus, any changes that occur in egg white during drying are apparently caused by changes in these proteins. Chromatographic and electrophoretic studies have indicated that only minor changes occur in egg-white protein when it is spray dried, with globulin proteins most sensitive to spray drying (Galyean and

Cotterill 1979). The denaturation and coagulation of the proteins are considered to be chemical changes, since they involve the unfolding of protein, which exposes certain chemical groups, such as the sulfhydryl group, and thus alters the chemical reactivity of the proteins. Studies of the heat denaturation of egg white by differential scanning endotherms at pH 7 have indicated major endotherms at 65°C produced by the denaturation of conalbumin and at 84°C by ovalbumin. At pH 9.0, the conalbumin endotherm is increased to 70°C, whereas the ovalbumin endotherm remains the same (Donovan et al. 1975). Since water is an integral part of the protein molecule, its removal may cause certain changes to occur in the properties of egg white. It is difficult to give a chemical explanation for the loss in whipping properties of egg white when small amounts of yolk are present. The effect of yolk contamination is greatly magnified when water is removed from egg white. Minute quantities of fats from the contaminating yolk are probably changed from a highly emulsified state to a free state during drying, as has been observed in whole egg. These free fats could spread and coat the dried particles, thus reducing the foaming ability of the white when reconstituted.

The presence of natural glucose in eggs can cause chemical changes during drying, as well as during storage after drying. In egg white, the reaction involves the reducing or aldehyde groups of glucose and the amino groups of the proteins (the Maillard reaction) (Stewart and Kline 1948; R. W. Kline and Stewart 1948). This reaction results in the development of a brown color, fluorescence, and a reduction in solubility. The reaction is minimized by drying to a low moisture content and the rate proceeds more slowly at lower pHs. Because glucose constitutes about 4.0% of the solids in egg white, almost all egg white that is dried has had the glucose removed before drying. Egg white can be pretreated in several different ways before drying. The primary purpose of pretreatment is to remove the glucose (this is covered in Chapter 13).

The changes that occur in egg yolk and yolk containing egg products are apparently even more complex than those in egg white. These changes can also be caused by the reaction of naturally occurring glucose. In this case, glucose can react with the amino groups not only in protein, but also in cephalin (Kline et al. 1951A, B). The reaction with cephalin which causes the development of off-flavors and off-odors during storage, can be prevented by the removal of glucose or inhibited by the addition of carbohydrates.

Half the water in yolk is distributed as free water and the rest is bound to three distinct fractions: the water-soluble protein, the high-density lipoprotein, and the low-density lipoprotein (Schultz and Forsythe 1967). Most of the lipids are in the low-density lipoprotein fraction, which in unaltered egg yolk liquid is in a finely dispersed state. These workers postulate that the triglycerides make up the inner core of the highly emulsified low-density lipoprotein, which is surrounded by a phospholipid shell. The protein molecules are wrapped around the shell. The high-density lipoproteins are contained in granules that can be removed by centrifugation.

Actually, all the lipids of unaltered egg yolk are associated with lipoproteins, and they cannot be easily extracted with a nonpolar solvent, such as hexane. On the other hand, when plain egg yolk is spray dried, a large percentage of the fat can be extracted by hexane (Kline *et al.* 1964). This is apparently caused by the release of free fat from the highly emulsified state, which, in turn, reduces the whipping ability of the egg yolk or whole egg products. This change can be prevented by the addition of carbohydrates. In dried-egg products to which carbohydrates have been added at a high enough level, fats are again difficult to extract with hexane.

In whole egg liquid heated at 57° to 87°C for 3.5 min, livetins and globulins were found to be the most heat sensitive as measured by electrophoresis and chromatography (Woodward and Cotterill 1983). Conalbumin and ovalbumin were the most stable. The addition of sugar and salt improved stability. Changes as indicated by disc-gel electrophoretic patterns have been noted in liquid egg yolk heated at 63° or 65°C for 4 min, but not at 61°C for 4 min. Also, spray-drying was noted to alter several egg yolk components (Cunningham and Varadarajulu 1973).

The changes in viscosity of plain egg yolk and whole egg, as indicated previously, are apparently due to changes in the lipoproteins. The lipids make up approximately 45% of whole egg solids and 60% of yolk solids, and thus play a predominant role in changes during drying. Of the lipids in egg yolk, 62% are glycerides, 33% phospholipids, and 5% cholesterol. Of the phospholipids, lecithin comprises 73% and cephalin 15%. The oxidation rate of cephalin is extremely rapid, being approximately 100 times that of lecithin (Lea 1957). These phospholipids are, of course, bound together with protein, and water is an essential part of this association. When water is removed, the balance is changed. Thus, it is difficult to remove water from lipoproteins without causing changes in their properties (Parkinson 1975). Changes have also been noted in proteins and lipoproteins on storage (Parkinson 1977).

Gelation, which occurs when yolk is frozen and thawed, is apparently due to the aggregation of yolk lipoproteins because of the imbalance and shift in water when yolk is frozen (Powrie *et al.* 1963). The carbohydrates that protect yolk from gelation during freezing also protect yolk that is spray dried. Carbohydrates prevent an increase in viscosity of yolk both during and after drying and during storage.

Microbiological Properties

Some changes may occur in the microbiological populations of eggs during drying. For example, the total bacteria count can drop considerably, depending on the type of microorganism present in the liquid, as well as on the conditions used in drying. Some bacteria may be quite sensitive to drying, whereas others exhibit strong resistance. Drying cannot be relied on as a method of pasteurizing eggs.

Because of pasteurization and better sanitary practices, the microbiological quality of dried eggs has greatly improved through the years. Whole egg and yolk products are heat pasteurized in the liquid state prior to drying, while egg-white products are usually heat treated after drying to reduce bacteria counts. Egg white can also be pasteurized in the liquid state before drying. A combination of pasteurization or heat treatment of the liquid before drying and heat treatment of the finished product ensures very low bacterial populations (Bergquist 1961A).

Other noncommercial processing methods for reducing bacteria counts include chemical treatment of the dried product, for example, with ethylene or propylene oxide and irradiation of the liquid or dried product with ultraviolet light and ionizing irradiation (Ijichi *et al.* 1964; Grim and Goldblith 1965). A decided disadvantage of irradiation methods is off-flavor development.

The practice of heat-treating dried egg-white products has been used for some time (Ayres and Slosberg 1949; Banwart and Ayres 1956; Smith 1964). Here, advantage is taken of the fact that dried egg white with glucose removed is very stable to heat, and can be subjected to relatively high temperatures. Good kills of the total bacteria, including *Salmonella*, can be achieved. The kill of bacteria seems to be related to the prior state of the bacteria. For example, if *Salmonella* had been subjected to heat treatment in the liquid state prior to drying, they are very susceptible to heat treatment of the dried product.

ROLE OF CHEMICAL ADDITIVES

Carbohydrates

The loss of foaming power of whole egg and yolk when they are dried can be prevented by the addition of certain carbohydrates to the liquid before drying. The amount of protection given by the carbohydrate depends on the percentage of carbohydrate used. The higher the level, the greater the protection. For example, 10% sugar added to whole egg will preserve almost all of its whipping properties. Carbohydrates react with the proteins and phospholipids of the low-density lipid fraction of the yolk, an interaction which stabilizes the low-density lipoprotein whose micellar structure would break down upon removal of water from the micelle. Researchers postulate that carbohydrates participate in the hydrogen bonds through the hydroxyl group in which the water of hydration has participated (Schultz and Forsythe 1967).

There is a rather complex relationship between the amount and kind of carbohydrate added, the degree of protection given to the beating power of egg by the carbohydrate, and the rate of development of off-odors and off-flavors. Studies have been conducted in which sucrose, 24 dextrose equivalent (DE) corn syrup solids, and 42 DE corn syrup solids were added to whole egg at different levels (Kline and Sugihara 1964; Kline *et al.* 1964). As the percentage of carbohydrate was increased, flavor stability improved until a maximum was reached. Then, further addition of carbohydrate caused a sharp decrease in flavor stability. This peak was reached in whole egg at about 5% with sucrose, 10% with 24 DE corn syrup solids, and 7.5% with 42 DE corn syrup solids. This variation is believed to be related to the average molecular size of the carbohydrate. Foaming power of the whole egg is approximately proportional to the carbohydrate content at each of the above levels.

It was also observed that fat that can be extracted with a nonpolar solvent decreases as the carbohydrate level is increased. Also, the phospholipid content of the extracted fat sharply decreases. The off-flavor associated with decreased fat extractability has been shown to be accompanied by an increase in oxidative rancidity, as measured by peroxide value. The off-flavor noted is usually described as "fishy."

Although the addition of a carbohydrate such as sucrose helps to preserve the whipping properties of whole egg and yolk, it has little or no effect in preserving the whipping power of egg white during drying. Sucrose does retard heat denaturation of egg white (Slosberg *et al.* 1948) and does stabilize the egg white against the Maillard reaction if the whites are dried with the natural glucose present (Kline *et al.* 1965A).

Salts

Although salt (sodium chloride) is used to stabilize much of the egg yolk that is frozen commercially, it is not used extensively in dried-egg products. It does have a stabilizing effect on whipping properties, similar to that of carbohydrates, but the levels where it would be effective cannot be tolerated because of flavor. Salt is thought to have some beneficial effects when small percentages are added to products containing carbohydrate.

Whipping Aids

Whipping aids are added to egg white products to give them more uniformity and to compensate for any changes that might occur during processing and drying. The most common whipping aids used commercially and approved by FDA include sodium lauryl sulfate (Mink 1939), triethyl citrate (Kothe 1953), triacetin (Maturi *et al.* 1960), and sodium desoxycholate (Kline and Singleton 1959). These additives are used at about 0.1%, based on egg white solids, and depending on the type of additive. There are also other additives which have been reported to improve whipping properties, including acetyl methyl carbinol, salts of capric acid, sodium diacetate, calcium stearyl-2 lactylate, and polyphosphates. Sodium hexametaphosphate has been shown to have a protective effect on the foaming properties of pasteurized egg white (Chang *et al.* 1970) and has been reported to increase the sensitivity of bacteria to destruction by heat (Kohl *et al.* 1970).

Whipping aids are needed because the whipping ability of egg white is often adversely affected by pasteurization, drying, and pumping. These aids are capable of giving to the egg-white product whipping properties that are comparable to, or even better than, the natural untreated product. Several chemicals are effective as whipping aids, but their action is not fully understood. All we know is that certain chemical compounds work, while others do not. For example, anionic surface-active agents (e.g., sodium lauryl sulfate) improve the whipping ability of dried egg white, whereas nonionic and cationic surface-active agents do not. Other chemical compounds that are effective are certain organic esters (e.g., triethyl citrate), bile salts (e.g., sodium desoxycholate), and salts of a fatty acid (e.g., sodium oleate).

Selection of the whipping aid depends on the type of egg product and its use. Whipping aids are used in liquid and frozen egg whites that have been pasteurized. The ester-type whipping aid, such as triethyl citrate, seems to be preferred. Sodium lauryl sulfate does not work as well for liquid and frozen products, but is the additive most commonly used for dried egg whites. It is used at a level of less than 0.1% (on a solids basis) and is the additive usually specified by cake-mix manufacturers who prefer to standardize on one compound.

The chemical additives that enhance the whipping properties of egg white have little or no effect on the whipping properties of whole egg and yolk when used at the same levels. However, some improvement in foaming ability and in foam stability has been noted when sodium lauryl sulfate and gum stabilizers are added at higher levels (Kim and Setser 1982). Polyphosphates have also been reported to have a beneficial effect on the beating properties of whole-egg solids (Lewis *et al.* 1953).

Other Additives

Several other additives are used in egg products to improve one or more properties of the final product. Thus, the flow properties of spray-dried whole egg and spray-dried egg yolk can be improved by adding low percentages of a colloidal silicon dioxide preparation or sodium silicoaluminate. The improvement in flowability apparently results from coating the surfaces of the spray-dried particles by the finely divided, free-flowing agent (Forsythe *et al.* 1964). U.S. Standards of Identity for dried-egg products permit the use of less than 2% sodium silicoaluminate and 1% silicon dioxide.

Addition of emulsifiers to eggs apparently is somewhat beneficial for their functional properties (Spicer and Sly 1966; Liot and Liot 1975). Emulsifiers added to egg before drying prevent removal of the water of crystallization during drying and thus promote more rapid reconstitution of the dried product. Examples of such emulsifiers are glyceryl monostearate, glyceryl dioleate, polyoxyethylene, and sorbitan mono-oleate. The only problem with the use of emulsifiers in eggs is that when the product is used, say in baked goods, other ingredients also have emulsifiers and recipes often call for the addition of emulsifiers. Cakes can become overemulsified, with deterimental effects on volume, texture, and other cake characteristics.

The direct addition of color to egg products is not permitted by the U.S. Standards of Identity. Indirect addition of color through feed is apparently permitted by FDA only if the pigmentation material is natural; if synthetic materials are used, they come under the color additive law. At the present time, only a few producers are feeding chickens for improved color in eggs.

DRYING EQUIPMENT

Spray Drying

Spray drying is the most important method of producing dried-egg products. In spray drying, the liquid is finely atomized into a stream of hot air. Because of the enormous surface area created by atomization, evaporation of water is very rapid. Although the principle of this operation is simple, the complete system of component parts is rather complex.

A typical spray dryer for drying of egg products is shown in Fig. 14.3, and a schematic diagram of a spray drying system is shown in Fig. 14.4. The air used for drying is filtered by a suitable means to remove undesirable dust. It is then heated to a temperature of between 121° and

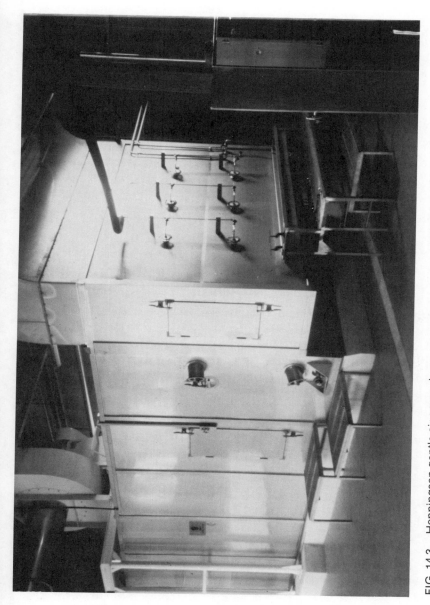

FIG. 14.3. Henningsen gentle air spray dryer.
Courtesy of Henningsen Foods, Inc.

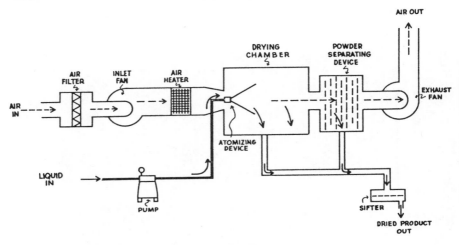

FIG. 14.4.   Schematic diagram of spray-drying system.

232°C and moved to the spray drying chamber by a fan. The liquid to be dried is pumped to various atomizing devices and sprayed into the hot-air stream. The water evaporates instantly. The powder so formed separates from the drying air in the drying chamber and also in a separating device which is part of, or separate from, the spray drying chamber. The air is then removed from the system by an exhaust fan. The dried product removed from the dryer is sometimes cooled and is usually sifted before packaging. The design of spray drying equipment varies considerably, and different types of spray dryers have been used for egg products.

The volume of air required to dry egg products depends on the overall efficiency of the spray drying system, the inlet air temperature and humidity, temperature of the liquid being fed, degree of atomization, moisture level desired in the dried product, as well as other factors. A rule of thumb is that 450 ft³ of air (at standard conditions, 21°C and atmospheric pressure) is needed to evaporate 1 lb of water when the inlet air is heated to 155°C. For example, a typical dryer with an evaporative capacity of 25 lb water per minute would need 11,000 ft³ of air per minute heated to 155°C.

Air Filtration.   Filtration of the air used for drying is an important consideration. The air is usually drawn from the outside through a duct which leads into the filter house. Filtration is necessary to remove dust and other comtaminants. The types of filters commonly used are listed below:

1. *Viscous filters.* These are coated with a viscous material to catch

and retain the dust. The disposable fiberglass filter is an example. Also, there are metallic filters which are coated with oil. These can be washed or steamed and then re-oiled. These filters usually have a low efficiency, and efficiency decreases with increased dust load. They are generally used for roughing filters to catch the larger particles before the air goes into a filter of higher efficiency.

2. *Dry filters.* These use various filter media and the passages for air flow are generally smaller than in the viscous-type filter. This type of filter can have very high filtering efficiency, as high as 99.97% using a dioctylphthalate test. High-efficiency filters are generally used after a roughing filter. Their efficiency usually increases with increased dust loads, but the pressure drop is greater than with the viscous-type filter and increases considerably as the dust load increases.

3. *Electrostatic filters.* These filters use a high voltage to create an electrostatic charge for removing particles from the air. They generally have high efficiency and are used in conjunction with other types of filters.

Heating of Air. The air for a spray dryer may be heated by an indirect heater, such as steam coils, or by a direct-fired furnace, using natural gas, propane, or other suitable fuel. Electrical heating is also possible, but this is usually restricted to small spray dryers where economics is not a consideration. In the past, there has been a question as to whether direct-fired furnaces are responsible for off-flavors and off-odors of whole egg and yolk products produced by certain spray dryers. It is possible that incomplete combustion could result in products such as formaldehyde, which could be absorbed by the powder and contribute to off-flavors. Also, nitrogen oxides can be generated with high flame temperatures and result in nitrites in the spray-dried products. Considerable progress has been made in direct-fired furnace design, however. Burners can now be designed to give 100% combustion and low levels of nitrogen oxides. With a direct-fired furnace, all products of combustion go through the drying system with the air. One of these products is water. A direct-fired furnace operated on natural gas will add 0.007 lb water per pound of drying air when the air is heated from 21° to 177°C. This additional water in the air requires that the drying temperatures, both inlet and outlet air temperatures, be somewhat higher than when drying with an indirect heater. The advantage of a direct-fired furnace over an indirect heater is higher thermal efficiency.

Indirect heating of the drying air has the advantage that no products of combustion, such as water, enter the air, although thermal efficiency is lower (60–90%) than with the direct-fired heating. Outlet temperatures required to achieve the same moisture level of the finished product are less due to the lower water content of the drying air. The most common type of indirect heat is the steam coil. Thermal efficiency of steam heaters can usually be improved by using devices such as an economizer on the boiler and a flash steam-recovery system on the steam coils. Indirect furnaces are also used in which heat is transferred from the combustion chamber to the drying air through a metal wall, and the products of combustion are vented to the outside. Improvements in design have resulted in thermal efficiencies as high as 90%.

With higher fuel costs, heat-recovery systems are now being installed on spray dryers. In these systems heat from the exhaust air is picked up and transferred to the inlet air. Ten to thirty percent of the heat required for drying can be recovered in this way. One of three different types of heat-recovery systems are used:

1. *Air-to-air heat exchangers.*
2. *Run-around system* where heat-transfer coils are in both the exhaust air and inlet air and are connected by piping through which a heat-transfer medium such as propylene glycol is circulated.
3. *Q-Dot system* which consists of a common coil, part of which is in the inlet air and part in the exhaust air. The coil contains a heat-transfer medium, such as one of the Freons, which vaporizes at the exhaust air temperatures and condenses at the inlet temperatures.

Atomization.   There are three basic types of atomization devices: pressure nozzles, centrifugal atomizers, and two-fluid nozzles. These are illustrated in Fig. 14.5.

Pressure nozzles are the most common type used for egg drying. These nozzles require pressures from 500 to 6,000 psig. High-pressure pumps of the same basic design as three-piston homogenizers are generally used to achieve these pressures. The pressure nozzle contains an inner chamber in which the liquid is given a swirl, and an orifice through which the liquid emerges as a hollow cone, which is unstable and breaks up into spherical particles.

The centrifugal atomizer consists of a rotating or spinning disk or device turning at a very high speed, from 3,500 to 50,000 rpm, depending on the diameter of the device.

Although two-fluid nozzles are generally limited to pilot-plant or laboratory spray dryers, some large commercial applications have been made (Bergquist 1963). The atomizing fluid is usually air, with pressures from 5 to 100 psig, depending on the type of atomizer and the capacity of the nozzle. The disadvantages of two-fluid nozzles are higher energy requirement and less uniform particle-size distribution. One advantage is that less severe physical forces are involved when atomizing a sensitive product such as egg white.

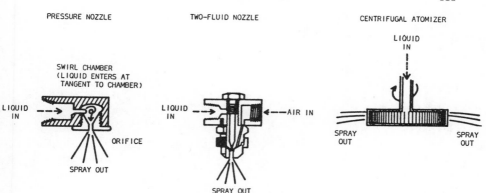

FIG. 14.5.   Atomization devices for spray drying.

Drying Chamber.   There are many different types of spray drying chambers. They may be classified as vertical or horizontal. Air flow may be cocurrent or countercurrent with the atomized product, or may have a complex mixing with the spray, as in the case of a cyclone-type dryer. A portion of the powder may be separated from the air in the drying chamber and devices such as screw conveyors, drags, vibrators, and air sweeps are used to remove the powder from the drying chamber. In some dryers, all the product leaves with the air and is collected in another part of the system. Most dryers require a secondary collection system to remove the particles entrained in the exhaust air. Examples of drying chambers used for egg products are shown in Fig. 14.6.

Secondary Collection.   Various types of secondary collectors are satisfactory for egg products. Well-designed cyclone collectors will recover almost 100% of the plain whole egg and yolk product. However, recovery is generally poor with products that contain carbohydrates because of their lower particle density. Egg-white powder is even more difficult to collect with cyclone collectors. Cloth collectors or bag filters are commonly used for egg products because of their high collection efficiency. The bag material may be cotton, wool, or a synthetic fiber, and the cloth area may vary from 1 to 0.1 ft^2 per cubic foot per minute of air flow. This depends on the product-entrainment load and type of collector. Bag collectors with a compressed-air-puff-back arrangement or with a blowring can generally carry a heavier load per square foot of cloth area than collectors where bags are mechanically shaken. Most operations require continuous function of the bag collectors. Bag-shaking collectors require that several chambers of bags be used and that air flow in a chamber be cut off when the bags are shaken; otherwise, the powder cannot be released from the bags. On the other hand, the compressed-air-puff-back

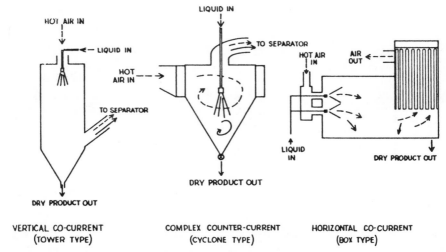

VERTICAL CO-CURRENT   COMPLEX COUNTER-CURRENT   HORIZONTAL CO-CURRENT
   (TOWER TYPE)   (CYCLONE TYPE)   (BOX TYPE)

FIG. 14.6.   Examples of spray-drying chambers used for egg products.

and the blow-ring types will operate in most cases without cutting off
the air flow.

The wet scrubber-type collector is not known to be used for egg
products. Although the wet scrubber can provide for a high recovery of
product and heat from the exhaust air, there are problems with sanitation and also with deterioration of the product.

Secondary Drying.   Some spray drying systems use a secondary
drying chamber to reduce moisture to very low levels. In these systems,
the product is picked up by a hot-air stream and then collected after a
certain holding period. Secondary dryers were common during the
1940s, when the Quartermaster Corps and the USDA had specifications
for whole egg requiring moisture levels below 2%. Most spray dryers
then were not able to produce the desired moisture level in one step.
Conrad *et al.* (1948) then developed a pneumatic drying system, which
suspended the powder in warm air and reduced its moisture content
from 3 or 4% to less than 2%.

With egg products, it is important to remove the powder from the air
stream as quickly as possible to minimize the effect of heat damage. It is
generally considered necessary to cool whole egg and yolk products to
temperatures below 30°C, as soon as the powder has been removed from
the drying system. Many different cooling systems have been devised.
The amount of cooling required depends on the temperature of the
powder coming out of the dryer. Basically three cooling systems can be
used:

1. *Intimate mixing of cooling air with the powder.* This is usually done by conveying the powder pneumatically with cooling air and then separating the powder from the air.

2. *Addition of a coolant, such as carbon dioxide.* Although cooling costs are relatively high, the temperature of the product can be closely controlled to almost any degree desired.

3. *Cooling by direct contact with a cold surface.* This can be used only for certain applications as condensation on the contact surface is sometimes a problem. Moisture in the warm powder will increase the humidity of the air in the cooling chamber, and if the cooling surface is at a temperature below the dew point of this air, condensation will result.

Pan Drying

Pan drying is still used for producing egg-white products. Flake-type egg-white products are made by this means. Drying on pans to a moisture level of 12 to 16% will produce a flake-type material. Products with dimensions of 1.5 to 12.5 mm are generally considered flakes. Material finer than this is sometimes called granular egg white. Pan-dried egg white can also be milled to a fine powder.

There are two types of dryers used for pan-drying egg white: air ovens and water-jacketed pans. Heat for evaporation of water is supplied by air in the case of the air oven dryer. In the water-jacket-type dryer, heat is supplied from warm water that flows beneath the pans. This has the advantage of uniform drying and good control of the temperature of the product in the pans. The latter is quite essential, since egg white is heat sensitive and coagulates rapidly if its temperature exceeds 54°C.

Freeze Drying

In freeze drying, water is removed from the product while it is in the frozen state. This is accomplished by freezing the product and then subjecting it to a very high vacuum. Heat must be supplied to the product while it is drying. For example, heated shelves are commonly used in the freeze-drying chamber. Although the principle of freeze drying is quite simple, the equipment needed to accomplish this operation is complex.

The cost of freeze drying is generally much higher than for other methods of drying. Thus, commercial applications of this method have been rather limited. Freeze drying has been used commercially in England for producing a dried whole-egg product containing emulsifiers primarily for bakery use (Spicer 1969). Freeze-dried precooked scrambled eggs have been produced and have found some application in the United States.

Other Drying Methods

Belt drying is a method that was used in China for making dried whole egg and dried yolk. The liquid was spread as a thin film on a continuous aluminum belt moving through a hot-air stream.

Another form of belt drying, "fluff" or "foam" drying, has been employed in the United States for commercial production of egg white solids. This method has also been tested with good results for whole egg and yolk products, but there has been no commercial application. In foam drying, the product is whipped into a stable foam, which is spread in a relatively thin layer on a continuously moving belt passing through heated air. In a modification of this method, foam is spread on a perforated belt; air blowing up through the belt creates craters which aid in the drying. The dried foam produced by these methods is broken up to the desired granulation. The particle characteristic of product produced in this way is usually quite desirable, being relatively easy to reconstitute in water.

Foam spray drying is a method being used for dairy products, but thus far, it has not found widespread use for egg products. It has been tested for egg products, however, and has been found to offer some advantages. Foam spray drying is accomplished by incorporating a gas, such as carbon dioxide, nitrogen, or air, into the liquid before it is atomized. The gas is mixed with the liquid ahead of the high-pressure pump, or is forced into the liquid after the pump (Kline *et al.* 1963A). Foam spray drying can be done on most conventional spray-drying equipment. Egg products produced by this method have a lower bulk density and a somewhat different particle characteristic than regular spray dried material. Whole egg and yolk products prepared by this method seem to dissolve faster.

Agglomeration of egg powders can be accomplished by the same methods that are used for producing instant skim milk. This is usually done by adding back moisture under controlled conditions and then redrying the particles that agglomerate. Agglomeration of egg white is aided by incorporating sugar, such as sucrose (Sugihara *et al.* 1962) or lactose (Peebles 1960). Egg white agglomerated by itself will break apart quite easily.

Concentration before Drying

Concentration of liquid egg products before drying is a means of improving thermal efficiency, increasing the capacity of a dryer, and changing product characteristics such as higher bulk density. Thus far, commercial applications have been mostly in Europe. For example, whole egg has been concentrated to about 40% solids and egg white to about 20% solids before spray drying by using vacuum-type evaporation equipment. However, there have been problems in concentrating products by this means. For example, there is severe damage to the functional properties of egg white even when using temperatures below those at which heat damage occurs. Apparently the physical action and shear forces within the concentrator have damaging effects.

Reverse osmosis and ultrafiltration have also been studied as means of concentrating liquid egg products (Lowe *et al.* 1969; Payne and Hill 1973; Tsai *et al.* 1977) and have been commercially applied to the concentration of egg white before drying. With these methods, energy requirements for water removal are relatively low, but there are some losses of low-molecular-weight substances such as salts. However, there is little or no loss of functional properties such as whipping ability. Sanitary improvements have been made in this equipment so that it can be cleaned satisfactorily using proper in-place cleaning procedures. Although a concentration of egg white to 24% solids is possible, concentration to 18–20% solids is more practical. The viscosity increases, and consequently flux drops off severely when going above 20% solids.

ECONOMIC CONSIDERATIONS

Conversion of Liquid Eggs to Solids

Figure 14.7 demonstrates the amount of water and solids in the liquid and dried forms of egg white, whole egg, and yolk. The moisture level of spray-dried egg white is less than 8%, and of pan-dried egg white, less than 16%. The moisture content of both plain dried whole egg and plain dried yolk is less than 5.0%. The solids content of the liquid egg products can vary somewhat; an average for egg white is 11.3%, for whole egg 24.5%, and for yolk 44.0%. Yolk solids can be adjusted in the breaking room and controlled by the amount of adhering whites staying with the yolk. By definition, yolk cannot have a solids content of less than 43.0%.

Approximately 8 lb of egg-white liquid are required to produce 1 lb of

FIG. 14.7.   Water, solids content, and comparable weights of liquid and dried egg products.

dried egg-white product. In the case of whole egg, it takes approximately 3.9 lb of liquid for 1 lb of solids and for yolk, 2.2 lb of liquid per pound of dried product.

Storage

Egg White. The shelf life of all dried glucose-free egg-white products is almost infinite. They are stable under any storage conditions with no deterioration anticipated with age. Dry storage is usually suitable.

Plain Whole Egg and Yolk. Plain whole unstabilized egg solids have a shelf life of only about 1 month at room temperature, but under refrigeration they can be held for about a year. Possible deterioration could be browning or a change in color with development of off-flavors. Also, solubility could decrease with age and thus the ability to function in various products would change. On the other hand, stabilized whole egg, with glucose removed, might be expected to have a shelf life up to 1 year at room temperature. Packing the product in an inert atmosphere, such as carbon dioxide or nitrogen, helps stability. Because glucose has been removed, the browning reaction, the principal deterioration occurring in unstabilized plain whole egg, would not occur. Oxidative changes may take place slowly, and there is evidence that oxidation of egg products is slower at higher moisture levels.

Plain unstabilized egg yolk solids have a shelf life of about 3 months at room temperature and can be stored over a year under refrigeration. The deterioration that takes place with age is change in color and flavor, due to the browning reaction and to oxidation involving the lipids. Actually, the browning reaction is slower than in whole egg because the glucose level is relatively low (0.4%); the pH is also lower. The reconstituted viscosity of egg yolk solids increases with age; that of freshly dried product is 1,000 to 3,000 cP. Storage at temperatures higher than 38°C will cause the viscosity to increase rapidly; for example it will increase to greater than 10,000 cP in 1 week at 43°C.

The removal of glucose from egg yolk improves the shelf life of the dried product to some degree. The shelf life of stabilized egg yolk is estimated to be about 8 months at room temperature and considerably more than a year under refrigeration. The browning reaction, of course, would not occur, but flavor changes due to oxidation involving the lipids could take place. Increase in viscosity takes place when the product is stored under elevated temperature, in the same way as with the egg yolk without glucose removed.

Egg with Carbohydrates. Egg products containing carbohydrates generally keep very well for 1 month at room temperature and over a year under refrigeration. Oxidative flavor changes develop before any others. Sometimes the product develops a "fishiness," or a "stale" or "burnt" taste. Surprisingly, the browning reaction does not occur very rapidly, even though these products contain relatively high percentages of glucose. This deterioration is apparently prevented by a separation of the two reaction groups in this type of product. Actually, these products are considered to be quite stable.

For commercial use, dried-egg products are generally packed in fiber drums with a polyethylene liner, or its equivalent. In this type of package, the rate of moisture transfer through the liner is relatively slow, and there is little moisture pickup when the product is stored.

Transportation

One advantage of dried-egg products over their liquid or frozen counterparts is low transportation costs. Only the solids with a small amount of moisture are shipped. Also, dried eggs do not require refrigeration when being shipped except under very special circumstances.

REFERENCES

Anon. 1969. Egg pasteurization manual. U.S., Agric. Res. Serv. ARS *ARS 74*-78, Mar.

Anon. 1975. Regulations Governing the Inspection of Eggs and Egg Products, 7CFR, Part 59, June 30. U.S. Dep. Agric., Washington, DC.

Ayres, J. C., and Slosberg, H. M. 1949. Destruction of *Salmonella* in egg albumen. Food Technol. *3*, 180–183.

Banwart, G. J., and Ayres, J. C. 1956. The effect of high temperature storage on the content of *Salmonella* and on functional properties of dried egg white. Food Technol. *10*, 68–73.

Bate-Smith, E. C., and Hawthorne, J. R. 1945. Dried egg. X. The nature of the reaction leading to loss of solubility of dried egg protein. J. Soc. Chem. Ind. *64*, 297–302.

Bergquist, D. H. 1961A. Method of producing egg albumen solids with low bacteria count. U.S. Pat. 2,982,663.

Bergquist, D. H. 1961B. Improved egg solids. Candy Ind. Confect. J. Apr. 11.

Bergquist, D. H. 1963. Method of preparing powdered egg albumen. U.S. Pat. 3,082,098, Mar. 19.

Bergquist, D. H. 1964. Eggs. *In* Food Dehydration. W. B. Van Arsdel and M. J. Copley (Editors), Vol. 2. AVI Publishing Co., Westport, CT.

Bergquist, D. H. 1979. Sanitary processing of egg products. J. Food Prot. *42*, 591–595.

Bergquist, D. H. 1980A. Measurable difference in quality of egg products. Proc. Eur. Poult. Conf. World Poult. Sci. Assoc., 6th, 1980 Vol. 1, pp. 232–243.

Bergquist, D. H. 1980B. New development in drying egg products. Proc. Eur. Symp. Qual. Egg Prod., 1st, 1980 (sponsored by Spelderholt Institute of Poultry Research, Beekbergen, Holland), pp. 7–14.

Bergquist, D. H., and Stewart, G. F. 1952A. Atomization as a factor affecting quality in spray dried albumen. Food. Technol. *6*, 201–203.

Bergquist, D. H., and Stewart, G. F. 1952B. Surface formation and shear as factors affecting the beating powder of egg white. Food Technol. *6*, 262–264.

Bergquist, D. H., and Wells, F. 1956. The monomolecular surface film method for determining small quantities of yolk or fat in egg albumen. Food Technol. *10*, 48–50.

Bergquist, D. H., *et al.* 1978. Instant dissolving egg white. U.S. Pat. 4,115,592, Sept. 19.

Brooks, J., and Hawthorne, J. R. 1943. Dried egg. IV. Addition of carbohydrates to egg pulp before drying. J. Soc. Chem. Ind. *62*, 165–167.

Brown, S. L., and Zabik, M. E. 1967. Effect of heat treatment on the physical and functional properties of liquid and spray dried egg albumen. Food Technol. *21*, 87–92.

Bull, H. B. 1938. Studies on surface denaturation of egg albumin. J. Biol. Chem. *123*, 17–30.

Chang, P. K. *et al.* 1970. Sodium hexametaphosphate effect on the foam performance of heat-treated and yolk-contaminated albumen. Food Technol. *24*, 63–67.

Chapin, R. B. 1951. Some factors affecting the emulsifying properties of hen's egg. Ph.D. Thesis, Iowa State Coll., Ames.

Clinger, C. *et al.* 1951. The influence of pasteurization, freezing, and storage on the functional properties of egg white. Food Technol. *5*, 166–170.

Conrad, R. M. *et al.* 1948. Improved dried whole egg products. Kans. State Coll., Tech. Bull. *64*.

Cotterill, O. J., and Funk, E. M. 1963. Effect of pH and lipase treatment on yolk-contaminated egg white. Food Technol. *17*, 1183–1188.

Cotterill, O. J. *et al.* 1965. Improving yolk-contaminated white by heat treatments. Poult. Sci. *44*, 228–235.

Cotterill, O. J. *et al.* 1976. Emulsifying properties of salted yolk after pasteurization and storage. Poult. Sci. *55*, 544–548.

Cotterill, O. J. *et al.* 1978. Nutrient composition of commercially spray-dried egg products. Poult. Sci. *57*, 439–442.

Cunningham, F. E., and Lineweaver, H. 1965. Stabilization of egg white proteins to pasteurizing temperatures above 60°C. Food Technol. *19*, 1442–1447.

Cunningham, F. E., and Varadarajulu, P. 1973. Electrophoretic study of fresh, heated, spray dried, and sonicated egg yolk. Poult. Sci. *52*, 365–369.

Davis, J. G., and Kline, L. 1965. Vapor pressure of water and carbon dioxide over whole egg solids containing added carbohydrates. J. Food Sci. *30*, 673–679.

Donovan, J. W. *et al.* 1975. A differential scanning calorimetric study of the stability of egg white to heat denaturation. J. Sci. Food Agric. *26*, 73–83.

Everson, G. J., and Souder, H. J. 1957. Composition and nutritive importance of eggs. J. Am. Diet. Assoc. *33*, 1244–1254.

Forsythe, R. H., and Bergquist, D. H. 1951. The effect of physical treatment on some properties of egg white. Poult. Sci. *30*, 302–311.

Forsythe, R. H., and Bergquist, D. H. 1973. Functionality of eggs in the modern bakery. Baker's Dig. *47*, 84–87, 90, 131.

Forsythe, R. H. *et al.* 1964. Improvement in flow properties of egg yolk solids. Food Technol. *18*, 153–157.

Froning, G. W. *et al.* 1981. Factors affecting puffing and sensory characteristics of extruded egg products. Poult. Sci. *60*, 2091–2097.

Galyean, R. D., and Cotterill, O. J. 1979. Chromatography and electrophoresis of native and spray-dried egg white. J. Food Sci. *44*, 1345–1349.

Gane, R. 1943. Dried egg. VI. The water relations of dried egg. J. Soc. Chem. Ind. *62*, 185–187.

Garibaldi, J. A. *et al.* 1968. Heat denaturation of ovomucin-lysozyme electrostatic complex—a source of damage to whipping properties of pasteurized egg white. J. Food Sci. *33*, 514-524.

Grim, A. C. and Goldblith, S. A. 1965. The effect of ionizing radiation on flavor of whole-egg magma. Food Technol. *19*, 1594-1596.

Hill, W. M. *et al.* 1965. Spray drying egg white at various pH levels. Poult. Sci. *44*, 1155-1163.

Ijichi, K. *et al.* 1964. Effects of ultraviolet irradiation of egg liquid on *Salmonella* destruction and performance quality with emphasis on egg white. Food Technol. *18*, 1628-1632.

Imai, C. 1976. Effects of bacteria fermentation and lipase treatment on the whipping properties of spray-dried egg white. Poult. Sci. *55*, 2409-2414.

Joslin, R. P., and Proctor, B. E. 1954. Some factors affecting the whipping characteristics of dried whole egg powders. Food Technol. *8*, 150-154.

Kandatsu, M., and Yamaguchi, M. 1973. Process for the manufacture of granular or powdery purified whole egg protein. U.S. Pat. 3,778,425, Dec. 11.

Kim, K., and Setser, C. S. 1982. Foaming properties of fresh and of commercially dried eggs in the presence of stabilizers and surfactants. Poult. Sci. *61*, 2194-2199.

Kline, L., and Singleton, A. D. 1959. Egg whites. U.S. Pat. 2,881,077, Apr. 7.

Kline, L., and Sonoda, T. T. 1951. Role of glucose in the storage deterioration of whole egg powder. I. Removal of glucose from whole egg by yeast fermentation before drying. Food Technol. *5*, 90-94.

Kline, L., and Sugihara, T. 1964. Drying of yolk-containing egg liquid. U.S. Pat. 3,162,540, Dec. 22.

Kline, L. *et al.* 1951A. Role of glucose in the storage deterioration of whole egg powder. II. A browning reaction involving glucose and cephalin in dried whole egg. Food Technol. *5*, 181-187.

Kline, L. *et al.* 1951B. Role of glucose in the storage deterioration of whole egg. III. Effect of glucose removal before drying on organoleptic, baking, and chemical changes. Food Technol. *5*, 323-331.

Kline, L. *et al.* 1954. Comparison of the quality and stability of whole egg powders desugared by yeast and enzyme methods. Food Technol. *8*, 343-349.

Kline, L. *et al.* 1963A. Process of spray-drying eggs. U.S. Pat. 3,115,413, Dec. 24.

Kline, L. *et al.* 1963B. Some characteristics of yolk solids affecting their performance in cake doughnuts. II. Variability in commercial yolk solids. Cereal Chem. *40*(1), 38-50.

Kline, L. *et al.* 1964. Properties of yolk-containing solids with added carbohydrates. J. Food Sci. *29*, 693-709.

Kline, L. *et al.* 1965A. Preparation of dried egg white. U.S. Pat. 3,170,804, Feb. 23.

Kline, L. *et al.* 1965B. Heat pasteurization of raw liquid egg white. Food Technol. *19*, 1709-1718.

Kline, L. *et al.* 1966. Further studies on heat pasteurization of raw liquid egg white. Food Technol. *20*, 1604-1606.

Kline, R. W., and Stewart. G. F. 1948. Glucose-protein reaction in dried egg albumen. Ind. Eng. Chem. *40*, 919-922.

Klose, AA. *et al.* 1943. Vitamin content of spray-dried whole egg. Ind. Eng. Chem. *35*, 1203-2105.

Kohl, W. F. *et al.* 1970. Process for the pasteurization of egg white. U.S. Pat. 3,520,700, July 14.

Kothe, H. J. 1953. Egg white composition. U.S. Pat. 2,637,654, May 5.

Koudele, J. W., and Heinsohn, E. C. 1960. The egg products industry of the United States. Part I. Historical highlights, 1900-59. Kansas State Univ. Agric. Exp. Stn., Bull. *423* (North Cent. Reg. Publ. *108*).

Larsen, J. E., and Froning, G. W. 1981. Extraction and processing of various components from egg yolk. Poult. Sci. *60*, 160-167.

Lea, C. H. 1957. Deteriorative reactions involving phospholipids and lipoproteins. J. Sci. Food Agric. *8*, 1-13.

Levin, E. 1975. Reconstituted egg product and method of preparing. U.S. Pat. 3,881,034, Apr. 29.

Lewis, M. A. *et al.* 1953. Stable foams from food proteins with polyphosphates. Food Technol. *7*, 261-264.

Lieu, E. H. *et al.* 1978. Effect of storage on lipid composition and functional properties of dried egg products. Poult. Sci. *57*, 912-923.

Lightbody, H. D., and Fevold, H. L. 1948. Biochemical factors influencing the shelf life of dried whole egg and means for their control. Adv. Food Res. *1*, 149-202.

Liot, R. and Liot, E. 1975. Method of making improved dried egg product. U.S. Pat. 3,930,054, Dec. 30.

Lowe, E. *et al.* 1969. Egg white concentrated by reverse osmosis. Food Technol. *23*, 753-762.

MacDonnell, L. R. *et al.* 1950. Shear—not pressure—harms egg white. Food Ind. *22*, 273-276.

Makower, B. 1945. Vapor pressure of water adsorbed on dehydrated eggs. Ind. Eng. Chem. *37*, 1018-1022.

Maturi, V. F. *et al.* 1960. Egg white composition. U.S. Pat. 2,933,397, Apr. 19.

Melnick, D. 1967. Low cholesterol dried egg yolk and process. U.S. Pat. 3,563,765, Aug. 31.

Mink, L. D. 1939. Egg material treatment. U.S. Pat. 2,183,516, Dec. 12.

Miyahara, T., and Bergquist, D. H. 1961A. Modern egg solids for the baker. I. Use of egg white solids. Baker's Dig. *35*, (1) 52-55.

Miyahara, T., and Bergquist, D. H. 1961B. Modern egg solids for the baker. II. Use of plain whole egg and yolk solids. Baker's Dig. *35*, (2), 71-73, 97.

Miyahara, T., and Bergquist, D. H. 1961C. Modern egg solids for the baker. III. Use of blends of egg and carbohydrates. Baker's Dig. *35* (3), 66-69, 86.

Parkinson, T. L. 1975. Effects of spray drying and freezing on the proteins of liquid whole egg. J. Sci. Food Agric. *25*, 1625-1637.

Parkinson, T. L. 1977. The effect of storage on the proteins of frozen liquid egg and spray dried egg. J. Sci. Food Agric. *28*, 811-821.

Passy, N., and Mannheim, C. H. 1982. Flow properties and water sorption of food powders. II. Egg powders. Lebensm.-Wiss. Technol. *15*, 222-225.

Payawal, S. R. *et al.* 1946. Pasteurization of liquid-egg products. II. Effect of heat treatment on appearance and viscosity. Food Res. *11*, 246-260.

Payne, R. E., and Hill, C. G., Jr. 1973. Concentration of egg white by ultrafiltration. J. Milk Food Technol. *36*, 359-363.

Peebles, D. D. 1960. Dried egg product and process of manufacture. U.S. Pat. 2,950,204, Aug. 23.

Powrie, W. D. *et al.* 1963. Gelation of egg yolk. J. Food Sci. *28*, 38-46.

Privett, O. S. *et al.* 1964. Ultraviolet absorbancy of volatiles as a measure of oxidative flavor deterioration in egg powder. Food Technol. *18*, 1485-1488.

Rolfes, T. *et al.* 1955. The physical and functional properties of lyophilized whole egg, yolk, and white. Food Technol. *9*, 569-572.

Schultz, J. R., and Forsythe, R. H. 1967. The influence of egg yolk lipoprotein-carbohydrate interaction on baking performance. Baker's Dig. *41* (1), 56-57, 60-62.

Slosberg, H. M. 1946. Some factors affecting the functional properties of liquid egg albumen. Ph.D. Thesis, Iowa State Coll., Ames.

Slosberg, H. M. *et al.* 1948. Factors influencing the effects of heat treatment on the leavening power of egg white. Poult. Sci. *27*, 294-301.

Smith, C. F.   1964.   Treatment of egg albumen, U.S. Pat. 3,161, 527, Dec. 15.

Spicer, A.   1969.   Freeze drying of foods in Europe. Food Technol. *23*, 1272–1274.

Spicer, A., and Sly, W.   1966.   Preparation of dried egg. U.S. Pat. 3,268,346, Aug. 23.

Stewart, G. F., and Kline, R. W.   1941.   Dried egg albumen. I. Solubility and color denaturation. Proc. Inst. Food Technol. *2*, 48–56.

Stewart, G. F., and Kline, R. W.   1948.   Factors influencing rate of deterioration in dried egg albumen. Ind. Eng. Chem. *40*, 916–919.

Sugihara, T. F. *et al.*   1962.   Application of instantizing methods to egg solids. Inst. Food Technol., 22nd Ann. Meet., Miami Beach, FL.

Sugihara, T. F. *et al.*   1966.   Heat pasteurization of liquid whole egg. Food Technol. *20*, 1076–1083.

Toney, J. W., and Bergquist, D. H.   1983.   Functional egg products for the cereal foods industries. Cereal Foods World *28*, 445–447.

Tsai, L. S. *et al.*   1977.   Concentration of egg white by ultrafiltration. J. Food Prot. *40*, 449–455.

Woodward, S. A., and Cotterill, O. J.   1983.   Electrophoresis and chromatography of heat-treated plain, sugared and salted whole egg. J. Food Sci. *48*, 501–506.

Quality Control and Product Specifications

James M. Gorman*
Hershell R. Ball, Jr.

DEFINITION AND PURPOSE

The purpose of the quality control (QC) division of an egg-products processing operation is to guide and monitor the production of safe and salable egg products. By serving this purpose well, it meets its responsibility to management and to users. If egg products are to be safe and wholesome, they must be prepared from shell eggs of acceptable edible quality and processed, packaged, transported, and stored under sanitary conditions. If the finished products are to be salable, they must meet the bacteriological, chemical, functional, and physical requirements of the customer.

ORGANIZATION

In the organizational structure of an egg-products company the QC department should share equal rank with other major organizational departments or divisions. This means primary responsibility directly to top management.

---

* Deceased.

EGG SCIENCE & TECHNOLOGY,
3rd Edition

Relationship to Other Divisions of Company

The production of safe, wholesome, and salable egg products to meet today's standards requires a team-effort approach. Good communication and working relationships with other divisions of the organization are therefore mandatory rather than optional.

The relationship of the QC department with the production division should include the joint development of processing and sanitation instructions for all items processed. The more critical instructions should be checked during daily inspections to maximize compliance and to assure the production of acceptable products. QC's greatest support for success comes from capable and knowledgeable production employees.

The relationship of the QC department with the sales division should emphasize the development of product specifications that will meet the approval of major customers. In developing meaningful specifications, it is highly important to maintain good communications with the QC departments of important customers.

Personnel

In meeting its responsibility to management, QC must be provided with laboratory facilities that are equipped and staffed to perform the examinations deemed important to the operation. Additionally, it should oversee product sampling and the reporting of findings that denote acceptance or rejection of production lots. The reports should be channeled to all company divisions involved in the production and movement of products to customers.

The number of employees needed for the QC function will depend largely on the productive capacity of the operation and the number of products processed. A small, single-plant operation might need only one individual having sound practical knowledge in bacteriology and chemistry. On the other hand, large operations processing a variety of products and composed of multiplant operations are likely to require a staff comprised of many chemists and microbiologists and supporting personnel. When large QC staffs are necessary, they are frequently structured to operate under the guidance of a QC, or technical, director.

Whatever the size of the staff, however, the individual charged with directing the QC function should possess a broad technical understanding of egg products and have appreciable skill in the processes required for their manufacture.

LABORATORY DESIGN AND EQUIPMENT

A suggested laboratory floor plan is shown in Fig. 15.1. The available floor space and the arrangement of QC laboratories will vary according to the needs of the processing plant and design of the building housing the laboratory. The arrangement of the laboratory in Fig. 15.1 suggests how multiple functions can be carried out in a relatively small space. Notice that the equipment is grouped according to the tasks that will be performed in the laboratory, i.e., chemical/physical, functional, and microbiological with open work space. Also note that the design calls for a common wall bearing most of the water and major electrical service. For safety considerations the laboratory should have two entrances, a fume hood, a solvent-storage cabinet, and other required safety equip-

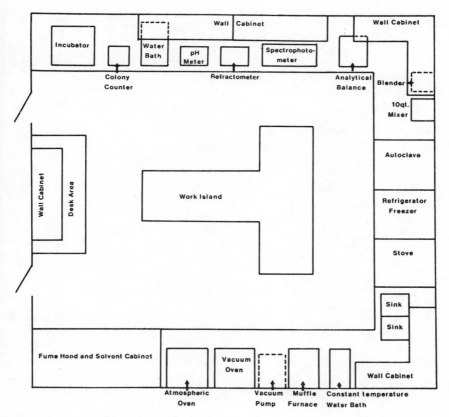

FIG. 15.1.   Laboratory floor plan.

ment such as an eye-wash station, fire blanket, fire extinguisher, and simple first-aid kit.

The following list describes some of the major and minor equipment and supplies that would be required in a QC laboratory:

Main Equipment

For Chemical, Functional, and Physical Testing

Analytical balance
Balance table
Atmospheric oven
Vacuum oven
Vacuum pump
Desiccators
Refractometer
Constant-temperature water bath with circulating pump
Spectrophotometer
Hobart Mixer, 10 qt
Home-type mixer
Range for baking

pH meter and electrodes with pH 5, 7, and 9 buffers
Planimeter
Muffle furnace
Quartz crucibles
Viscosimeter
Portable UV lamp
Microscope
Slides
Glassware (graduated cylinders, volumetric flasks, funnels, and beakers)
Thermometers, various types

For Microbiological Testing

Refrigerator with freezer compartment
Top-loading electronic scale
Dilution bottles
Autoclave
Incubators
Bunsen burners
Quebec colony counter
Timer
Glassware (petri dishes, 2.2- and 10-ml pipettes, disposable or glass)

Test tubes with racks, baskets, and brushes
Forceps
Dissecting scissors
Jars, 4 oz
Pipette cans, stainless steel
Blender with jars and lids
Chromel inoculating needles
Inoculating needle holder
Hand tally counter

Safety Equipment

Fume hood
Eye-wash station
First-aid kit
Fire extinguisher

Solvent-storage cabinet
Fire blanket
Chemical spill kit

CHEMICAL AND MICROBIOLOGICAL METHODS OF ANALYSIS

The following brief annotated bibliography of methods presents the most commonly used procedures. Earlier editions of this book cited methods published by the Institute of American Poultry Industries (IAPI) and by the Poultry and Egg Institute of America (PEIA). Unfortunately, those publications are no longer in print and are difficult to obtain. Since those publications are not readily available and since many of the methods they presented are outdated or now published in other manuals, references to IAPI and PEIA procedures have been omitted. "Laboratory Methods for Egg Products" (USDA) 1984 may be especially useful because of the extensive explanations of the procedures being described. [See References Added in Proof, No. 1 (p. 343).]

Chemical Methods of Analysis

Method	Basis of Method	References
Total Solids or Moisture		
1. Vacuum oven (official)	Moisture removal at 100°C and 28–29 in. vacuum for 5 hr [b–d]	AOAC[a] (1985); USDA (1984);
2. Atmospheric oven (alternate)	Moisture removal from liquid and defrosted frozen samples overnight at 105°–107°C from dried samples for 1.75 hr	USDA (1984)
3. Microwave oven (alternate)	Moisture removal by microwave	Cotterill and Delaney (1959)
4. Refractometer (Abbé) (alternate)	Solids of liquid eggs by index or sugar-scale correlation. Electrolytes facilitate whole egg and yolk readings. Prism temperature control important	Cahn and Epstein (1936); Triebold (1946)
5. Moisture balances (alternate)	Moisture removal by infra-red	Kramer and Twigg (1962)

[a] Association of Official Analytical Chemists.
[b] Heat set liquid and defrosted samples to prevent "boil over" during vacuum drying. Place in steam bath or in atmospheric oven for 1 hr.
[c] Blends containing carbohydrates should be dried in vacuum oven.
[d] Pulverize flake or granular albumen.

Color

AOAC method	Yolk pigments extracted with acetone. Correlated to β-carotene standards using spectrophotometer. Reported as μg β-carotene per gram of sample	AOAC (1985); USDA (1984); also see Fletcher (1980)

Salt

AOAC method	Titration of excess silver nitrate with standard thiocyanate using ferric ammonium sulfate indicator	AOAC (1985); USDA (1984)

Method	Basis of Method	References
Sugar AOAC-Munson-Walker general method	Gravimetric method. Involves inversion of dextrose and reduction of copper sulfate	AOAC (1985); USDA (1984)
Percentage of Yolk in Egg Whites Monomolecular film method	Monolayer films utilized for detection of trace levels of yolk in egg whites	Bergquist and Wells (1956); Colburn *et al.* (1964)
Total Fat AOAC-acid hydrolysis	Ethyl and petroleum ether extraction following acid hydrolysis Also see	AOAC (1985); USDA (1984) Fletcher *et al.* (1984)
Protein AOAC-Kjeldahl	Total nitrogen $\times$ 6.25 for protein	AOAC (1985)
Ash AOAC	Sample is dried and then ashed in muffle furnace at about 525°C	AOAC (1985)
Pesticide Residues 1. Chlorinated hydrocarbons	Sample cleanup followed by electron-capture gas chromatography	Stemp *et al.* (1964); AOAC (1985)
2. Organic phosphorous compounds	Sample cleanup and procedures need modifications for adaption to egg products	AOAC (1985)
pH 1. Dried eggs	Reconstitution of specified amounts of dried egg with distilled water and pH determination using a pH meter	AOAC (1985); USDA (1984)
2. Liquid eggs	Direct determination with pH meter	

Microbiological Methods of Analysis

Total Count Aerobic plate count	Viable count determined using appropriate dilutions and making poured plates with Standard Methods agar. Reported as standard plate count per milliliter or gram after 48-hr incubation at 32°C	Speck (1984); USDA (1984)
Coliforms Agar plate method	Appropriate dilutions plated using desoxychocholate or violet red bile agar	Am. Public Health Assoc. (APHA) (1978)

Method	Basis of Method	References
Salmoneliae		
Salmonella detection and identification	Following lactose pre-enrichment selenite and tetrathionate enrichments streaked on differential media. Suspicious colonies picked to triple sugar iron and lysine iron agar slants. *Salmonella*-indicative cultures confirmed serologically and biochemically	APHA (1978); USDA (1984)
Yeasts and Molds		
Agar plate method	Appropriate dilutions plated with potato dextrose agar acidified to pH 3.5 with 10% sterile tartaric acid solution. Incubated at 23°C for 5 days. Reported as yeasts and/or molds per milliliter or gram	APHA (1978); USDA (1984)
Escherichia coli		
1. Differential tests	I.M.V.C. Tests (indole, methyl red. Voges-Proskauer, and sodium citrate)	APHA (1975)
2. Fermentation tubes	Lactose broth fermentation tubes used. Incubation 48 hr at 35°C Tubes showing gas streaked on Levine's EMB agar. Incubation 24 hr at 35°C	Speck (1984)
Coagulase-Positive *Staphylococcus aureus*	Baird-Parker agar used to isolate suspect colonies and coagulase test for confirmation	AOAC (1985); USDA (1984)

FUNCTIONAL PERFORMANCE EVALUATION

Whip Tests

Liquid Whites.   Measure 450 ml of liquid egg whites into 10-qt Hobart electric mixer bowl. Use wire whip, and whip on speed no. 2 for 90 sec, then on speed no. 3 for 90 sec. Remove wire whip, level foam, and measure foam depth with 12-in. rule. Foam depth should measure 5.5 to 6.5 in. (14.0–16.3 cm.)

Dried Albumen.   Dissolve 43 g of pan- or tray-dried albumen in 430 ml of water at 21°C. Large crystals may require overnight soaking. Transfer rehydrated product into 10-qt Hobart electric mixer bowl; whip and measure foam depth as above. Whipping performance of spray-dried albumen may be evaluated in the same manner. Good rehydration of spray-dried albumen can usually be achieved by mixing the albumen and water in a Hobart electric mixer bowl on speed no. 1 (low) for 5 min prior to whipping on speeds no. 2 and 3.

Cake Performance

Angel Cake Preparation

Angel cake albumen	43.00 g	5.40%
Monocalcium phosphate (V-90)	4.60	0.60
Salt	3.80	0.50
Granulated sugar	96.50	12.60
Water at 21°C	288.00 ml	37.60

Pour water into 5-qt Hobart electric mixer bowl. Add dry-blended albumen, monocalcium phosphate, salt, and sugar. Use wire whip, and whip on speed no. 1 about 5 min for reconstitution. Whip on speed no. 2 about 9 min to desired peak or foam gravity, 0.15 to 0.17 gm/ml

Fold in blend of:

Powdered sugar	210.00g	27.72%
Cake flour	93.60	12.20
Wheat starch	25.00	3.25
Bicarbonate of soda	1.00	0.13
	765.50 g	100.00%

Scale 709 g (25 oz) of batter into a 10-inch aluminum tube pan and bake at 190.5°C about 33 min. Invert for cooling. Cake height should measure 110 to 120 mm. For evaluation of pH 8.5 to 9.0 liquid whites, replace albumen and water with 329.3 g of liquid whites and increase monocalcium phosphate level to 9.2 g.

Sponge Cake Preparation

For liquid whole eggs and whole egg and egg yolk products co-dried with carbohydrates.

Sugar	284.0 g	33.90%
Dried egg-CHO product	95.0	11.30
Salt	3.8	0.45
NF milk solids	10.6	1.30

Blend dry ingredients by sifting into 5-qt Hobart mixer bowl.

Add water at 54°	274.0 g	32.75%

Whip on speed no. 2 for 11 min. A teaspoon of vanilla may be added during whipping for flavor.

Fold in cake flour	170.0 g	20.30%
	837.4 g	100.00%

Scale 340 g (12 oz) of batter into round 8- × 2-in. pans and bake at 191°C for about 22 min. Invert for cooling. Cake height should measure 4.5 –5.0 cm.

Salad Dressing Preparation (for Yolk Solids)

Yolk solids	21.0 g	2.3%
Vegetable oil	340.0	37.5
Starch paste	546.0	60.2
	907.0 g	100.0%

Starch Paste

Sugar	90.0 g	15.0%
Cornstarch (salad-dressing type)	46.2	7.7
Salt	24.0	4.0
Dry mustard	6.6	1.1
Water	330.0	55.0
Vinegar (5% HOAC)	103.2	17.2
	600.0 g	100.00%

Cooks down in prescribed time to quantity needed for 907 g of salad dressing.

1. Make starch paste by transferring blended dry ingredients to 2-qt stainless steel pan. Add vinegar and water. Place over medium high heat and bring to boiling point with constant stirring with wire whip. Boil for 1 min. Mixture should become clear and shiny.

2. Cool one-third of starch paste (182 g) to 49°C in 5-qt Hobart electric mixer bowl. With timer set at 2 min, add yolk solids rapidly while agitating with dough paddle on speed no. 1. Yolk solids should go into solution readily. If specks of yolk are visible at the end of 1 min of mixing, stop mixer and scrape down bowl and paddle. Continue mixing on speed no. 1 for the remaining minute. Examine for specks of yolk at the end of the 2-min mixing period. If specks of yolk are still visible, continue mixing for 1 but not more than 2 additional min.

3. Add remaining starch paste. Set timer for 2 min. Start paddle agitator on speed no. 2. Add oil during first minute of mixing. At completion of 2 min of mixing, remove paddle and examine finished emulsion for specks of yolk and consistency. Proper consistency is smooth and spoonable. It should not be pourable or stringy.

4. Store product in glass jars in refrigerator. Observe for separation.

Solubility of Dried Products

Method	Basis of Method	References
1. Solubility of dried whole eggs	Product dissolved in 5% sodium chloride solution. Refractive index determined and solubility reported as Haenni value below: $(N_D^{25}$ of sample solution $-N_D^{25}$ of solvent) $\times 1000$	Hawthorne (1944)
2. Solubility of dried whole eggs	Aqueous solution filtered. Filtrate treated with $0.1/M$ sodium acetate buffer solution at pH 4.6 in 15-ml centrifuge tube. Protein precipitated in boiling water bath. Centrifuged precipitate read to nearest 0.1 ml and reported as Stuart Solubility Index.	Stuart et al. (1942)
3. Percentage insolubles in dried albumen	Aqueous solution centrifuged. Insolubles transferred to tared aluminum moisture dish for drying. Residue calculated as percentage of insolubles.	

SPECIFICATIONS AND STANDARDS

Specifications and standards are published by governmental regulatory agencies and by egg-products producers and customers. They are the basis for the consistent manufacturing and sales of egg products. *Standards* for eggs and egg products published in the *Federal Register*, Vol. 50, pp. 14462–14464, March 15, 1977. The F.D.C. regulations Part 160 define egg products, aspects of their composition, how they may be prepared, and how they should be labeled. They are the legal definitions of eggs and egg products. *Specifications* differ from standards in several important aspects. They are not legal definitions except as they apply to specific contractual agreements between a saler and a buyer. Specifications are generally more specific than the legal standards for egg products and are used to more precisely define the product being sold. Typical specifications for various egg products are given below, as well as Part 160 of the F.D.A. Regulations.

Eggs and Egg Products Definitions and Standards (Title 21 Code of Federal Regulations. U.S. Department of Health, Education and Welfare Food and Drug Administration F.D.C. Regs., Part 160, March 1977).

Eggs and Egg Products

§ 160.100 Eggs.

(a) Dried eggs, dried whole eggs are prepared by drying liquid eggs that conform to § 160.115, with such precautions that the finished food is free of viable *Salmonella* microorganisms. They may be powdered. Before drying, the glucose content of the liquid eggs may be reduced by one of the optional procedures set forth in paragraph (b) of this section. Either silicon dioxide complying with the provisions of § 172.480 of this chapter or sodium silicoaluminate may be added as an optional anticaking ingredient, but the amount of silicon dioxide used is not more than 1 percent and the amount of sodium silicoaluminate used is less than 2 percent by weight of the finished food. The finished food shall contain not less than 95 percent by weight total egg solids.

(b) The optional glucose-removing procedures are:

(1) *Enzyme procedure.* A glucose-oxidase-catalase preparation and hydrogen peroxide solution are added to the liquid eggs. The quantity used and the time of reaction are sufficient to substantially reduce the glucose content of the liquid eggs. The glucose-oxidase-catalase preparation used is one that is generally recognized as safe within the meaning of section 201(s) of the Federal Food, Drug, and Cosmetic Act. The hydrogen peroxide solution used shall comply with the specifications of the United States Pharmacopeia, except that it may exceed the concentration specified therein and it does not contain a preservative.

(2) *Yeast procedure.* The pH of the liquid eggs is adjusted to the range of 6.0 to 7.0, if necessary, by the addition of dilute, chemically pure hydrochloric acid, and controlled fermentation is maintained by adding food-grade baker's yeast (*Saccharomyces cerevisiae*). The quantity of yeast used and the time of reaction are sufficient to substantially reduce the glucose content of the liquid eggs.

(c) The name of the food for which a definition and standard of identity is prescribed by this section is "Dried eggs" or "Dried whole eggs" and if the glucose content was reduced, as provided in paragraph (b) of this section, the name shall be followed immediately by the statement "Glucose removed for stability" or "Stabilized, glucose removed."

(d) (1) When either of the optional anticaking ingredients specified in paragraph (a) of this section is used, the label shall bear the statement "Not more than 1 percent silicon dioxide added as an anticaking agent" or "Less than 2 percent sodium silicoaluminate added as an anticaking agent," whichever is applicable.

(2) The name of any optional ingredient used, as provided in paragraph (d) (1) of this section, shall be listed on the principal display panel or panels of the label with such prominence and conspicuousness as to render such statement likely to be read and understood by the ordinary individual under customary conditions of purchase.

§ 160.110 Frozen eggs.

(a) Frozen eggs, frozen whole eggs, frozen mixed eggs is the food prepared by freezing liquid eggs that conform to § 160.115, with such precautions that the finished food is free of viable *Salmonella* microorganisms.

(b) Monosodium phosphate or monopotassium phosphate may be added either directly or in a water carrier, but the amount added does not exceed 0.5 percent of the weight of the frozen eggs. If a water carrier is used, it shall contain not less than 50 percent by weight of such monosodium phosphate or monopotassium phosphate.

(c) When one of the original ingredients specified in paragraph (b) of this section is used, the label shall bear the statement "Monosodium phosphate (or monopotassium phosphate) added to preserve color", or in case the optional ingredient used is added in a water carrier, the statement shall be "Monosodium phosphate (or monopotassium phosphate), with — — — percent water as a carrier, added to preserve color", the blank being filled in to show the percent by weight of water used in proportion to the weight of the finished food. The statement declaring the optional ingredient used shall appear on the principal display panel or panels with such prominence and conspicuousness as to render it likely to be read and understood under customary conditions of purchase.

§ 160.115 Liquid eggs.

Liquid eggs, mixed eggs, liquid whole eggs, mixed whole eggs are eggs of the domestic hen broken from the shells and with yolks and whites in their natural proportion as so broken. They may be mixed, or mixed and strained, and they are pasteurized or otherwise treated to destroy all viable *Salmonella* microorganisms. Pasteurization or such other treatment is deemed to permit the adding of safe and suitable substances (other than chemical preservatives) that are essential to the method of pasteurization or other treatment used. For the purposes of this paragraph, safe and suitable substances are those that perform a useful function in the pasteurization or other treatment to render the liquid eggs free of viable *Salmonella* microorganisms, and that are not food additives as defined in section 201(s) of the Federal Food, Drug, and Cosmetic Act; or, if they are food additives, they are used in conformity with regulations established pursuant to section 409 of the act.

§ 160.140 Egg whites.

(a) Egg whites, liquid egg whites, liquid egg albumen is the food obtained from eggs of the domestic hen, broken from the shells and separated from yolks. The food may be mixed, or mixed and strained, and is pasteurized or otherwise treated to destroy all viable *Salmonella* microorganisms. Pasteurization or such other treatment is deemed to permit the adding of safe and suitable substances (other than chemical preservatives) that are essential to the method of pasteurization or other treatment used. Safe and suitable substances that aid in protecting or restoring the whipping properties of liquid egg whites may be added. For the purposes of this paragraph, safe and suitable substances are those that perform a useful function as whipping aids or in the pasteurization or other treatment to render liquid egg whites free of viable *Salmonella* microorganisms and that are not food additives as defined in section 201(s) of the Federal Food, Drug, and Cosmetic Act; or, if they are food additives, they are used in conformity with regulations established pursuant to section 409 of the act.

(b) Any optional ingredients used as whipping aids, as provided for in paragraph (a) of this section, shall be named on the principal display panel or panels of labels with such prominence and conspicuousness as to render such names likely to be read and understood by ordinary individuals under customary conditions of purchase.

§ 160.145 Dried egg whites.

(a) The food dried egg whites, egg white solids, dried egg albumen, egg albumen solids is prepared by drying liquid egg whites conforming to the requirements of § 160.140 (or deviating from that section only by not being *Salmonella* free). As a preliminary step to drying, the glucose content of the liquid egg whites is reduced by adjusting the pH where necessary, with food-grade acid and by following one of the optional procedures set forth in paragraph (b) of this section. If the food is prepared from liquid egg whites conforming in all respects to the requirements of § 160.140, drying shall be done with such precautions that the finished food is free of viable *Salmonella* microorganisms. If the food is prepared from liquid egg whites that are not *Salmonella* free, the dried product shall be so treated by heat or otherwise as to render the finished food free of viable *Salmonella* microorganisms. Dried egg whites may be powdered.

(b) The optional glucose-removing procedures are:

(1) *Enzyme procedure.* A glucose-oxidase-catalase preparation and hydrogen peroxide solution are added to liquid egg whites. The quantity used and the time of reaction are sufficient to substantially reduce the glucose content. The glucose-oxidase-catalase preparation used is one that is generally recognized as safe within the meaning of section 201(s) of the Federal Food, Drug, and Cosmetic Act. The hydrogen peroxide solution used shall comply with the specifications of the United States Pharmacopeia, except that it may exceed the concentration specified therein and it does not contain a preservative.

(2) *Controlled fermentation procedures—(i) Yeast procedure.* Food-grade baker's yeast (*Saccharomyces cerevisiae*) is added to the liquid egg whites and controlled fermentation is maintained. The quantity of yeast used and the time of reaction are sufficient to substantially reduce the glucose content.

(ii) *Bacterial procedure.* The liquid egg whites are subjected to the action of a culture of glucose-fermenting bacteria either generally recognized as safe within the meaning of section 201(s) of the Federal Food, Drug, and Cosmetic Act or the subject of a regulation established pursuant to section 409 of the act, and the culture is used in conformity with such regulation. The quantity of the culture used is sufficient to predominate in the fermentation and the time and temperature of reaction are sufficient to substantially reduce the glucose content.

(c) When the dried egg whites are prepared from liquid egg whites containing any optional ingredients added as whipping aids, as provided for in § 160.140(a), the common names of such optional ingredients shall be listed on the principal display panel or panels of the label with such prominence and conspicuousness as to render the names likely to be read and understood by ordinary individuals under customary conditions of purchase.

§ 160.150   Frozen egg whites.

(a) Frozen egg whites, frozen egg albumen is the food prepared by freezing liquid egg whites that conform to § 160.140, with such precautions that the finished food is free of viable *Salmonella* microorganisms.

(b) When frozen egg whites are prepared from liquid egg whites containing any optional ingredients added as whipping aids, as provided for in § 160.140(a), the common names of such optional ingredients shall be listed on the principal display panel or panels of the label with such prominence and conspicuousness as to render such names likely to be read and understood by ordinary individuals under customary conditions of purchase.

§ 160.180   Egg yolks.

Egg yolks, liquid egg yolks, yolks, liquid yolks are yolks of eggs of the domestic hen, so separated from the whites thereof as to contain not less than 43 percent total egg solids, as determined by the method prescribed in "Official Methods of Analysis of the Association of Official Agricultural Chemists," 10th edition, 1965, p. 247, sections 16.002 and 16.003, under "Total Solids." They may be mixed, or mixed and strained, and they are pasteurized or otherwise treated to destroy all viable *Salmonella* microorganisms. Pasteurization or such other treatment is deemed to permit the adding of safe and suitable substances (other than chemical preservatives) that are essential to the method of pasteurization or other treatment used. For the purposes of this paragraph, safe and suitable substances are those that perform a useful function in the pasteurization or other treatment to render the egg yolks free of viable *Salmonella* microorganisms, and that are not food additives as defined in section 201(s) of the Federal Food, Drug, and Cosmetic Act; or, if they are food additives, they are used in conformity with regulations established pursuant to section 409 of the act.

§ 160.185   Dried egg yolks.

(a) Dried egg yolks, dried yolks is the food, prepared by drying egg yolks that conform to § 160.180, with such precautions that the finished food is free of *Salmonella* microorganisms. Before drying, the glucose content of the liquid egg yolks may be reduced by one of the optional procedures set forth in paragraph (b) of this section. Either silicon dioxide complying with the provisions of § 172.480 of this chapter or sodium silicoaluminate may be added as an optional anticaking ingredient, but the amount of silicon dioxide used is not more than 1 percent and the amount of sodium silicoaluminate used is less than 2 percent by weight of the finished food. The finished food shall contain not less than 95 percent by weight total egg solids.

(b) The optional glucose-removing procedures are:

(1) *Enzyme procedure.* A glucose-oxidase-catalase preparation and hydrogen peroxide solution are added to the liquid egg yolks. The quantity used and the time of reaction are sufficient to substantially reduce the glucose content of the liquid egg yolks. The glucose-oxidase-catalase preparation used is one that is generally recognized as safe within the meaning of section 201(s) of the Federal Food, Drug, and Cosmetic Act. The hydrogen peroxide solution used shall comply with the specification of the United States Pharmacopeia, except that it may exceed the concentration specified therein and it does not contain a preservative.

(2) *Yeast procedure.* The pH of the liquid egg yolks is adjusted to the range of 6.0 to 7.0, if necessary, by the addition of dilute, chemically pure hydrochloric acid, and controlled fermentation is maintained by adding food-grade baker's yeast (*Saccharomyces cerevisiae*). The quantity of yeast used and the time of reaction are sufficient to substantially reduce the glucose content of the liquid egg yolks.

(c) The name of the food for which a definition and standard of identity is prescribed by this section is "Dried egg yolks," or "Dried yolks," and if the glucose content was reduced, as provided in paragraph (b) of this section, the name shall be followed immediately by the statement "Glucose removed for stability" or "Stabilized, glucose removed."

(d) (1) When either of the optional anticaking ingredients specified in paragraph (a) of this section is used, the label shall bear the statement "Not more than 1 percent silicon dioxide added as an anticaking agent" or "Less than 2 percent sodium silicoaluminate added as an anticaking agent," or whichever is applicable.

(2) The name of any optional ingredient used, as provided in paragraph (d) (1) of this section, shall be listed on the principal display panel or panels of the label with such prominence and conspicuousness as to render such statement likely to be read and understood by the ordinary individual under customary conditions of purchase.

§ 160.190   Frozen egg yolks.

Frozen egg yolks, frozen yolks is the food prepared by freezing egg yolks that conform to § 160.180, with such precautions that the finished food is free of viable *Salmonella* microorganisms.

Food Dressings

Two or more food products which contain eggs are mayonnaise and salad dressing and possibly French dressing. These are generally known as "standard of identity" products. Many people have asked us to include these regulations, see FDA 21 CFR Ch. 1 (4-1-85 Edition) for further details (Subpart B-Requirements for specific standardized food dressings and flavorings). In the case of French dressing, egg products may or may not be used.

§ **169.140 Mayonnaise.** [see References Added in Proof, No. 2, p. 343]

(a) *Description.* Mayonnaise, mayonnaise dressing, is the emulsified semi-solid food prepared from vegetable oil(s), one or both of the acidifying ingredients specified in paragraph (b) of this section, and one or more of the egg yolk-containing ingredients specified in paragraph (c) of this section. One or more of the ingredients specified in paragraph (d) of this section may also be used. The vegetable oil(s) used may contain an optional crystallization inhibitor as specified in paragraph (d) (7) of this section. All the ingredients from which the food is fabricated shall be safe and suitable. Mayonnaise contains not less than 65 percent by weight of vegetable oil. Mayonnaise may be mixed and packed in an atmosphere in which air is replaced in whole or in part by carbon dioxide or nitrogen.

(b) *Acidifying ingredients.* (1) Any vinegar or any vinegar diluted with water to an acidity, calculated as acetic acid, of not less than 2½ percent by weight, or any such vinegar or diluted vinegar mixed with an optional acidifying ingredient as specified in paragraph (d) (6) of this section. For the purpose of this paragraph, any blend of two or more vinegars is considered to be a vinegar.

(2) Lemon juice and/or lime juice in any appropriate form, which may be diluted with water to an acidity, calculated as citric acid, of not less than 2½ percent by weight.

(c) *Egg yolk-containing ingredients.* Liquid egg yolks, frozen egg yolks, dried egg yolks, liquid whole eggs, frozen whole eggs, dried whole eggs, or any one or more of the foregoing ingredients listed in this paragraph with liquid egg white or frozen egg white.

(d) *Other optional ingredients.* The following optional ingredients may also be used:

(1) Salt.

(2) Nutritive carbohydrate sweeteners.

(3) Any spice (except saffron or turmeric) or natural flavoring, provided it does not impart to the mayonnaise a color simulating the color imparted by egg yolk.

(4) Monosodium glutamate.

(5) Sequestrant(s), including but not limited to calcium disodium EDTA (calcium disodium ethylenediaminetetraacetate) and/or disodium EDTA (disodium ethylenediaminetetraacetate), may be used to preserve color and/or flavor.

(6) Citric and/or malic acid in an amount not greater than 25 percent of the weight of the acids of the vinegar or diluted vinegar, calculated as acetic acid.

(7) Crystallization inhibitors, including but not limited to oxystearin, lecithin, or polyglycerol esters of fatty acids.

(e) *Nomenclature.* The name of the food is "Mayonnaise" or "Mayonnaise dressing."

(f) *Label declaration of ingredients.* Each of the ingredients used in the food shall be declared on the label as required by the applicable sections of Part 101 of this chapter.

§ 169.150 Salad dressing.

(a) *Description.* Salad dressing is the emulsified semisolid food prepared from vegetable oil(s), one or both of the acidifying ingredients specified in paragraph (b) of this section, one or more of the egg yolk-containing ingredients specified in paragraph (c) of this section, and a starchy paste prepared as specified in paragraph (e) of this section. One or more of the ingredients in paragraph (e) of this section may also be used. The vegetable oil(s) used may contain an optional crystallization inhibitor as specified in paragraph (e) (8) of this section. All the ingredients from which the food is fabricated shall be safe and suitable. Salad dressing contains not less than 30 percent by weight of vegetable oil and not less egg yolk-containing ingredient than is equivalent in egg yolk solids content to 4 percent by weight of liquid egg yolks. Salad dressing may be mixed and packed in an atmosphere in which air is replaced in whole or in part by carbon dioxide or nitrogen.

(b) *Acidifying ingredients.* (1) Any vinegar or any vinegar diluted with water, or any such vinegar or diluted vinegar mixed with an optional acidifying ingredient as specified in paragraph (e) (6) of this section. For the purpose of this paragraph, any blend of two or more vinegars is considered to be a vinegar.

(2) Lemon juice and/or lime juice in any appropriate form, which may be diluted with water.

(c) *Egg yolk-containing ingredients.* Liquid egg yolks, frozen egg yolks, dried egg yolks, liquid whole eggs, frozen whole eggs, dried whole eggs, or any one of more of the foregoing ingredients listed in this paragraph with liquid egg white or frozen egg white.

(d) *Starchy paste.* It may be prepared from a food starch, food starch-modified, topioca flour, wheat flour, rye flour, or any two or more of these. Water may be added in the preparation of the paste.

(e) *Other optional ingredients.* The following optional ingredients may also be used:

(1) Salt.

(2) Nutritive carbohydrate sweeteners.

(3) Any spice (except saffron or turmeric) or natural flavoring, provided it does not impart to the salad dressing a color simulating the color imparted by egg yolk.

(4) Monosodium glutamate.

(5) Stabilizers and thickeners. Dioctyl sodium sulfosuccinate may be added in accordance with §172.810 of this chapter.

(6) Citric and/or malic acid may be used in an amount not greater than 25 percent of the weight of the acids of the vinegar or diluted vinegar calculated as acetic acid.

(7) Sequestrant(s), including but not limited to calcium disodium EDTA (calcium disodium ethylenediamine-tetraacetate) and/or disodium EDTA (disodium ethylenediamine-tetraactetate), may be used to preserve color and/or flavor.

(8) Crystallization inhibitors, including but not limited to oxystearin, lecithin, or polyglycerol esters of fatty acids.

(f) *Nomenclature.* The name of the food is "Salad dressing."

(g) *Label declaration of optional ingredients.* Each of the ingredients used in the food shall be declared on the label as required by the applicable sections of Part 101 of this chapter.

(Sec. 401, 701, 52 Stat. 1046, as amended, 1055—1056, as amended by 70 Stat. 919 and 72 Stat. 948 (21 U.S.C. 341, 371))
(42 FR 14481, Mar. 15, 1977, as amended at 42 FR 25325, May 17, 1977)

CONTROLLING PRODUCT QUALITY

The QC unit of the organization should possess a broad technical understanding of egg products and the processess required for their manufacture. Important technical aspects and processing requirements are well covered in the other chapters of this publication. Additionally both the production and QC departments should be thoroughly familiar of FDA Current Good Manufacturing Practice (Sanitation) in Manufacturing, Processing, Packing, or Holding Human Foods (21 CFR, Part 110) and USDA Regulations for the Inspection of Eggs and Egg Products (7 CFR, Part 2859). Knowledgeable and responsible production employees in critical processing areas offer keystone support to any QC program.

Following the production of pasteurized dried or frozen egg products, representative samples should be drawn on a daily lot or sublot basis for the chemical, microbiological, and physical examinations required to assure customer acceptance. A noteworthy classification system and sampling plan was described by Foster (1971). See Tables 15.1, 15.2, and 15.3.

TABLE 15.1. Categories of Risk for Formulated Foods

| Type of food | Hazard characteristic[a] | | | Category |
	A Ingredient	B Process	C Abuse	
Food intended for infants, the aged, and the infirm	+[b]	+ or 0	+ or 0	I
Food intended for general use	+	+	+	II
	+	+	0	III
	+	0	+	III
	0	+	+	III
	+	0	0	IV
	0	+	0	IV
	0	0	+	IV
	0	0	0	V

SOURCE: Foster (1971).
[a] See text for definition of hazard characteristics. Heat-sterilized canned foods for infants offer no significant Salmonella hazard and should be assigned to Category V.
[b] + = Hazard present; 0 = hazard not present.

TABLE 15.2. Acceptance Plan for 25-g Test Units

Product category	No. of units tested[a]		Significance: 95% probability no more than one organism[b] (g)
	None positive	No more than one positive	
I	60 (1,500)	95 (2,375)	500
II	30 (750)	48 (1,200)	250
III, IV, V	15 (375)	24 (600)	125

SOURCE: *Foster (1971).*
[a] Numbers in parentheses indicate total grams of sample tested.
[b] Confidence limits in Attribute Sampling Tables are commonly expressed in terms of the proportion of samples defective. However, at very low levels of contamination it may be more meaningful to speak in terms of concentration per unit. Thus, an average proportion of 0.05, 0.1, and 0.2 defective (positive) 25-g test units corresponds respectively to one organism in 500, 250, and 125 g of product.

TABLE 15.3. Acceptance Plan for 50- and 100-g Test Units

Product category	No. 50-g units tested[a]		No. 100-g units tested		Significance: 95% probability no more than one organism (g)
	None positive	No more than one positive	None positive	No more than one positive	
I	30 (1,500)	63 (3,150)	15 (1,500)	46 (4,600)	500
II	15 (750)	32 (1,600)	8 (800)	23 (2,300)	250
III, IV, V	8 (400)	16 (800)	4 (400)	12 (1,200)	125

SOURCE: *Foster (1971).*
[a] Numbers in parentheses indicate total grams of sample tested.

SUMMARY

A QC program involves:

1. Inspection of raw materials, ingredients, and packaging supplies
2. Inspection of operating procedures
3. Inspection of sanitation conditions
4. Chemical, microbiological and physical examinations of finished products
5. Waste-disposal control
6. Warehousing control
7. Compliance with customer, Federal, and State requirements
8. Record keeping and reporting
9. Assistance to customers in handling and use of products

SELECTED REFERENCES

American Public Health Association (APHA)   1975.   Standard Methods for the Examination of Water and Waste Water, 14th Edition. APHA, Washington, DC.

American Public Health Association (APHA)   1978.   Standard Methods for the Examination of Dairy Products, 14th Edition. APHA, Washington, DC.

Association of Official Agricultural Chemists   (1985).   "Official Methods of Analysis," 10th Ed., p. 247, Sects. 16.002 and 16.003. AOAC, Washington, DC.

Bergquist, D. H., and Wells, F.   1956.   The monomolecular surface film method for determining small quantities of yolk or fat in egg albumen. Food Technol. *10*, 48–50.

Cahn, F. J., and Epstein, A. K.   1936.   Methods of analysis. U.S. Pat. 2,065,114.

Colburn, J. T., Cotterill, O. J., and Funk, E. M.   1964.   A modified monomolecular film test for micro-quantities of lipids in foods. Res. Bull.—Mo., Agric. Exp. Stn. *856*.

Cotterill, O. J., and Delaney, I.   1959.   Microwave heating for the determination of total solids in liquid egg products. Food Technol. *13*, 476.

Fletcher, D. L.   1979.   An evaluation of the AOAC method of yolk color analysis. Poult. Sci. 59, 1059–1066.

Fletcher, D. L., Britton, W. M., and Cason, J. A. A comparison of various procedures for determining total yolk lipid content. Poult. Sci. 63, 1759–1763.

Foster, E. M.   1971.   The control of *Salmonella* in processed foods: A classification system and sampling plan. J. Assoc. Off. Anal. Chem. *54* (2), 259–266.

Hawthorne, J. R.   1944.   Dried egg. VII. Methods for the determination of the solubility of dried whole eggs. J. Soc. Chem. Ind. *63*, 6.

Kramer, A., and Twigg, B. A.   1962.   Fundamentals of Quality Control for the Food Industry, 3rd Edition, Vol. 1. AVI Publishing Co., Westport, CT.

Sawyer, A. D.   1966.   Extraction procedure for chlorinated organic pesticides in fresh eggs. J. Assoc. Off. Anal. Chem. *49*, 643.

Speck, M. L. (Editor)   1984.   Compendium of Methods for the Microbiological Examination of Foods. American Public Health Association, Washington, DC.

Stemp, A. R., Liska, B. J., Langlios, B. E., and Stadelman, W. J.   1964.   Analysis of egg yolk and poultry tissues. Poult. Sci. *43*, 273–275.

Stuart, L. S., Grewe, E., and Dick, E. E.   1942.   Solubility of spray dried whole egg powder. U.S. Egg Poult. Mag. *42*, 488.

Triebold, H. O.   1946.   Quantitative Analysis of Agricultural and Food Products. Van Nostrand-Reinhold, Princeton, New Jersey.

U.S. Department of Agriculture (USDA)   1984.   Laboratory Methods for Egg Products. U.S. Department of Agriculture, Agricultural Marketing Service, Poultry Division, Grading Branch, Washington, DC.

U.S. Department of Health, Education and Welfare/Food and Drug Administration   1977.   Title 21 Code of Federal Regulations. F.D.C. Regulations, Part 160. U.S. Government Printing Office, Washington, DC.

United States Federal Food and Drug Administration, 21 CFR, Ch. 1, dated 4-1-85.

REFERENCES ADDED IN PROOF:

1. Froning, G. W., Acton, J. C., Ball, J. R., Jr., *et al.*   1986.   Recommended methods of analysis of eggs and poultry meat. No. Central Regional Bull. *307*.

For copies contact:

> I. T. Omtvedt, Dean and Director
> Agriculture Research Division
> Institute of Agriculture and Natural Resources
> University of Nebraska
> Lincoln, NE 68583

2. Harrison, L. J., and Cunningham, F. E.   1985.   Factors influencing the quality of mayonnaise: A review. J. Food Quality *8*, 1–20.

16

Functional Properties of Eggs in Foods

Ruth E. Baldwin

The term "functional properties" refers to the attributes of eggs which make them a useful ingredient in foods such as noodles, mayonnaise, cakes, and candy (Forsythe 1963). The major functions have been classified as follows: (1) coagulating, (2) foaming, (3) emulsifying, and (4) contributing nutrients. In addition, eggs serve as coloring and flavoring ingredients, and in some instances they are used to control growth of sugar crystals. The adequacy of performing these functions determines the value of eggs in food products. The emphasis in this chapter is on the inherent functional properties of the egg in foods; the effects of processing on functional performance is covered elsewhere.

Although many foods are capable of foaming, eggs are especially effective in this capacity. Under proper conditions, they produce large-volume, stable foams which coagulate during heating. This is of particular value in the production of cakes and certain confections. The ability of the protein to coagulate on heating is utilized when eggs are used as binding agents and when they are ingredients in custards and pie fillings. Egg yolk is an excellent emulsifier for fats or oils and water. Emulsification is their major function in mayonnaise and one of their functions in shortened (fat-containing) cakes.

Although attempts have been made to simulate the functions of eggs, no other food or combination of ingredients has completely duplicated these properties. Eggs are polyfunctional, in that they usually perform in more than one capacity, whereas materials that have been considered as possible substitutes are monofunctional. For

EGG SCIENCE & TECHNOLOGY,
3rd Edition

example, the substitute may serve as a foaming agent but lack co-agulative properties. The usefulness of eggs as a food ingredient is evident when the large number of food products that contain eggs, either fresh, frozen, or as egg solids is considered.

COAGULATION

Changes in the structure of the egg-protein molecules resulting in the loss of solubility and thickening, or change from the fluid (sol) to the solid or semisolid (gel) state, may be brought about by heat, mechanical means, salts, acids, alkalies, and other reagents such as urea. This change from the sol to the gel state is known as coagulation. Strictly speaking, coagulation designates a loss of solubility, whereas gelation refers to thickening and loss of fluidity (Lowe 1955). However, Lepeschkin (1922) considered coagulation as synonomous with agglutination.

The success of many cooked foods is dependent on the coagulation of proteins, especially the irreversible heat coagulation of egg proteins. Both the white and the yolk are utilized because of their ability to coagulate and bind food materials together. The coagulum or gel formed in some mixtures may be so firm (overcoagulated) that it squeezes liquid out, so that liquid and curd phases form. This phenomenon is referred to as syneresis.

Mechanism of Coagulation

Chick and Martin (1910) concluded that heat coagulation is a reaction between proteins and water, followed by separation or agglutination of the protein. This conclusion was based on observations of crystalline albumen that remained soluble even after heating at 120°C (248°F) for 5 hr, whereas it was rendered insoluble in a few minutes when heated in the presence of steam. In 1982, Ma and Holme postulated that thermo-coagulation requires a balanced electrostatic attraction between protein molecules and presented evidence of hydrophobic interaction during gel formation. Marked oxidation of sulfhydryl groups during coagulation was not indicated.

The mechanism of coagulation has been studied with individual proteins under a limited range of conditions. Smith (1964) investigated the influence of urea on ovalbumin and found, first, an unfolding of the molecule followed by aggregation. Second, a secondary unfolding may occur during or after aggregation. Kauzman (1959) suggested that formation of intermolecular hydrophobic bonds, hydrogen bonds, and disulfide bonds causes the protein to become insoluble. The extent of the unfolding of the molecule is related to the particular protein and various conditions of the system: the higher the concentration of unfolded molecules in the system and the faster the rate of unfolding, the finer the gel network. Since a temperature increase accelerates the first step in coagulation more than the second, an increase in temperature also contributes to a fineness of the gel network (Ferry 1948).

A strong electrostatic force plays a role in keeping proteins in solution. When the coulombic repulsion is lessened by the presence of a neutral salt, the polypeptide chains tend to associate and form a gel network. It is possible for gelation to occur before a large number of free chains have formed. When this happens, a coarse network results (Ferry 1948). Conversely, in 7 M urea at pH 3, the polypeptide chain of albumin unfolds rapidly, but aggregation does not ensue (McKenzie et al. 1963).

Coagulable Components

Early researchers believed that egg white was a single protein. However, protein fractions with varying properties were found as a result of investigations on the solubility of albumen. These fractions, in turn, were found to contain more than one protein (Fevold 1951). Feeney (1964) listed the proteins in percentage of the egg white solids as follows: ovalbumin, 54; conalbumin, 13; ovomucoid, 11; lysozyme, 3.5; ovomucin, 1.5; flavoprotein-apoprotein, 0.8; "proteinase inhibitor," 0.1; avidin, 0.05; and unidentified proteins, 8. Of these, ovomucoid and ovomucin are noncoagulable by heat (Parkinson 1966; Johnson and Zabik 1981A, B). Conalbumin is especially heat sensitive unless it is complexed with a metal ion such as iron or aluminum (Azari and Feeney 1958; Cunningham and Lineweaver 1965). Ovalbumin ranks next to conalbumin in the tendency to denaturation by heat (Johnson and Zabik 1981B).

Other than a greater concentration of ovomucin in the thick white, no marked differences have been found between the composition of the thick and thin layers of egg white (Forsythe and Foster 1949; Steven 1961; Parkinson 1966). The mucin content is the major factor responsible for differences between the coagulation properties of the thick and thin white. The chalazas are high in mucin content, and their resistance to coagulation is readily observable.

In contrast to egg white, egg yolk contains considerable proportions of nonprotein components. Expressed as percentage of yolk solids, they are as follows: protein (livetins), 4–10; phosphoproteins, 4–15; lipoproteins, 16–18; lipids, 46; carbohydrates, 2; minerals, 2; and some vitamins (Parkinson 1966). Although most of the proteins of the yolk are subject to heat coagulation, phosvitin and livetin are not (Mecham and Olcott 1949; Tsutsui and Obara 1980A).

Measurement of Coagulation

Three tests for the properties of coagulated egg white were devised by Barmore (1936). The first of these was based on measuring the stress required to pull apart coagulated samples of standard size and shape. The second test was the determination of the depth of penetration of a standard steel ball into the coagulum. A similar type of data was collected by Merrill (1963) by using a penetrometer on coagulated albumen. Also, an improvised device was described by Hanning (1945), which operated on a principle similar to that of the penetrometer. Although this device was used with puddings, it would be equally suitable for stirred custard or partially coagulated egg protein. It consisted of a tapered graduated centrifuge tube which was released above the product. The time required for the tube to penetrate to a specified depth was measured. The third of Barmore's (1936) tests consisted of introducing a plunger into the coagulum and measuring the force required to cause penetration. The principle of this device is similar to that of the Bloom gelometer used by Baldwin et al. (1967) in measuring the firmness of coagulated egg white. The Delaware Jelly Strength tester described by Baker (1938) also is similar in that a plunger is pushed through the surface of the gel. These instruments differ in the source of the force and the manner in which the force is measured. With a curd tension meter, the resistance of the coagulum to a cutting edge can be measured (Carr and Trout 1942).

Estimation of the amount of sag is useful in investigations of baked custards. Griswold (1962) stated that percentage of sag can be calculated by measuring the height of the food before and after removal from the cooking container, or by using the Exchange Ridgelimeter, which was designed for determining the sag of fruit jellies. The ratio of height to diameter of custards was called the "standing index" by Carr and Trout (1942).

The reduction of the tendency of egg white to flow was measured as gelation score by Seideman et al. (1963). McKenzie et al. (1963) also considered loss of the tendency to flow an index of gelation and, in turn, an index of coagulation. The line-spread test (Grawemeyer and Pfund 1943) would be useful as an indicator of flow properties for soft custards or partially coagulated egg. For this test, a diagram with concentric circles is covered with a glass plate. A cylinder (sample container) is placed above the center circle showing through the glass. The cylinder is filled with the food, is then lifted, and the product is allowed to flow for a controlled length of time. An average distance of flow can be calculated by observing the circles covered by the food.

Viscosity measurements have been reported by many researchers. Changes in viscosity indicate the tendency for ovalbumin to aggregate and for the polypeptides to change in configuration. Viscosity increases when the polypeptide chains of the protein become loosely coiled (Smith 1964). This change in configuration is associated with a loss of solubility and thus with coagulation.

Optical rotation, along with viscosity, was used by Smith (1964) in studying the extent of aggregation of the protein. An increase in levorotation was attributed to an unfolding of albumin molecules (McKenzie *et al.* 1963). Cunningham and Cotterill (1962) and Seideman *et al.* (1963) determined absorbance and electrophoretic mobility, together with gelation, in egg-white systems adjusted to different pH levels. The maxima of absorbance represent the pH levels at which components in the protein are the least stable, and the minima indicate the greatest stability. The maxima, no doubt, coincide with the isolectric pH of the individual proteins in the system.

Solubility is probably one of the most widely used and sensitive indexes of coagulation. Changes in solubility may be estimated by measuring the rate of formation and the amount of precipitate, coagulum, or sedimentation under given conditions (Chick and Martin 1910, 1911, 1912A,B; Galston and Kaur 1962; Smith and Back 1965). Some investigators recovered the precipitated protein and determined the nitrogen content (Cunningham and Lineweaver 1965). Titration with ammonium sulfate to turbidimetric end points was a method employed by Smith and Back (1965) to compare solubilities of two proteins. Melnick and Oser (1949) suggested supplementing the solubility test with information on the salting-out characteristics of proteins.

Factors Affecting Coagulation

A summary of the major factors influencing coagulation of eggs is shown in Table 16.1. Under the usual conditions of food preparation, several factors exert an influence simultaneously, and some of these forces may be in opposition to one another.

Temperature. Chick and Martin (1910) emphasized that heat coagulation of albumen is influenced by a variety of conditions. It does not occur instantaneously at a given temperature, but is a time process in which heat is the accelerator. The average speed of coagulation of albumen is increased 191 times with a rise in temperature of 1°C (1.8°F) and approximately 635 times with a 10°C (18°F) increase in temperature. Lowe (1955) pointed out that at high temperatures, coagulation occurs almost instantaneously.

Coagulation of albumen begins at about 62°C (144°F), and the coagulum ceases to flow when a temperature of about 65°C (149°F) is reached. At 70°C (158°F) the coagulum is fairly firm, but tender, and it becomes

TABLE 16.1. Factors Influencing Coagulation of Eggs

Factor	Influence
Temperature	Speed of coagulation influenced by temperature (Chick and Martin 1910; Lowe 1955). White begins to coagulate at 62° C (144° F) (Romanoff and Romanoff 1949; Beveridge *et al.* 1980). Yolk begins to coagulate at 65° C (149° F) (Romanoff and Romanoff 1949). Overcoagulation with extended time of heating (Lowe 1955). High temperature reduces the difference between temperature at which optimum and overcoagulation occur (Andross 1940)
Dilution	Raises temperature required for coagulation (Andross 1940). Decreases firmness of coagulum (Beveridge *et al.* 1980)
Salts	Naturally occurring salts essential to coagulation (Lepeschkin 1922). Added salts promote coagulation (Lowe 1955). $CuSO_4$ and $AlCl_3$ decrease firmness (Beveridge *et al.* 1980)
Sugar	Raises temperature of coagulation (Barmore 1936).
Acid	Lowers temperature of initial coagulation (Barmore 1936)
Alkali	Causes translucent gel above pH 11.9 (Cunningham and Cotterill 1962)

very firm at higher temperatures (Romanoff and Romanoff 1949; Beveridge *et al.* 1980). When heated by exposure to microwaves, albumen was observed to coagulate at 57.2°C (135°F) (Baldwin *et al.* 1967). The white from duck egg coagulates at 55°C (131°F) if held at this temperature for 10 min (Rhodes *et al.* 1960). Egg yolk begins to coagulate at 65°C (149°F) and ceases to flow when it reaches a temperature of about 70°C (158°F) (Romanoff and Romanoff 1949).

The coagulation reaction is endothermic; heat is absorbed. Hence foods with a high egg content may remain at the same temperature during the period of time that coagulation is occurring (Griswold 1962). The composition of the food product determines the temperature at which thickening occurs. For example, custards depending only on egg white for thickening, coagulate at a lower temperature than those containing egg yolk (Andross 1940).

Successful egg cookery is dependent upon the relation between time and temperature. Too much heat results in overcoagulation, regardless of whether the excess heat is the result of too high a temperature or exposure to heat for too long a time. It is possible to achieve good results with high temperatures if the time is controlled in relation to the temperature. Heating beyond that required to cause optimum firmness results in undesirable syneresis (Lowe 1955). The syneresis observed by Chen and Hsu (1981) when whole egg or albumen was cooked by microwaves may have been due to overcoagulation. Both the speed of cooking induced by microwaves and the tendency for carry-over heating make it difficult to prevent overcooking with some foods.

When custards are cooked at high temperatures, there is very little difference between the point of optimum thickening and that of undesirable curdling. When cooked over boiling water, custards thicken at 87°C (189°F) and curdle at 90°C (194°F). When cooked over water brought to a boil during the cooking process, the custard thickens at 82°C (180°F) and curdles at 87°C (189°F). The low-temperature cookery allows ease in handling the product, since there is 5°C difference between optimum coagulation and curdling, whereas with the high temperature there is only a 3°C margin (Andross 1940). The practice of cooking custards in a double boiler or setting the container in a pan of water during baking is a protection against excessive heating.

When hard-cooking eggs in the shell, both the temperature and the proportion of water to eggs are essential considerations. Griswold (1962) recommended starting the cooking process with cold water and bringing it to a boil, or starting with boiling water and turning off the heat while the eggs cook. Andross (1940) reported that when shell eggs are cooked at temperatures as low as 85°C (185°F), the yolk does not completely coagulate in 30 min. Further, the white does not coagulate firmly in 1.5 hr when the cooking water is held at 72°C (162°F).

If the temperature for frying eggs is as low as 115°C (239°F), the egg white is likely to spread excessively, and at a temperature of 146°C (295°F), overcoagulation occurs. Andross (1940) recommended 126° to 137°C (259° to 270°F) for frying eggs.

Dilution. The temperature required for coagulation of eggs is elevated by dilution, and the firmness of the coagulum decreases with increased dilution (Beveridge *et al.* 1980). The amount of liquid combined with eggs for scrambling determines whether a small, firm mass or a large, soft one is obtained. If excessive liquid is added to the egg, there is likely to be a curdlike product which may separate during cooking. If too little liquid is added and the eggs are cooked too long, a rubbery mass results (Griswold 1962). According to Andross (1940), 10 to 25 ml of milk per egg is suitable for scrambled eggs of optimum consistency. With this amount of milk, there is a sufficient margin of safety between the coagulation and curdling temperatures to allow ease in handling the product. Dilution is also a factor in coagulation of custards. Custards containing a high proportion of eggs thicken at a lower temperature than those with a lesser egg content.

Firmness of the coagulum decreases less with increasing dilution with water when coagulation is brought about by exposure to microwaves rather than to heat in a conventional oven at 163°C (325°F). Also, the final temperature of the coagulum that has been exposed to microwaves is lower when the amount of dilution is increased. Conversely, as dilution is increased in conventionally heated albumen, higher final temperatures are attained. Within each method of cookery, the amount of evaporation from the product remains constant, but there is less evaporation with microwave cookery than with conventional heating (Baldwin et al. 1967).

Salts. Removal of naturally occurring salts from egg white impairs its ability to coagulate. This property can be restored by adding back the salts in optimum amounts (Lepeschkin 1922). Similarly, when the salt concentration of a custard mix is lowered by substitution of water for milk, gelation does not occur. Coagulation may be accomplished in this custard by adding salts to it, but curdling rather than smooth coagulation results if the salts are added after heating is completed. Lactates, chlorides, sulfates, phosphates and combinations of $MgCl_2$ and NaSCN, and NaCl, Na_2SO_4, and $CaCl_2$ promote coagulation, some more than others (Lowe 1955). Beveridge et al. (1980) noted that $AlCl_3$ and $CuSO_4$ decreased firmness of coagula. The gel strength is a function of the balance of the cation and anion activity of the salt. Curdling is caused by high salt concentration. However, when $FeCl_3$ is used, the curds can be dispersed by heating. Decreasing amounts of salts are required to promote coagulation as the valence of the cation increases (Lowe 1955). Addition of NaCl to the water in which eggs are poached improves the appearance of the finished product due to its tendency to promote coagulation (Andross 1940; St. John and Flor 1931).

Sugar. Sucrose raises the temperature required to cause coagulation of albumen in proportion to the amount added (Lowe 1955). This effect is greater when the pH of the albumen is above 8.5 (Seideman et al. 1963). However, Barmore (1936) observed that sugar elevates the temperature of coagulation of egg white both in the presence and in the absence of acid. The acid lowers the temperature of initial coagulation both in the presence and absence of sugar. Addition of sugar and potassium acid tartrate to egg white in the proportion used in cake batter results in elevation of the initial coagulation temperature.

The effect of sucrose on the quality of baked custards was investigated by Wang et al. (1973). Sucrose contributed to tenderness of the crust of the custard. However, it did not influence the strength of the gel provided that the egg and milk solids were kept constant.

Acidity or Alkalinity. The effect of acid or alkali on albumen depends upon the pH and its relation to the isoelectric point of the proteins.

Seideman *et al.* (1963), in investigating heat coagulation, found maxima in absorbance when albumen was adjusted to pH levels near the iso-electric pH of the proteins in the system.

Seideman *et al.* (1963) and Cunningham and Lineweaver (1965) determined that the stability of conalbumin to heat is minimal near pH 6.0. At pH 9 the heat stability of this protein is markedly increased. In contrast, the heat stability of ovalbumin is greatest at a neutral pH (Lewis 1926).

When the pH of albumin is in the range from 5.3 to 5.7, pectins inhibit its coagulation by heat. However, at pH levels below this, coagulation of albumin is promoted by pectins. Boiled pectins lack the ability to affect the coagulation of albumin (Galston and Kaur 1962).

Coagulation is favored and the shape of poached eggs is improved by adding acid to the water in which they are cooked (Andross 1940). Also, a soft gel can be attained in a custard containing distilled water, and lacking the salts of milk, if the pH is lowered before the mixture is heated. Coagula resulting from exposure to microwaves or to heat are more tender when the albumen is acidified (Baldwin *et al.* 1967).

Although no practical application has been found for alkaline coagulation of egg white, it is of interest in relation to the functional properties of eggs. When the pH of albumen is adjusted to 11.9 or above, it forms a translucent gel which in time undergoes self-liquefaction. The time required for gelation depends upon the strength of the alkali, the rate at which it is added to the albumen, the volume ratio of the two, and the temperature (Cunningham and Cotterill 1962). For further information on textural properties of egg white, whole egg, and yolk gel properties. Please review Woodward (1984) and the resulting papers from this thesis published in *Food Science* (1985, 1986). See Woodward and Cotterill (1985) and Woodward and Cotterill (1986), etc.

FOAMING

A foam is a colloidal dispersion in which a gaseous phase is dispersed in a liquid phase. When egg white is beaten, air bubbles are trapped in the liquid albumen, and a foam is formed. Although both phases of the foam do not meet the classical description for colloidal particle size, the continuous film of albumen does. During the beating of egg white, the air bubbles decrease in size and increase in number, and the translucent albumen takes on an opaque but moist appearance. As increased amounts of air are incorporated, the foam becomes stiff and loses its flow properties. If beating is continued, the foam becomes friable and loses its moist, glossy appearance (Lowe 1955).

The foaming power of albumen makes it useful as a carrier of air for leavening, and thus it contributes lightness to certain food products. Essentially, meringues are highly aerated egg whites containing sugar (Godston 1950). Angel cakes are little more than meringues with flour added to contribute to the structure. Although souffles and foamy omelets contain both yolk and white of egg, their lightness is due chiefly to beaten white. Beaten yolks, either alone or together with beaten whites, form the basis of sponge cakes.

Mechanism of Foam Formation

Griswold (1962) described the mechanism of foam formation as the unfolding of the protein molecules so that the polypeptide chains exist with the long axes parallel to the surface. This change in molecular configuration results in loss of solubility or coagulation of some of the albumen, which collects at the liquid-air interface. This adsorption film is essential to the stability of the foam.

Also essential to formation of a stable foam is a liquid with high tensile strength or elasticity. Elasticity is lost if too much air is beaten into the foam, and the albumen is stretched to such an extent that it can expand no further. This is especially important for egg foams subjected to heat. As the air in the cells becomes hot it expands, and the albumen surrounding it must either stretch or break. If the latter occurs, the foam cannot contribute maximum leavening to the food.

Foaming power is often attributed to low surface tension. This quality allows the creation of a large surface which is essential to foaming. However, other characteristics, such as high viscosity and low vapor pressure, are more important than low surface tension. A viscous liquid has little tendency to drain away from the air cells, and a liquid with low vapor pressure has little tendency to evaporate from around the bubbles.

Foaming Components

Ovomucin is the component of egg white that forms a film of insoluble material which stabilizes the foam (MacDonnell et al. 1955). Johnson and Zabik (1981A) stated that conalbumin, lysozyme, ovomucin, and ovomucoid alone have little or no foaming capacity but confirmed that interaction of lysozyme and globulins is important in the foaming process.

Overwhipping insolubilizes too much of the ovomucin and lowers the elasticity of the bubbles (MacDonnell et al. 1955). Egg white drained from foams is capable of repeated foaming provided that too much ovomucin has not been coagulated and that the initial foam is not too stable. If the first foam is beaten for a long period, the drainage liquid exhibits poor whipping properties (Forsythe and Bergquist 1951). According to MacDonnell *et al.* (1955) and Cunningham (1976), each successive whip of liquid drained from foams requires longer beating and results in a less stable product. The greatest amount of ovomucin is removed with the first whipping (MacDonnell *et al.* 1955), but about one-third of the ovomucin can be recovered from the drainage after the third whipping (Cunningham 1976). In addition to ovomucin, lysozyme, globulins A_1 and A_2, and conalbumin are retained in the foam (Cunningham 1976). MacDonnell *et al.* (1955) and Cunningham (1976) found that albumin is not removed by the foaming process. The albumin that is collected in the drip is identical with that of fresh egg white.

Globulins contribute to high viscosity and decrease the tendency for liquid to drain away from the air bubbles of a foam. Globulins also lower the surface tension, which is helpful especially in the initial stages of foaming. Low surface tension also promotes small bubble formation and smooth texture (MacDonnell *et al.* 1955).

Meringues and angel cake batter can be made from globulins and ovomucin without the other components of egg white. However, when the heat-coagulable protein ovalbumin is not present, the cake batter collapses after rising during the baking period. MacDonnell *et al.* (1955) observed that when ovalbumin alone is used for making angel cakes, long whipping is required to form a foam, and a coarse-textured product is obtained. Egg white from which the globulins and ovomucin are removed requires long whipping, and when used in cakes it causes reduced volume. Returning globulins to the egg white improves the volume of cakes in which it is used, but whipping time remains long. When ovomucin is replaced, the time required for whipping is shortened, but the volume of cakes made with this combination is not improved. Return of both globulins and ovomucin restores the whipping and cake-making quality to the egg white. Addition of extra ovomucin results in increased speed of foam formation and excess coagulation of protein in the bubble surface, together with reduced elasticity of the film surrounding the bubble. Therefore, the volume of cakes made from the product is low. Addition of excess globulins to egg white results in quick foaming properties and increased volume in cakes. Stability in this case is achieved by the normal amount of ovomucin.

The possible role of lysozyme in foaming was investigated by Sauter and Montoure (1972). Smaller volume was evident in foams from eggs with high lysozyme content than from those which were low in lysozyme. This difference may have been due to the greater ratio of thick to thin white in high lysozyme eggs. The amount of drainage from the foams did not differ significantly.

Measuring the Properties of Foams

As indicated by the foregoing discussion, angel cakes have long been used as a critical test of the foaming properties of egg whites. Although most attention has centered on the volume of the cakes, other factors such as tenderness, texture and grain, and elasticity of the crumb also reflect the quality of the albumen foam (Smith 1960).

Generally, the volume of angel cake is measured by the seed displacement procedure described by Binnington and Geddes (1938). If the baking pan is slightly deeper than the baked cake, volume can easily be determined using the baking pan. The difference between the volume of seed held by the pan with and without the cake represents the volume of the cake. Volume can also be estimated by polar planimeter measurements of the area covered by representative slices of cakes.

Since photographs provide permanent records, they can be used as a means for comparing texture, as well as volume, of cakes. Ink prints also serve as reasonably accurate representations of the texture and volume of cakes. Barnard (1966) described materials for preparing ink prints. If desired, the ink prints can be used as a basis for the polar planimeter measurements mentioned above.

Barmore (1936) described in detail an apparatus for testing the toughness of angel cakes. For this test, the weight required to break the lower portion of an hour glass-shaped sample from the upper part is determined and reported in relation to the area of the break.

Although the volume, texture, and tenderness of cake can be measured objectively, it is generally conceded that acceptability can be assessed only by sensory evaluation. Sauter and Montoure (1975), however, obtained values with the Instron Universal testing machine for shear compression force for angel cakes that agreed closely with sensory scores for tenderness.

For information on the methodology of sensory testing, the reader is referred to the comprehensive treatment of the subject by Amerine et al. (1965). Harns et al. (1952) reported that angel cakes can be held frozen prior to sensory testing without introducing appreciable error. This procedure can ease the pressure due to timing of sensory evaluations.

Whip time is indicative of the ease with which foaming occurs, but it does not describe the foam. It is defined as the time required to beat a foam to a specified stage of aeration. The increase in specific volume of foam per unit of time of beating is the rate of beating and is expressed in milliliters per gram per second. The specific volume represents the relationship between the volume of the foam and its weight and is found by dividing the milliliters of foam by the weight in grams (Slosberg et al. 1948).

Volume of foam is somewhat difficult to measure accurately since air pockets are easily trapped. Attempts to remove the air pockets may result in damage to the foam and loss of air from the product. Depth of foam in the bowl in which it is whipped can be measured (Carney 1938), or depth can be measured and the volume of the bowl at this depth reported in milliliters (Baldwin et al. 1968). McKeller and Stadelman (1955) designed a device with which volume can be read in milliliters on a calibrated handle when a plunger is brought into contact with the surface of the foam.

Specific gravity is an indirect measure of the volume of a foam: the lower the specific gravity, the larger the volume. Barmore (1934) stated that the specific gravity and stability of albumen foams are proportional to the viscosity of the liquid. Therefore, the stability of foams is proportional to the specific gravity. Although Barmore (1936) found maximum volumes in angel cakes made from foams with specific gravities between 0.150 and 0.170, other researchers have shown that angel cake volumes cannot be predicted on the basis of specific gravity (Smith 1960; Slosberg et al. 1948; MacDonnell et al. 1950).

In general, beating different batches of egg white to the same degree of lightness is estimated by visual observation. Niles (1937) described the characteristics of egg-white foams beaten to different stages and listed the kinds of food products that required foams beaten to these stages. Specific gravity, of course, would be an objective way of determining the stage of aeration. However, during the time this measurement is being made on a small portion of the foam, changes in the state of foam continue (Brooks and Taylor 1955). The Brookfield viscometer (Brookfield Engineering Laboratories, Inc., Stoughton, MA) equipped with a helipath stand has potential as a rapid and objective test of stage of whipping (Cloninger 1967).

Most tests of the stability of egg white foams are based on the amount of liquid released from the foam in a given time. The drainage is reported on a weight, volume, or a rate-of-drainage basis. Barmore (1934) described a simple method of measuring drainage from foams by weighing. Instead of a continuous collection procedure, some researchers (Cotterill et al. 1963B) pour the liquid away from the foam for measurement. Relative stability of meringues was demonstrated by Godston (1950) by placing the meringues in closed jars and allowing them to separate into a liquid and a foam phase. Screens can be used as supports for meringues while they drain (Gillis and Fitch 1956; Baldwin et al. 1968).

Cotterill et al. (1965) collected the insoluble material, rather than the drainage from the foam, and washed, dried, and weighed it. This procedure is similar to Barmore's (1934) analysis for insoluble protein. The insolubilized protein also can be compared by drying foams and observing the volume after drying (Henry and Barbour 1933).

Factors Affecting Foams

Egg foams are influenced by method and extent of beating, pretreatments (Table 16.2), and by added ingredients (Table 16.3). Comparison of data from one laboratory to another is difficult due to differences in equipment, techniques, and methods of expressing results.

Methods of Beating.   The stage or extent of beating is the major factor influencing the characteristics of egg foams. Barmore (1936) found that when specific gravities of foams are above 0.170, there is insufficient air for adequate leavening of cakes. When values are below 0.150, the foams lack sufficient stability for suitable leavening. The stage of maximum

TABLE 16.2.   Influence of Method of Beating and Pretreatments on Egg Foams

Variable	Influence
Method of beating	Greater volume with thin than with thick white when hand-operated beater used (St. John and Flor 1931)
	Greater volume with thin than with thick white when beaten with electric beater, but loss of volume with continued beating (Henry and Barbour 1933)
	Greater volume with thick than with thin white when beaten with hypocycloidal action (Bailey 1935)
	Foam immobilizes around beaters operating on stationary axes (Bailey 1935)
	Maximum stability attained before maximum volume in egg white foam (Barmore 1936)
	Increased time of beating improves volume of whole egg foams, but volume of cakes made from these foams is not increased (Briant and Willman 1956)
Blending	Blending (until ovomucin fibers are 300 μm in length) increases beating rate and volume of cakes made with blended whites (Forsythe and Bergquist 1951).
Homogenizing	Reduced whip time and volume of cakes made with homogenized egg white (Forsythe and Bergquist 1951)
Temperature	Quicker foaming at room temperature than at refrigerator temperature (St. John and Flor 1931)
	Stability of egg-white foam similar at 20°C (68°F) and 34°C (93°F) (Barmore 1934).
	Holding albumen at 58°C (136°F) not detrimental to foaming properties (Slosberg et al. 1948)
	Adjusting albumen of pH 6.5 increases stability to heat (Slosberg et al. 1948).
	Heating albumen, pH 8.75, at 58°C (136°F) for 3 min improves foaming (Seideman et al. 1963)
	Pasteurization (60°C or 140°F, 3.5 min) increases whip time (Cunningham and Lineweaver 1965; Garibaldi et al. 1968)

TABLE 16.3. Influence of Added Ingredients on Egg Foams

Ingredient	Influence
Acid or base	Increased whiteness of cake with added acid (Grewe and Child 1930)
	Improved stability of foams with acids and acid salts at specific pH levels, but no effect due to selected bases (Barmore 1934)
	Improved stability of egg white to heat by adjusting to pH 6.5 with selected acidulants (Slosberg *et al.* 1948)
	No decrease in beading of meringues with potassium acid tartrate (Hester and Personius 1949)
	Discoloration of meringues prevented by citric acid (Forsythe 1957)
	Improved whipping properties of duck eggs with addition of lemon juice (Rhodes *et al.* 1960)
	Improved whip time and volume of cakes with albumen adjusted to pH 8.75 and heated 3 min at 58°C (136°F) (Seideman *et al.* 1963)
	Increased heat stability of ovalbumin, lysozyme and ovomucoid at pH 7 (Cunningham and Lineweaver 1965)
Water	Increased volume and decreased stability of foams with dilution of 40% or more (Henry and Barbour 1933)
	Increased volume and decreased stability of egg-yolk foams (Cunningham 1972)
NaCl	Decreased stability in egg-white foams (Hanning 1945) Contributes to soft whole-egg foams (Briant and Willman 1956)
	Increased whip time in egg white (Smith 1960)
Sugar	Delayed foam formation and decreased expansion (Hanning 1945; Gillis and Fitch 1956)
	Functional properties of albumen protected against detrimental effects of heat if sucrose, lactose, dextrose, or maltose added prior to heating (Slosberg *et al.* 1948)
Egg yolk	Reduced volume in egg-white foams (St. John and Flor 1931)
	Triglycerides of egg yolk more detrimental to albumen foams than cholesterol and phospholipid fractions (Smith, 1959)
	Selected heat treatments beneficial to foaming of egg white containing yolk (Cunningham and Cotterill 1964)
	Detrimental effects of yolk counteracted by addition of freeze-dried egg white (Sauter and Montoure 1975)
Oil	Reduced volume in egg-white foams and decreased stability with (≥5%) cottonseed oil (Henry and Barbour 1933)
	Stiff egg-white foams not prevented by cottonseed, corn, or coconut oils (Dizmang and Sunderlin 1933)
	Stability of egg-white foams inhibited by butterfat, cream, or raw whole milk (Dizmang and Sunderlin 1933)

(Continued)

TABLE 16.3. (Continued)

Ingredient	Influence
Surfactant	Olive oil more inhibitory to albumen foams than egg yolk (Bailey 1935)
	Anionic surfactants improve albumen foams (Cotterill et al. 1963B)
	Cationic surfactants improve yolk-free albumen foams (Cotterill et al. 1963B)
	Nonionic surfactants detrimental to albumen foams (Cotterill et al. 1963B)
Ester	Triethyl citrate reduces whip time of egg white with or without added yolk and improves volume of cakes containing systems with added yolk (Cotterill et al. 1963B)
	Triethyl citrate restores speed of foam formation in pasteurized egg white (Cunningham and Lineweaver 1965; Garibaldi et al. 1968)
Chemical modifier	3,3-Dimethylglutaric anhydride increases resistance of albumen to heat damage (Gandhi et al. 1968)
	Succinic anhydride improves heat stability and foaming of albumen (Ball and Winn 1982)
	Acetic anhydride improves foaming, but at high levels causes reduced volume of angel cakes (Ball and Winn 1982)
Emulsifier	Volume of sponge cakes is reduced when lecithin is added to the yolks used for foam (Cunningham 1975)
	Some reduction in synersis of souffles when lecithin:cholesterol ratios are 10:1 (Shemwell and Stadelman 1976)
Stabilizer	Hydroxymethyl cellulose prevents collapse of frozen souffles and meringues (Cimino et al. 1967)
	Hydrocolloid stabilizers recommended to reduce synersis in meringues (Godston 1950)
	Guar gum improves meringues cooked by microwaves (Baldwin et al. 1968)

stability for egg-white foams is reached before maximum volume is attained (Barmore 1936). With whole eggs, Briant and Willman (1956) found that increasing the time of beating caused improved volume in foams but did not influence the volume of the batter or of the sponge cakes made with the foams.

Both the rate and motion of beaters can influence the incorporation of air into foams. St. John and Flor (1931) used a hand beater and obtained greater volume with thin rather than with thick egg white. This was also true for Henry and Barbour (1933), who used an electric mixer for beating. Bailey (1935) produced larger, more stable foams with thick whites than with thin when using a mixer with a hypocycloidal motion and found that the foam became immobilized around beaters with fixed axes. The ease with which thin egg-white foams with some beaters is due to the ease with which the thin egg white can be spread into thin layers which trap air quickly.

Blending or Homogenizing.   Forsythe and Bergquist (1951) reported that blending causes a decrease in the ovomucin fiber length until a constant length of about 200 μm is attained. Increases are found in both beating rate of eggs and volume of cakes when the egg white is blended so that ovomucin fibers are about 300 μm in length. Homogenization also exerts an influence on the physical state of the ovomucin, which causes reduced whip time and lower volume in cakes containing homogenized albumen.

Temperature.   Elevation of temperature results in lowering surface tension. Thus, albumen foams more easily and attains greater volume at room temperature than at refrigerator temperature (St. John and Flor 1931). Foam stability, however, is affected little by a change in temperature from 20°C (68°F) to 34°C (93°F) (Barmore 1934).

Slosberg *et al.* (1948) heated egg white for very short periods at temperatures up to 58°C (136°F) without detrimental effects on the subsequent rate of beating or upon the volume of cakes made with the heat-treated albumen. Holding at temperatures higher than 58°C (136°F) adversely affected the beating characteristics of albumen. This finding confirmed Barmore's (1934) conclusion that egg white can be heated to 50°C (122°F) for 30 min without inhibiting foam-forming properties, but heating for 15 min at 65°C (146°F) causes decreased stability of foams.

Pasteurized egg white requires longer whipping to attain a foam comparable in specific gravity to foam from unpasteurized albumen. Garibaldi *et al.* (1968) attributed this to an irreversably denatured ovomucin-lysozyme network, the removal of which restores normal foaming ability to egg white.

Acidity or Alkalinity.   When the pH of egg white is adjusted to 8.75, improved performance, reflected by volume of cakes and whip time, results from heating albumen for 3 min at 58°C (136°F). This does not hold for other pH levels (7.0 to 9.5) or for more rigorous heat treatments (Seideman *et al.* 1963). Also, an increase in the stability of egg white to heat was shown by Slosberg *et al.* (1948) when the pH was 6.5 and when the acidulant was lactic or hydrochloric acid or potassium acid tartrate. However, beading and leakage of meringues are not reduced by addition of potassium acid tartrate (Hester and Personius 1949).

Consideration of the influence of pH on the heat stability of specific proteins was essential to Cunningham and Lineweaver's (1965) recommendations to adjust albumen to pH 7 before pasteurization, thus protecting the ovalbumin, lysozyme, and ovomucoid against damage by heat. Conalbumin is stabilized by the addition of aluminum. Although whip time is increased by this treatment, excellent volume and texture is attained in angel cakes made from albumen so treated.

Rhodes *et al.* (1960) improved the whipping properties of albumen from duck eggs by adding lemon juice. The acid apparently exerts an influence on the ovomucin which reduces the time required for whipping. Angel cakes made with the acidified duck egg white compare favorably with those made from chicken egg white.

Barmore (1934) concluded that $Ca(OH)_2$, NaOH, or Na_2SO_4 had little, if any, effect on albumen foams. However, acids increase whiteness of cake crumb (Grewe and Child 1930), and acids and acid salts improve stability of albumen foams. This can be attributed to a change in the protein concentrated at the liquid-air interface of the foam. Potassium acid tartrate affects the foam more favorably than acetic or citric acid when the pH is about 6.0. At pH 8.0 little difference can be found among the effects of the three acidulants. Increased stability of foam allows time for heat to penetrate cakes and causes coagulation without collapse of air cells, and thus prevents shrinkage of the cake during the last part of the baking period. Acid hydrolysis, however, plays no role in maintaining volume during baking of angel cakes (Barmore 1934).

Forsythe (1957) found that citric acid protects meringues against discoloration due to a red conalbumin-iron complex. The acid sequesters the iron thus preventing it from complexing with conalbumin.

Dilution. Henry and Barbour (1933) found that the volume of foams can be increased by adding water to the albumen before beating. The foams are almost as stable as those made from whites to which no water is added. Addition of 5 ml of water per 10 g of egg white not only increases the volume of foams but imparts a fluffy, less-pasty quality (St. John and Flor 1931). When more than 40% water is added, however, liquid separates from the foam on standing (Henry and Barbour 1933).

Cunningham (1972) reported that dilution permitted formation of egg-yolk foam with increased volume. However, the foams are less stable, and volume is diminished in sponge cakes made with the diluted yolk. When albumen is used to dilute the yolk, the foaming properties are superior to those of yolk containing water.

Sodium Chloride. Salt generally is believed to affect the stability of albumen foams adversely. Hanning (1945) illustrated this with foams beaten 6 min and containing 2 g of NaCl per 80 g of egg white. Smith (1960) indicated that NaCl increases whip time, but protects angel cakes against the detrimental influences of high humidity and low baking temperature. However, products containing NaCl exhibit an "open" grain. Conversely, a "close" grain is found in cakes containing $CaCl_2$. This may be due to the presence of calcium or to the increased whipping required when $CaCl_2$ is present. With whole eggs, adding NaCl before beating causes soft foams which are small in volume. Sponge cakes prepared with these foams are small and tough (Briant and Willman 1956).

Sugar. Delay of foam formation is apparent when sugar is incorporated into egg white. According to Hanning (1945), this effect is especially noted in the first part of the beating period. With 50% sugar, more than 9 min of beating is necessary to incorporate all liquid into the foam, whereas this requires only 3 to 4 min when no sugar is present. In addition, to attain comparable stiffness in foams, 32 min of beating is required for albumen containing 50% sugar, as opposed to 16 min without. At every period of beating, Hanning observed less expansion in the foams containing sugar than in those without. The relationship is linear when expansion factors versus the logarithms of the beating periods are plotted.

Gillis and Fitch (1956) viewed sugar content and beating as opposing forces, in that increased length of beating is required when the amount of sugar is increased. When sugar is added prior to heating for a short period (58°C or 136°F), the rate of beating is improved (Slosberg *et al.* 1948). The favorable effects are caused by addition of sucrose, lactose, dextrose, and maltose, especially the last three.

Soft meringues for pie toppings have been used extensively as a test medium for the interrelation among levels of sugar, extent of beating, and temperature of baking. Control of degree of coagulation appears to be the key to regulating undesirable leakage and beading in soft meringues. Hester and Personius (1949) found incomplete coagulation when meringues were spread on a cold base and baked a short time at a high temperature (218°C or 425°F, 4.5 min). Increased tendency for meringues to slip and greater amounts of liquid to collect also occurs when the temperature of the base decreases. When hot, rather than cold, syrup is used in preparation, less liquid is released from meringues (Felt *et al.* 1956). Beads of liquid form on the surface of meringues placed on a hot base and baked 10 min at 163°C (325°F) or 7 min at 191°C (375°F). Gillis and Fitch (1956) attributed this to overcoagulation of the egg white protein. When levels of sugar are the same, the greater leakage with meringues baked at 204°C (400°F) rather than 163°C (325°F) may be due to inadequate coagulation in the center of the meringues and overcoagulation in the outer areas. Increasing the amount of sugar retards overcoagulation at the surface. High levels of leakage due to inadequate coagulation occur in meringues containing high levels of sugar if they are baked at a low temperature (163°C or 325°F).

Souffles may or may not contain sugar. In either case overcoagulation accompanied by loss of fluid from the foam may occur. To lessen this tendency, souffles are generally baked in a container surrounded by hot water. This permits sufficient coagulation for the product to hold shape without overcoagulation occurring.

Egg Yolk. The influence of egg yolk on the performance of egg white is of special concern because it is practically impossible to produce completely yolk-free white on a commercial basis. St. John and Flor (1931) stated that as little as one drop of yolk causes a reduction from 135 to 40 ml in volume of egg-white foam.

Smith (1959) found that the triglyceride fraction of egg yolk is more detrimental to foaming of egg white than the cholesterol and phospholipid fractions. Pancreatic lipase, but not wheat-germ lipase, causes hydrolysis of the glyceride fractions of yolk to less inhibitory substances, thus counteracting the adverse effects of egg yolk on foaming of egg white (Cotterill and Funk 1963).

Cunningham and Cotterill (1964) attributed the detrimental influence of yolk on albumen to a complexing of a yolk component with ovomucin. Since heat treatments are beneficial to the foaming properties of egg white containing yolk, the complex may dissociate under the influence of heat. The time and temperature required to achieve this is 10 hr at 38°C (100°F), 1.5 hr at 43°C (110°F), or 15 min at 54°C (130°F). Improvement in function of the egg white when it is adjusted in either direction away from pH 8.5 indicates that dissociation occurs more easily at extremes of pH.

Sauter and Montoure (1975) successfully conteracted the detrimental effects of 0.4% yolk in albumen utilized for angel cakes by adding 2% freeze-dried egg white. The volume of cakes was similar to that of the control, and, with the exception of decreased tenderness, only minor differences in sensory properties were noted. It was postulated that the beneficial effects of freeze-dried egg white may be due to the replacement of protein fractions complexed by egg yolk with a lysozyme-ovomucin complex and/or to the additional ovoglobulin and ovomucin supplied.

Pankey and Stadelman (1969) altered the yolk composition and foaming properties of whole eggs by manipulating the composition of the dietary fat of laying hens. They ascribed little practical significance to the differences in volume of sponge cakes made with eggs representing different dietary treatments. Addition of 1, 2, or 4% lecithin to yolk, however, exerted a marked detrimental influence, which was reflected by diminished volume of sponge cakes made with the eggs containing these amounts of lecithin (Cunningham 1975).

Oil.   Henry and Barbour (1933) reported a reduction in volume of egg white foam and a tendency for foam structure to break down during continued beating when 0.01 to 1.0% refined cottonseed oil was added. However, stability of the beaten foam was not affected unless the amount of oil exceeded 0.5%. Contrary to these findings of Henry and Barbour, cottonseed was listed by Dizmang and Sunderlin (1933), along with corn and coconut oils, as fats lacking the power to inhibit stiff foam formation, whereas butterfat, cream, or unhomogenized raw whole milk were all classified as having pronounced inhibitory effects. Forsythe (1957) stated that even the small amounts of fat in flour are detrimental to the stability of egg-white foams.

Because olive oil reduces the foaming power of egg white more than the same amount of fat in the form of egg yolk, Bailey (1935) postulated that olive oil has greater foam inhibition properties than yolk, or that substances in the yolk have the power to counteract foam inhibition. However, movement of lipid material from the yolk to the albumen is the likely reason that albumen from mottled eggs has less foaming capacity than that from normal eggs (Cunningham 1977).

Surfactants, Modifiers, Emulsifiers, or Stabilizers.   Gardner (1960) reported on the effects of a wide variety of chemical additives on the physical and functional properties of egg white. He concluded that the complexing of an anionic surfactant (such as sodium lauryl sulfate) with lysozyme, coupled with a reduction in surface tension of the egg white system, are responsible for the improved performance caused by this group of compounds. Increased volume of angel cakes results when small quantities of sodium lauryl sulfate are added to egg white, but when concentrations are 0.48% or higher, coarseness results in the cakes in which the egg white foam is used. Sodium lauryl sulfate (0.03%) plus triethyl citrate (0.03%) increase the volume of meringues, but are not useful in improving the quality of meringues cooked by microwaves (Baldwin *et al.* 1968).

Cotterill *et al.* (1963B) reviewed the use of chemical additives for the purpose of improving the foaming of egg white which contains yolk, and conducted an investigation to clarify the role of these additives in improving the performance of egg white. Their study suggested that the ionic character of the additive may have significance in relation to altered performance in both yolk-free and yolk-containing egg-white systems. Improved volume of angel cakes is obtained with both yolk-free and yolk-containing foams when an anionic surfactant is added to the albumen. With the system containing yolk, the lipid is retained in the foam with this type of additive. Cationic surfactants do not cause retention of yolk lipids by the foam, and improvement in performance occurs only with the yolk-free system. A nonionic surfactant causes an increase in the lipid content of the drip collected from albumen foams and is detrimental to the performance of both the yolk-free and the yolk-containing albumen in angel cakes.

Triethyl citrate counteracts the detrimental effects of yolk on egg-white foams. It decreases the time required to whip both yolk-free and yolk-containing systems, but is beneficial in improving cake volume only for egg white containing yolk (Cotterill et al. 1963B). Addition of triethyl citrate to pasteurized egg was recommended by Cunningham and Lineweaver (1965) and by Garibaldi et al. (1968) to restore foaming properties to pasteurized egg. The improvement in whipping properties of the pasteurized albumen was attributed by Cunningham (1963) to increased insolubilization of protein at the air-liquid interface of the foam. Garibaldi et al. (1968) suggested that triethyl citrate may alter surficial viscosity of the egg white, thus decreasing the susceptibility of the foam to mechanical damage.

Addition of 3, 3-dimethylglutaric anhydride protects albumen (pH 7.5–8.5) against heat damage. Gandhi et al. (1968) suggested use of this additive (<15 mol/mol egg-white protein) to counter effects of high-temperature pasteurization on egg white. Succinic anhydride (≤20 mol/50,000 g protein) also improves heat stability of albumen without detrimental effects on volume of cakes made with the succinylated product. In contrast, acetic anhydride (20 mol/50,000 g egg white) diminishes heat stability but improves foaming ability of albumen (Ball and Winn, 1982).

Although loss of fluid from souffles is due to overcoagulation of the egg foam, a lecithin: cholesterol ratio of 10:1 lessens fluid loss. Other lecithin: cholesterol ratios (8:1, 6.7:2, 6.7:4) are ineffective in reducing fluid loss from spinach souffles (Shemwell and Stadelman 1976).

Godston (1950) recommended the use of hydrocolloid stabilizers for meringues at a level of 1% based on the weight of the meringue, or 3% based on the amount of water added. With the exception of seaplant extractives, hydrocolloids compete for water, and thus do not function well in combination. Seaplant extractives cause improvement in meringues by reacting directly with the protein of the egg white in such a way that loss of liquid from the meringues is prevented.

Guar gum (0.06% based on weight of liquid egg white) is effective in improving the quality of meringues cooked by microwaves, but algin imparts a gel-like quality to the products (Baldwin et al. 1968). Guar gum, especially in combination with triacetin (0.1 to 0.3%), reduces the time required for whipping meringues and also reduces the amount of drip released. This combination of additives can be utilized in a meringue-type dessert which is cooked successfully by microwaves (Upchurch and Baldwin 1968).

In general, food products containing large proportions of beaten egg, and low levels of other ingredients that contribute to the structure, do not maintain their high quality when subjected to frozen storage. However, the stabilizer, carboxymethylcellulose, is effective in preventing collapse of frozen souffles and meringues. In addition to this stabilizer, slight increases in the flour and sugar in the formulas improve stability and texture (Cimino et al. 1967).

EMULSIFICATION

Egg yolk is itself an emulsion, a dispersion of oil droplets in a continuous phase of aqueous components. In addition, the yolk is an efficient emulsifying agent. For this reason, egg yolk, or whole egg, is an essential ingredient in mayonnaise and salad dressings, and plays an important role in foods such as cake batter containing shortening, cream puffs, and hollandaise sauce.

Mechanism of Emulsion Formation

Vincent et al. (1966) stated that reduction of interfacial tension is probably the first step in formation of an emulsion and that surface-active agents in egg yolk are essential to its function in emulsification. The surface-active agents form a film around the oil globules and prevent their coalescence. In surrounding the oil droplets, the emulsifier orients itself with the nonpolar part of the molecule extending into the oil and the polar portion into the aqueous phase. An emulsifier that is more attracted to one phase than the other lowers the surface tension in the liquid in which it is more soluble. Thus, this liquid forms the continuous phase of the emulsion.

Components Involved in Emulsification

The results of early research on emulsions suggest that there may have been some confusion in distinguishing among foaming, stabilizing, and emulsifying. For example, Clark and Mann (1922) stated that albumin was a good emulsifier in comparison with sucrose, starch, dextrin, and gum arabic. In the light of present understanding of emulsions, the only one of this series that could be expected to have even slight emulsifying action is albumin.

Sell et al. (1935) found that egg yolk oil had little emulsifying action, but when it was combined with lecithoprotein, excellent emulsifying ability was attained. The emulsifying efficiency of the lecithoprotein is reduced by addition of lecithin or cephalin.

Yeadon *et al.* (1958) concluded that efficient emulsifiers are mixtures or complexes. This conclusion was based on findings that highly purified lecithin lacks the emulsifying properties of unpurified phosphatides. Chapin (1951) identified the protein and lipoprotein (the ether-insoluble portion) as the most important emulsifying substances in whole egg and found that addition of phospholipids reduces the emulsifying ability. The emulsifying capacity of whole egg is unaltered by changes in fatty acid composition of egg yolk accomplished through dietary regimen of the laying hens (Pankey and Stadelman 1969).

Lecithin favors formation of an oil-in-water emulsion, whereas cholesterol tends to form a water-in-oil emulsion (Corran and Lewis 1924). The usual method of preparing mayonnaise by combining egg yolk and seasonings with part of the water phase, adding oil, and finally the remaining aqueous phase, is conducive to formation of a water-in-oil emulsion. The emulsifying agent is responsible for reversing this phenomenon. It appears, however, that egg yolk should promote a water-in-oil emulsion, because the lecithin:cholesterol ratio in fresh egg yolk is 6.7:1, and ratios lower than 8:1 are conducive to water-in-oil emulsions. Evidently, the finely divided solids in mayonnaise contribute significantly to the oil-in-water state. Sugar and spices tend to hold water, thus preventing excessive free water (Kilgore 1935).

For a stable emulsion, there must be a sufficient amount of emulsifier (King 1941), but an excess contributes to instability. Cunningham (1975) verified this detrimental effect of excess emulsifier and presented data illustrating decreased emulsifying capacity of egg yolk to which 2 to 4% lecithin was added.

In preparing emulsions such as mayonnaise, increasing the amount of egg yolk from 5 to 12% increases the thickness of the product. Amounts of yolk in excess of 12% are not associated with a change in consistency of the mayonnaise. It is essential that there be a sufficient amount of aqueous phase (usually vinegar) to surround the oil droplets as they become smaller and expose more surface. However, an excess of aqueous phase is not conducive to a good emulsion (Kilgore 1933, 1935).

Salad dressings are also oil-in-water emulsions, but the concentration of oil is only about 45% as compared to 65 to 75% in mayonnaise. In addition to egg yolk, stabilizers are permitted in this product. Hanson and Fletcher (1961) increased the egg-yolk level and found increased stability of salad dressing in frozen storage. Use of 8% fresh yolk prevented oil separation for 3 months in salad dressing stored at $-7°$ to $-34°C$ ($+20°$ to $-30°F$).

Measuring Emulsifying Properties

In studying the initial state of emulsions, Chapin (1951) measured viscosity by means of a "mobilometer." This instrument consists of a vertical tube in which the test sample is placed. A plunger of known weight travels through the sample, and the time required for this is recorded (Gray and Southwick, 1929). A yield pressure is no doubt required before the plunger begins its fall, because mayonnaise and similar foods exhibit non-Newtonian flow properties. For this reason, instruments are unsuitable for testing viscosity if they depend strictly upon the flow properties of the emulsion, or if they cut a path through the product which offers little if any resistance as the instrument retraces its course.

Chapin (1951) plotted viscosity against concentration of emulsifier to obtain an emulsifying index. This should not be confused with emulsifying capacity, which is a measure of grams of oil emulsified per milligram of protein nitrogen. To determine emulsifying capacity, Gandhi *et al.* (1968) prepared solutions containing 6.0 mg of protein nitrogen in 10 ml KCl (0.6 M) buffered at a specific pH. Oil was stirred in (0.6 ml/sec, 1500 rpm) until the emulsion broke.

Phase-inversion titrations were employed by Cotterill *et al.* (1976) to determine emulsifying capacity. Measurement of electrical resistance was employed by these researchers to determine the phase inversion point, but it was indicated that this point can be detected visually. Both monophasic and biphasic titrations were described, and the applicability of the latter method to both oil-in-water and water-in-oil emulsions was explained.

The distribution and size of oil droplets in mayonnaise can be studied by direct microscopic count of oil droplets (Finkle *et al.* 1923). Also, calculation of the distribution of oil particles by size can be derived based on the rate of change in density of the upper part of a diluted emulsion during standing. Further information concerning the dispersion of the oil droplets can be obtained by photographing the light-scattering pattern of diluted emulsions (Stamm 1925).

The stability of emulsions is determined by measuring changes in characteristics of the emulsion with time. These changes include resistance to coalescence of oil drops and the appearance of a water layer. The latter phenomenon is referred to as creaming. Kilgore (1933) tested the tendency toward creaming by shaking a sample of mayonnaise with an equal weight of water, pouring it into a graduated cylinder and allowing it to stand until separation occurred.

Centrifugation of emulsions hastens the formation of separate layers and is also a useful test of the stability. Yeadon *et al.* (1958) developed a test that requires homogenizing, adding oil, and centrifuging. The amount of oil remaining emulsified after a standing period following completion of centrifugation was calculated as percent and expressed as the stability index. The stability index for a good emulsion should be 94% or above, whereas it might be only 50% for a poor emulsion. Fifty-gram samples are adequate for performing this test.

Factors Affecting Emulsions

Egg yolk contributes viscosity to emulsions, and viscous emulsions tend to be stable. Also, a viscous continuous phase promotes stability in emulsions because it impedes movement and coalescence of the dispersed oil droplets (King and Mukherjee 1940).

The decrease in viscosity of egg yolk caused by dilution, may have influenced the lessened stability of emulsions observed by Varadarajulu and Cunningham (1972), but the reduction in solids content was probably the major factor. These researchers admonished processors to maintain sufficient solids content in their products to permit them to function effectively as emulsifiers. To do this, dilution of yolk with albumen should be maintained below 20%.

The stability of emulsions can be improved significantly by addition of 0.05% sodium-2-lactylate, but no effect is achieved with levels of 0.02% or less of this additive (Varadarajulu and Cunningham 1972). Stability of emulsions also can be improved by acetylation and succinylation of egg yolk lipoproteins (Tsutsui *et al.* 1980; Tsutsui and Obara 1980A). Another approach to improving stability of emulsion is by the fermentation of egg yolk. Dulith and Groger (1981) observed both improved emulsifying capacity and heat stability (100°C or 212°F, 30 min) in mayonnaise prepared with egg yolk fermented with pancreatic phospholipase A2.

CONTROL OF CRYSTALLIZATION

The major role of egg white in candy is to control the growth of sugar crystals. Other functions include promoting incorporation of air, imparting a short or friable quality, and binding water. The latter quality contributes to reduction or prevention of syneresis, excessive evaporation, and fluidity due to the inversion of sucrose (Cotterill *et al.* 1963A).

Solutions of sugar and water cannot entrap air when beaten. However, when albumen is dispersed in such a solution, it is possible to incorporate air by beating. The large protein molecules increase the viscosity of the solution and also agglomerate at the interface of the air cells, thus stabilizing the foam. The foam, as such, contributes to the quality of some candies, whereas the foam plus the supporting structure provided by heat-coagulated albumen are essential in other candies (Carlin 1948; Forsythe and Cunningham 1968).

Swanson (1929) conducted an extensive investigation of the effect of egg white on crystallization of sugar from syrups. The effectiveness of egg white in delaying crystallization was illustrated by contrasting the rapidity with which crystals formed in a sugar and water syrup heated to 113°C (236°F) before whipping, with the long time required for crystal formation when the same type of syrup was incorporated into stiffly beaten egg white. Her work indicated that adding both egg white and corn syrup results in formation of smaller crystals than adding corn syrup alone. Swanson (1933) recommended using 38 g of egg white per 454 g of sugar in divinity. Too little egg white allows growth of large sugar crystals, and too much results in a fluffy product which becomes powdery. In fondant, 3% egg white added after the cooked syrup has crystallized prevents excessive growth of crystals during storage.

Crystal size, specific volume, and solubility were the criteria used by Cotterill et al. (1963A) to evaluate the influence of pH, surface tension, and presence of yolk on performance of albumen in divinity candy. Crystal size was determined by microscopic observations of crystals. It was found that divinity made with egg white adjusted within the pH range of 4.0 to 5.5, exhibits greater specific volume and greater solubility than that prepared with egg white adjusted to pH levels of 8.0 to 10.0. Also, more beating is required and crystallization and "set-up" occur faster in candy prepared from albumen adjusted to the low pH range. Addition of yolk (up to 0.4%) to egg white results in increased crystal size, and a cationic surfactant causes a reduction in crystal size.

Cotterill et al. (1963A) suggested that since both egg white and corn syrup function as "doctors" and exert some control over crystal growth, the role of egg white might be delineated more clearly by altering the formula for candy. Reducing the level of corn syrup would place increased stress on the role of egg white in the product.

COLOR

According to Bornstein and Bartov (1966) the visual impression of egg yolk color determines the acceptability of eggs. Preference has been shown for gold- or lemon-colored yolks (Schroeder 1933).

The naturally occurring pigments in chicken egg yolk are mainly alcohol-soluble xanthophylls, lutein and zeaxanthin. Increased intensity of orange color can be achieved in the yolks of chicken eggs by

altering the diet of the laying hens (Angalet *et al.* 1976). Very little
β-carotene or crypotoxanthin are present in chicken egg yolks (Forsythe
1963), whereas β-carotene is one of the principal pigments in duck eggs.
The orange color, characteristic of yolk of Philippine-bred duck eggs, is
due to ketocarotenoids (Rodriquez *et al.* 1976).

Contribution of Color to Foods

Egg yolk contributes a pleasing yellow color to many foods, especially
baked products, noodles, ice creams, custards, sauces, and omelets.
However, it is rare for eggs to be used as an ingredient in food products
for their color contribution alone. A rich yellow color was used in the egg
substitutes marketed in the early part of the twentieth century, but aside
from the yellow color, which suggested the presence of eggs, these
substitutes did not resemble eggs in their functional properties (Romanoff and Romanoff 1949).

Deethardt and co-workers (1965A,B) conducted extensive investigations on the influence of egg yolk color on the quality of sponge cakes. By
controlling the diet of the hens, eggs were obtained in which the yolks
ranged from a light yellow to a deep orange color. Yolks from eggs laid
by hens fed a diet high in canthaxanthin caused an undesirable pink
color in cakes (Deethardt *et al.* 1965A). In general, cakes made with
light-colored yolks were preferred. When red lights were used for illumi-
nation in order to minimize color differences, 74% of the judges indicated
a preference for cakes containing light-colored yolks. This preference
increased to 96% when natural lighting was used (Deethardt *et al.*
1965B).

Sensory evaluations indicated that sponge cakes containing dark-
colored yolks were more moist than those made from light-colored yolks,
although moisture content did not differ significantly. Specific gravity
was higher for beaten yolk-sugar mixtures containing dark-colored
yolks than for those containing light-colored yolks. There appeared to be
a relationship between the specific gravity of the egg yolk-sugar mixture
and the moistness ratings by the sensory panel. Moistness ratings were
higher for cakes containing yolks darkened by natural xanthophyll
than for cakes containing yolks darkened by synthetic additives in the
diets of the hens. In qualities other than moistness, however, these cakes
were not preferred (Deethardt *et al.* 1965A,B).

Color Changes Due to Food Preparation Procedures

Xanthophylls, the major pigments in egg yolk, tend to be stable to the
conditions encountered in normal food preparation. The main color
problem is that of a greenish-black discoloration on the surface of the

yolk of hard-cooked eggs. This discoloration is due to formation of FeS and not to the pigments in the yolk (Tinkler and Soar 1920; Baker *et al.* 1967). Albumen is the source of the H_2S which enters into the FeS reaction during the heating of eggs in the shell. Baker *et al.* (1967) suggested that the enzyme cystine desulfhydrase may be responsible for the occurrence of H_2S in egg white. Duck eggs are less subject to this type of discoloration than chicken eggs, perhaps because duck eggs contain less potentially volatile sulfides (Rhodes *et al.* 1960). The longer the eggs are held in boiling water during cooking, the greater the discoloration. Cooling and peeling the eggs as soon as possible after cooking tends to retard green color formation. Baker *et al.* (1967) postulated that during heating the outward pressure of gases tends to prevent the reaction between H_2S from the albumen and the iron from the yolk. However, while the egg cools, it contracts and the H_2S moves toward the yolk. If the eggs are held for a period of time after cooking, the discoloration lessens due to oxidation of FeS. Increased alkalinity in the egg yolk favors FeS formation.

Formation of FeS is sometimes encountered on the underside of omelets. As a precaution against this, the egg white should be beaten well and thoroughly blended with the yolk, and the mixture should be cooked promptly.

Occasionally, the albumen of hard-cooked eggs is caramel-brown in color. This color is due to carbonyl-amine-type browning. To minimize this discoloration, Baker and Darfler (1969) recommended using fresh eggs for hard-cooking and cooling them immediately after cooking.

Hard-cooked egg rolls undergo a brownish discoloration when exposed to ultraviolet radiation. Schnell *et al.* (1969) attributed this to oxidation of tryptophan. The oxidative nature of the reaction was suggested by observations of a brown color in coagulated albumen heated with oxidizing agents.

Conalbumin complexes with iron to form a red product. The conalbumin-iron complex is not likely to influence the color of food products. However, it is possible for the reaction to occur in an acidic product high in content of egg white, if it is prepared in an iron container. This reaction is the basis for the test of conalbumin activity devised by Seideman *et al.* (1964) which utilizes the chemical assay technique of Fraenkel-Conrat and Feeney (1950).

FLAVOR

One of the most important factors influencing acceptability of eggs is flavor (Hard *et al.* 1963; Romanoff and Romanoff 1949). Because early researchers were concerned with the flavor and aroma of eggs, a committee was organized to study and develop a technique for evaluating their odor and flavor. Sharp (1937) summarized the recommendations of this committee as follows:

1. Egg yolks with a minimum of flavor should be used as standards. To develop a series of samples with a range of flavors, pure chemicals should be added to the yolks selected for minimum flavor.
2. Judges should evaluate eggs with a range of flavor from time to time, and, after agreeing on the flavor in the samples, should attempt to memorize the standards.

Perhaps, because the flavors and/or odors are seldom due to a single compound, little reference is made to the above recommendations in later work.

Off-flavors in eggs should be detected readily because the characteristic flavor is bland (Maga 1982). However, Miller *et al.* (1960) found that consumers did not distinguish between the flavor of eggs of different ages, and Holdas and May (1966) noted that some panel members consistently identified a fishy flavor in certain eggs, although no statistically significant off-flavor was established. Thus, some individuals may be sensitive to certain off-flavors, while others are unable to detect a difference. Another factor that has complicated some research on flavors and off-flavors is the fact that flavor differences among freshly laid eggs from hens fed different rations disappear during 3 months of storage under refrigeration (Jordan *et al.* 1960).

Gaebe (1940) tested only the yolk of eggs, because it tends to absorb flavors more readily than the white due to its high fat content. Egg white possesses a flavor and odor, but has little tendency to absorb flavors. In Gaebe's study, tests were conducted mainly on raw eggs in order to avoid changes in volatile components due to cooking. Some odor judgements were made on cooked eggs as soon as the tip was removed and before flavor was tested.

When flavor differences have been identified, descriptions of off-flavors have been somewhat vague and include terms such as "taint," "stale," "foreign," or "flat" (Maga 1982). The terms "fresh," "mild," and "sweet" were used to describe eggs that were rated as acceptable in the study by Hard and co-workers (1963).

Taste panel evaluations, conducted by Koehler and Jacobson (1966), indicated that richness, mustiness, and astringency of yolk are masked or diluted by the presence of egg white. The flavor described as "hydrolyzed protein" was apparent in whole egg and in certain fractions of yolk, but not in unfractionated yolk. Also, a chemical flavor was evident in certain yolk fractions, but not in whole egg or unfractionated yolk.

Koehler and Jacobson (1966) pointed out that most previous studies were concerned with flavor of eggs as influenced by diet of the hens, season of the year, or conditions of handling and storage. Continued interest in the relationship between the dietary regimen of laying hens and flavor of eggs is evident. Koehler and Bearse (1975) attributed stale, rancid, chemical, and fishy flavors to fish meals or oils in the feed provided for laying hens. This same condition of off-flavor can result from the inclusion of rapeseed in the diet of the hen (Pearson *et al.* 1979; Ahmed Shuaib *et al.* 1981).

Flavor Components

In an extensive review of the volatile constituents of eggs and egg products compiled by Maga (1982), both saturated and unsaturated hydrocarbons, phenols, indans, indoles, pyrroles, pyrazines, and sulfides were listed. It was pointed out that, although MacLeod and Cave (1975, 1976) failed to report some details of their procedures, they extracted more than 100 volatile components from eggs. Compounds found in highest concentrations were *n*-pentadecane, 5-heptadecene, toluene, and indole. Only a few aldehydes, ketones, and sulfur compounds were present. Dimethyl sulfide and dimethyl trisulfide, although occurring in small amounts, were credited with contributing to the characteristic flavor, but no one compound was determined as being responsible for the characteristic flavor of eggs.

Occurrence of decomposition in eggs is reflected by fluctuations of volatiles other than ethanol. Therefore, Rayner *et al.* (1980) proposed a gas chromatographic technique for monitoring the shelf life of eggs.

Flavor Changes during Cooking

Romanoff and Romanoff (1949) stated that the flavor of fresh eggs can be distinguished from that of stale eggs even when the eggs are combined with other ingredients. However, in an extensive investigation of flavor of eggs as related to their cooking quality, off-flavor was least apparent when the eggs were used in angel cakes as compared to custards or soft cooking. Off-flavors were most apparent in the soft-cooked eggs (Dawson *et al.* 1956). Banwart *et al.* (1957) reported that the off-flavor in stored oiled eggs after soft-cooking was more detectable in the white than in the yolk.

When eggs were evaluated after holding in cold storage (0°C or 32°F), soft-cooked eggs are comparable in flavor to high-quality eggs for approximately 15 weeks as compared to 13 weeks when stored in a refrigerator (8°C or 45°F), and only 1 week when held at room temperature. However, when eggs that are held under the above conditions are used in baked custards or angel cakes, the flavor score remains high for as long as 24 weeks. Thus, added ingredients minimize or mask flavor changes that occur in eggs during storage (Dawson *et al.* 1956). Yet Gaebe (1940) stated that cooking makes flavors more pronounced. It is understandable that if the off-flavor is a volatile substance, it would be more easily driven off during the baking of a custard or angel cake than during soft-cooking of eggs. Also, the type of cooking container may be an influential factor. Chen and Hsu (1981) found less H_2S, NH_3, and total carbonyls in eggs that were scrambled in a double boiler than in those prepared in pouches and cooked either in boiling water or by microwaves.

Flavor changes accompany the carbonyl-amine browning observed in some hard-cooked egg whites. The off-flavor, described as bitter by

TABLE 16.4.   Use and Function of Eggs in Foods

Food product	Functional property	Selected studies
Cakes, angel	Foaming and coagulation	Ball and Winn (1982); Barmore (1934, 1936); Cotterill and Funk (1963); Cotterill et al. (1963B); Cunningham and Cotterill (1962); Forsythe and Bergquist (1951); Gardner (1960); MacDonnell et al. (1955); Sauter and Montoure (1972); Seideman et al. (1963); Slosberg et al. (1948); Smith (1960)
	Foaming and coagulation of duck egg white	Rhodes et al. (1960)
	Flavor	Dawson et al. (1956)
Cakes, sponge	Foaming and coagulation of whole egg	Briant and Willman (1956); Cunningham (1972, 1975)
	Color of egg yolk	Deethardt et al. (1965A,B)
Candy	Inhibition of crystals	Cotterill et al. (1963A); Swanson (1929, 1933)
Custards	Coagulation	Andross (1940); Carr and Trout (1942); Lowe (1955); Wang et al. (1973)
	Flavor	Dawson et al. (1956)
Egg white	Coagulation	Barmore (1936); Baldwin et al. (1967); Beveridge et al. (1980); Cunningham and Cotterill (1962); Rhodes et al. (1960); Romanoff and Romanoff (1949); Seideman et al. (1963)
Eggs, cooked in shell, fried, scrambled, or poached	Coagulation	Andross (1940); Chen and Hsu (1981)
	Flavor	Chen and Hsu (1981); Dawson et al. (1956); Hard et al. (1963); Jordan et al. (1960); Koehler and Jacobson (1966); Miller et al. (1960)
Mayonnaise	Emulsifying	Kilgore (1933, 1935); Sell et al. (1935)
Meringues and soufflés	Foaming	Cimino et al. (1967); Godston (1950); Upchurch and Baldwin (1968)
	Foaming and coagulation	Baldwin et al. (1968); Gillis and Fitch (1956); Felt et al. (1956); Hester and Personius (1949); Shemwell and Stadleman (1976)
Salad dressing	Emulsifying	Hanson and Fletcher (1961)

some judges, increases if the cooked eggs are held at an elevated temperature (90°C or 176°F) up to 25 hr (Baker and Darfler 1969).

Data on methods of preparation of eggs indicate the following order of preference among consumers: fried, scrambled, soft-boiled, poached, and hard-boiled (Miller *et al.* 1960). Since fried and scrambled eggs have the highest preference rating, it is possible that the fat used in cooking may have some marked influence on the flavor components.

SUMMARY

The functional properties of eggs include coagulating, foaming, emulsifying, and contributing color and flavor. Nutritive value is also an important functional property; it is discussed in Chapter 7. Table 16.4 is a summary of the studies in which functional properties of eggs have been utilized in various ways.

REFERENCES

Ahmed Shuaib, A. C. A., Beswick, G., and Tomlins, R. I. 1981. The thiocyanate ion (SCN⁻) content of eggs from hens (*Gallus domesticus*) fed on a diet containing rapeseed meal. J. Sci. Food Agric. *32*, 347–352.

Amerine, M. A., Pangborn, R. M., and Roessler, E. B. 1965. Principles of Sensory Evaluation of Food. Academic Press, New York.

Andross, M. 1940. Effect of cooking on eggs. Chem. Ind. (London) *59*, 449–454.

Angalet, S. A., Fry, J. L., Damron, B. L., and Harms, R. H. 1976. Evaluation of waste activated sludge (citrus) as a poultry feed ingredient. 2. Quality and flavor of broilers, egg yolk color and egg flavor. Poult. Sci. *55*, 1219–1225.

Azari, P. R., and Feeney, R. E. 1958. Resistance of metal complexes of conalbumin and transferrin to proteolysis and to thermal denaturation. J. Biol. Chem. *232*, 293–302.

Bailey, M. I. 1935. Foaming of egg white. Ind. Eng. Chem. *27*, 973–976.

Baker, G. L. 1938. Improved Delaware jelly strength tester. Fruit Prod. J. *17*, 329–330.

Baker, R. C., and Darfler, J. 1969. Discoloration of egg albumen in hard-cooked eggs. Food Technol. *23*, 77–79.

Baker, R. C., Darfler, J., and Lifshitz, A. 1967. Factors affecting the discoloration of hard-cooked egg yolks. Poult. Sci. *46*, 664–672.

Baldwin, R. E., Matter, J. C., Upchurch, R., and Breidenstein, D. M. 1967. Effects of microwaves on egg whites. I. Characteristics of coagulation. Food Sci. *32*, 305–309.

Baldwin, R. E., Upchurch, R., and Cotterill, O. J. 1968. Effects of additives on meringues cooked by microwaves and by baking. Food Technol. *22*, 1573–1576.

Ball, H. R., Jr., and Winn, S. E. 1982. Acylation of egg white proteins with acetic anhydride and succinic anhydride. Poult. Sci. *61*, 1041–1046.

Banwart, S. F., Carlin, A. F., and Cotterill, O. J. 1957. Flavor of untreated, oiled, and thermostabilized shell eggs after storage at 34°F. Food Technol. *11*, 200–204.

Barmore, M. A. 1934. The influence of chemical and physical factors on egg-white foams. Colo., Agric. Exp. Stn., Tech. Bull. *9*.

Barmore, M. A.   1936.   The influence of various factors, including altitude, in the pro-
    duction of angel food cake. Colo., Agric. Exp. Stn., Tech. Bull. 15.
Barnard, K. M.   1966.   New materials aid comparisons of volume, texture, and contour
    of baked products. J. Home Econ. 58, 470–480.
Becher, P.   1957.   Emulsions: Theory and Practice. Van Nostrand-Reinhold, Princeton,
    New Jersey.
Beveridge, T., Arntfield, S., Ko, S., and Chung, J. K. L.   1980.   Firmness of heat induced
    albumen coagulum. Poult. Sci. 59, 1229–1236.
Binnington, D. S., and Geddes, W. G.   1938.   An improved wide-range volume measur-
    ing apparatus for small loaves. Cereal Chem. 15, 235–246.
Bornstein, S., and Bartov, I.   1966.   Studies on egg yolk pigmentation. Poult. Sci. 45,
    287–296.
Briant, A. M., and Willman, A. R.   1956.   Whole-egg sponge cakes. J. Home Econ. 48,
    420–421.
Brooks, J., and Taylor, D. J.   1955.   Eggs and egg products. Part II. Physico-chemical
    properties of the hen's egg. Dep. Sci. Ind. Res., Food Invest. Board, Spec. Rep. 60.
Carlin, G. T.   1948.   How to use egg albumen in candy. MC, Manuf. Confect. 28(10),
    30–32, 65–67.
Carney, R. I.   1938.   Determination of stability of beaten egg albumen. Food Ind. 10, 77.
Carr, R. E., and Trout, G. M.   1942.   Some cooking qualities of homogenized milk. I.
    Baked and soft custard. Food Res. 7, 360–369.
Chapin, R. B.   1951.   Some factors affecting the emulsifying properties of hen's egg.
    Ph.D. Dissertation, Iowa State Univ., Ames.
Chen, T. C., and Hsu, S. Y.   1981.   Quality attributes of whole egg and albumen mixtures
    cooked by different methods. J. Food Sci. 46, 984–986.
Chick, H., and Martin, C. J.   1910.   On the "heat coagulation" of proteins. J. Physiol.
    (London) 40, 404–430.
Chick, H., and Martin, C. J.   1911.   On the "heat coagulation" of proteins. Part II. The
    action of hot water upon egg albumen and the influence of acid and salts upon
    reaction velocity. J. Physiol. (London) 43, 1–27.
Chick, H., and Martin, C. J.   1912A.   On the "heat coagulation" of proteins. Part III.
    The influence of alkali upon reaction velocity. J. Physiol. (London) 45, 61–67.
Chick, H., and Martin, C. J.   1912B.   On the "heat coagulation" of proteins. IV. The
    conditions controlling agglutination of proteins already acted upon by hot water. J.
    Physiol. (London) 45, 261–295.
Cimino, S. L., Elliott, L. F., and Palmer, H. H.   1967.   The stability of souffles and
    meringues subjected to frozen storage. Food Technol. 21, 1149–1152.
Clark, G. L., and Mann, W. A.   1922.   A quantitative study of the adsorption in solution
    and at interfaces of sugars, dextrin, starch, gum arabic, and egg albumin, and the
    mechanism of their action as emulsifying agents. J. Biol. Chem. 52, 157–182.
Cloninger, M.   1967.   Salmonella typhimurum in egg white foam pie cooked by micro-
    wave. M.S. Thesis, University of Missouri-Columbia.
Corran, J. W., and Lewis, C. M.   1924.   Lecithin and cholesterol in relation to the
    physical nature of cell membranes. Biochem. J. 18, 1364–1370.
Cotterill, O. J.   1981.   A Scientist Speaks about Egg Products. American Egg Board,
    1460 Renaissance Drive Park Ridge, IL. 60068
Cotterill, O. J., and Funk, E. M.   1963.   Effect of pH and lipase treatment on yolk-
    contaminated egg white. Food Technol. 17, 1183–1188.
Cotterill, O. J., Amick, G. M., Kluge, B. A., and Rinard, V. C.   1963A.   Some factors
    affecting the performance of egg white in divinity candy. Poult. Sci. 42, 218–224.
Cotterill, O. J., Cunningham, F. E., and Funk, E. M.   1963B.   Effect of chemical addi-
    tives on yolk-contaminated liquid egg white. Poult. Sci. 42, 1049–1057.

Cotterill, O. J., Seideman, W. E., and Funk, E. M. 1965. Improving yolk-contaminated egg white by heat treatments. Poult. Sci. *44*, 228-235.

Cotterill, O. J., Glauert, J., and Bassett, H. J. 1976. Emulsifying properties of salted yolk after pasteurization and storage. Poult. Sci. *55*, 544-548.

Cunningham, F. E. 1963. Factors affecting the insolubilization of egg white proteins. Ph.D. Dissertation, Univ. of Missouri, Columbia.

Cunningham, F. E. 1972. Viscosity and functional ability of diluted egg yolk. J. Milk Food Technol. *35*, 615-617.

Cunningham, F. E. 1975. Influence of added lecithin on properties of hens' egg yolk. Poult. Sci. *54*, 1307-1308.

Cunningham, F. E. 1976. Properties of egg white foam drainage. Poult. Sci. *55*, 738-743.

Cunningham, F. E. 1977. Properties of albumen from eggs having mottled yolks. Poult. Sci. *56*, 1819-1821.

Cunningham, F. E., and Cotterill, O. J. 1962. Factors affecting alkaline coagulation of egg white. Poult. Sci. *41*, 1453-1461.

Cunningham, F. E., and Cotterill, O. J. 1964. Effect of centrifuging yolk-contaminated liquid egg white on functional performance. Poult. Sci. *43*, 283-291.

Cunningham, F. E., and Lineweaver, H. 1965. Stabilization of egg-white proteins to pasteurizing temperatures above 60°C. Food Technol. *19*, 1142-1447.

Dawson, E. H., Miller, C., and Redstrom, R. A. 1956. Cooking quality and flavor of eggs as related to candled quality, storage conditions, and other factors. Agric. Inf. Bull. (U.S., Dep. Agric.) *164*.

Deethardt, D. E., Burrill, L. M., and Carlson, C. W. 1965A. Relationship of egg yolk color to the quality of sponge cakes. Food Technol. *19*, 73-74.

Deethardt, D. E., Burrill, L. M., and Carlson, C. W. 1965B. Quality of sponge cakes made with egg yolks of varying color produced by different feed additives. Food Technol. *19*, 75-77.

Dizmang, V. R., and Sunderlin, G. 1933. The effect of milk on the whipping quality of egg white. U.S. Egg Poult. Mag. *39* (11), 18-19.

Dulith, C. E., and Groger, W. 1981. Improvement of product attributes of mayonnaise by enzymic hydrolysis of egg yolk with phospholipase A2. J. Sci. Food Agric. *32*, 451-458.

Feeney, R. E. 1964. Egg proteins. *In* Symposium on Foods: Proteins and Their Reactions. H. W. Schultz and A. F. Anglemier (Editors). AVI Publishing Co., Westport, CT.

Felt, S. A., Longree, K., and Briant, A. M. 1956. Instability of meringued pies. J. Am. Diet. Assoc. *32*, 710-715.

Ferry, J. D. 1948. Protein gels. Adv. Protein Chem. *4*, 2-78.

Fevold, H. L. 1951. Egg proteins. Adv. Protein Chem. *6*, 187-252.

Finkle, P., Draper, H. O., and Hildebrand, J. H. 1923. The theory of emulsification. J. Am. Oil. Chem. Soc. *45*, 2780-2788.

Forsythe, R. H. 1957. Some factors affecting the use of eggs in the food industry. Cereal Sci. Today *2*, 211-216.

Forsythe, R. H. 1963. Chemical and physical properties of eggs and egg products. Cereal Sci. Today. *8*, 309-310, 312, 328.

Forsythe, R. H., and Bergquist, D. H. 1951. The effect of physical treatments on some properties of egg white. Poult. Sci. *30*, 302-311.

Forsythe, R. H., and Cunningham, F. E. 1968. Expand use for egg albumen in candy products. Candy Ind. Confect. J. *130*, 85-91.

Forsythe, R. H., and Foster, J. F. 1949. Note on electrophoretic composition of egg white. Arch. Biochem. *20*, 161-163.

Fraenkel-Conrat, H. J., and Feeney, R. E. 1950. The metal binding activity of conalbumin. Arch. Biochem. Biophys. 29, 101–113.

Gaebe, O. F. 1940. A comparative odor and flavor study of eggs stored in avenized and unavenized fillers and flats. U. S. Egg Poult. Mag. 46(6), 346–349.

Galston, A. W., and Kaur, R. 1962. Interactions of pectin and protein in the heat coagulation of proteins. Science 138, 903–904.

Gandhi, S. K., Schultz, J. R., Boughey, F. W., and Forsythe, R. H. 1968. Chemical modification of egg white with 3,3-dimethylglutaric anhydride. Food Sci. 33, 163–169.

Gardner, A. F. 1960. The role of chemical additives in altering the functional properties of egg white. Ph.D. Dissertation, Univ. of Missouri, Columbia.

Garibaldi, J. A., Donovan, J. W., Davis, J. G., and Cimino, S. L. 1968. Heat denaturation of the ovomucin-lysozyme electrostatic complex: A source of damage to the whipping properties of pasteurized egg white. J. Food Sci. 33, 514–524.

Gillis, J. N., and Fitch, N. K. 1956. Leakage of baked soft-meringue topping. J. Home Econ. 48, 703–707.

Godston, J. 1950. The production of stable meringues, Baker's Dig. 110, 67–69.

Grawemeyer, E. A., and Pfund, M. D. 1943. Line-spread as an objective test for consistency. Food Res. 7, 105–108.

Gray, D. M., and Southwick, C. A., Jr. 1929. Scientific research affords a means of measuring stability of mayonnaise. Food Ind. 1, 300–304.

Grewe, E., and Child, A. M. 1930. The effect of acid potassium tartrate as an ingredient in angel cake. Cereal Chem. 7, 245–250.

Griswold, R. M. 1962. The Experimental Study of Foods. Houghton Mifflin Co., Boston, MA.

Hanning, F. M. 1945. Effect of sugar or salt upon denaturation produced by beating and upon the ease of formation and the stability of egg white foams. Iowa State Coll. J. Sci. 20, 10–12.

Hanson, H. L., and Fletcher, L. R. 1961. Salad dressings stable to frozen storage. Food Technol. 15, 256–262.

Hard, M. M., Spencer, J. V., Locke, R. S., and George, M. H. 1963. A comparison of different methods of preserving shell eggs. I. Effect on flavor. Science 42, 815–824.

Harns, J. V., Sauter, E. A., McLaren, B. A., and Stadelman, W. J. 1952. The use of angel food cake to test egg white quality. Poult. Sci. 31, 1083–1087.

Henry, W. C., and Barbour, A. D. 1933. Beating properties of egg white. Ind. Eng. Chem. 25, 1054–1058.

Hester, E. E., and Personius, C. J. 1949. Factors affecting the beading and leakage of soft meringues. Food Technol. 3, 236–240.

Holdas, A., and May, K. N. 1966. Fish oil and fishy flavor of eggs and carcasses of hens. Poult. Sci. 45, 1405–1407.

Johnson, T. M., and Zabik, M. E. 1981A. Egg albumen proteins interactions in an angel food cake system. J. Food Sci. 46, 1231–1236.

Johnson, T. M., and Zabik, M. E. 1981B. Gelation properties of albumen proteins, singly and in combination. Poult. Sci. 60, 2071–2083.

Jordan, R., Vail, G. E., Roger, J. C., and Stadelman, W. J. 1960. Functional properties and flavor of eggs laid by hens on diets containing different fats. Food Technol. 14, 418–422.

Kauzman, W. 1959. Some factors in the interpretation of protein denaturation. Adv. Protein Chem. 14, 1–63.

Kilgore, L. B. 1933. Observations of the proper amount of yolk in mayonnaise. U.S. Egg Poult. Mag. 39(3), 42–45, 63.

Kilgore, L. B. 1935. Egg yolk "makes" mayonnaise. Food Ind. 7, 229–230.

King, A. 1941. Some factors governing the stability of oil-in-water emulsions. Trans. Faraday Soc. *37*, 168–180.

King, A., and Mukherjee, L. N. 1940. The stability of emulsions. II. Emulsions stabilized by hydrophilic colloids. J. Soc. Chem. Ind., London, Trans. Commun. *59*, 185–191.

Koehler, H. K., and Bearse, G. E. 1975. Egg flavor quality as affected by fish meals or fish oils in laying rations. Poult. Sci. *54*, 881–889.

Koehler, H. H., and Jacobson, M. 1966. Flavor of egg yolk: Fractionation and profile analysis. Poult. Sci. *45*, 1371–1380.

Lepeschkin, W. W. 1922. The heat coagulation of proteins. Biochem. J. *16*, 678–701.

Lewis, P. C. 1926. The kinetics of protein denaturation. Part II. The effect of variation in the hydrogen ion concentration on the velocity of the heat denaturation of egg-albumin; the critical increment of the process. Biochem. J. *20*, 978–992.

Lowe, B. 1955. Experimental Cookery, 4th Edition. John Wiley and Sons Co., New York.

Ma, C. Y., and Holme, J. 1982. Effect of chemical modifications on some physiochemical properties and heat coagulation of egg albumen. J. Food Sci. *47*, 1454–1459.

MacDonnell, L. R., Hanson, J. L., Silva, R. B., Lineweaver, H., and Feeney, R. E. 1950. Shear—not pressure—harms egg white. Food Ind. *22*, 273–276.

MacDonnell, L. R., Feeney, R. E., Hanson, H. L., Campbell, A., and Sugihara, R. F. 1955. The functional properties of the egg white proteins. Food Technol. *9*, 49–53.

MacLeod, A. J., and Cave, S. J. 1975. Volatile flavour components of eggs. J. Sci. Food Agric. *26*, 351–360.

MacLeod, A. J., and Cave, S. J. 1976. Variations in the volatile flavour components of eggs. J. Sci. Food Agric. *27*, 799–806.

McCammon, R. B., Pittman, M., and Wilhelm, L. A. 1934. The odor and flavor of eggs. Poultry Sci. *13*, 95–101.

McKeller, D. M. B., and Stadelman, W. J. 1955. A method for measuring volume and drainage of egg white foams. Poult. Sci. *34*, 455–458.

McKenzie, H. A., Smith, M. B., and Wake, R. G. 1963. The denaturation of proteins. I. Sedimentation, diffusion, optical rotation, viscosity and gelation in urea solutions of ovalbumin and bovine serum albumin. Biochim. Biophys. Acta *69*, 222–239.

Maga, J. S. 1982. Egg and egg product flavor. J. Agric. Food Chem. *30*, 9–14.

Mecham, D. K., and Olcott, H. S. 1949. Phosvitin, the principal phosphoprotein of egg yolk. J. Am. Chem. Soc. *71*, 3670–3679.

Melnick, D., and Oser, B. L. 1949. The influence of heat-processing on the functional and nutritive properties of protein. Food Technol. *3*, 57–71.

Merrill, D. D. 1963. Effects of microwaves on egg white. Master's Thesis, Univ. of Missouri, Columbia.

Miller, C., Sanborn, L. D., Abplanalp, H., and Stewart, G. F. 1960. Consumer reaction to egg flavor. Poult. Sci. *39*, 3–7.

Niles, K. B. 1937. Egg whites on parade. U.S. Egg Poult. Mag. *43*(6), 337–340.

Pankey, R. D., and Stadelman, W. J. 1969. Effect of dietary fats on some chemical and functional properties of eggs. J. Food Sci. *34*, 312–317.

Parkinson, T. L. 1966. The chemical composition of eggs. J. Sci. Food Agric. *17*, 101–111.

Pearson, A. W., Butler, E. J., and Fenwick, G. R. 1979. Rapeseed meal goitrogens and egg taint. Vet. Rec. *104*, 168.

Rayner, E. T., Dupuy, H. P., Legendre, M. G., Schuller, W. H., and Holbrook, D. M. 1980. Assessment of egg flavor (odor) quality by unconventional gas chromatography. Poult. Sci. *59*, 2348–2351.

Rhodes, M. B., Adams, J. L., Bennett, N., and Feeney, R. E. 1960. Properties and food uses of duck eggs. Poult. Sci. *39*, 1473–1478.

Rodriquez, D. B., Arroyo, P. T., Bucoy, A. S., and Chichester, C. O. 1976. Identification of pigments of the egg yolk from Philippine-bred ducks. J. Food Sci. *41*, 1418–1420.

Romanoff, A. L., and Romanoff, A. J. 1949. The Avian Egg. John Wiley and Sons Co., New York.

Sauter, E. A., and Montoure, J. E. 1972. The relationship of lysozyme content of egg white to volume and stability of foams. J. Food Sci. *37*, 918–920.

Sauter, E. A., and Montoure, J. E. 1975. Effects of adding 2% freeze-dried egg white to batter of angel food cakes made with white containing egg yolk. J. Food Sci. *40*, 869–871.

Schnell, P. G., Vadehra, D. V., and Baker, R. C. 1969. The cause of discoloration of hard cooked egg rolls. Food Sci. *34*, 423–426.

Schroeder, C. H. 1933. Interior egg quality. U.S. Egg Poult. Mag. *39*(6), 44–46.

Seideman, W. C., Cotterill, O. J., and Funk, E. M. 1963. Factors affecting heat coagulation of egg white. Poult. Sci. *42*, 406–417.

Seideman, W. E., Hill, W. M., and Cotterill, O. J. 1964. Colorimetric determination of conalbumin activity in egg white. Food Science and Nutrition Dept., University of Missouri, Columbia (unpublished).

Sell, H. M., Olsen, A. G., and Kremers, R. E. 1935. Lecithoprotein: The emulsifying ingredient in egg yolk. Ind. Eng. Chem. *27*, 1222–1223.

Sharp, P. F. 1937. Report of committee studying odors and flavors in eggs. U.S. Egg Poult. Mag. *43*(6), 307.

Shemwell, G. A., and Stadelman, W. J. 1976. Some factors affecting weepage of baked spinach souffles. Poult. Sci. *55*, 1467–1472.

Slosberg, H. M., Hanson, H. L., Stewart, G. F., and Lowe, B. 1948. Factors influencing the effects of heat treatment on the leavening power of egg white. Poult. Sci. *27*, 294–301.

Smith, C. F. 1959. Shell egg deterioration. Diffusion of yolk lipids into albumen as the natural cause of failure in performance. Poult. Sci. *38*, 181–192.

Smith, C. F. 1960. Functional properties of dried egg albumen in baked products. H. J. Heinz Co., Pittsburgh, PA (personal communication).

Smith, M. B. 1964. Studies on ovalbumin. I. Denaturation by heat, and the heterogencity of ovalbumin. Aust. J. Biol. Sci. *17*, 261–270.

Smith, M. B., and Back, J. F. 1965. Studies on ovalbumin. II. The formation and properties of S-ovalbumin, a more stable form of ovalbumin. Aust. J. Biol. Sci. *18*, 365–377.

Stamm, A. J. 1925. An experimental study of emulsification on the basis of distribution of size particles. Colloid Symp. Monogr. *3*, 251–267.

Steven, F. 1961. Starch gel electrophoresis of hen egg white, oviduct white, yolk, ova and serum proteins. Nature (London) *192*, 972.

St. John, J. L., and Flor, I. H. 1931. A study of whipping and coagulation of eggs of varying quality. Poult. Sci. *10*, 71–82.

Swanson, E. L. 1929. The effect of egg albumen on the crystallization of sugar from sirups. M.S. Thesis, Iowa State Univ., Ames.

Swanson, E. L. 1933. Egg whites and sugar crystals. U.S. Egg Poult. Mag. *39*(9), 32–33, 56.

Tinkler, C. K., and Soar, M. C. 1920. The formation of ferrous sulphide in eggs during cooking. Biochem. J. *14*, 114–119.

Tsutsui, T., and Obara, T. 1980A. Studies of changes in egg yolk components during heating. J. Jpn. Soc. Food Sci. Technol. *27*, 7–13.

Tsutsui, T., and Obara, T.   1980B.   Effect of succinylation on the emulsifying properties of egg yolk protein components. J. Jpn. Soc. Food Sci. Technol. 27, 293–297.

Tsutsui, T., Matsumoto, S., and Obara, T.   1980.   Effect of acetylation on emulsifying properties of egg yolk protein components. J. Jpn. Soc. Food Sci. Technol. 27, 448–452.

Upchurch, R., and Baldwin, R. E.   1968.   Guar gum in meringues and a meringue product cooked by microwaves. Food Technol. 22, 107–108.

Varadarajulu, P., and Cunningham, F. E.   1972.   A study of selected characteristics of hens' yolk. I. Influence of albumen and selected additives. Poult. Sci. 51, 542–546.

Vincent, R., Powrie, W. D., and Fennema, O.   1966.   Surface activity of yolk, plasma and dispersions of yolk fractions. Food Sci. 31, 643–648.

Wang, A. C., Funk, K., and Zabik, M. E.   1973.   Effect of sucrose on quality characteristics of baked custards. Poult. Sci. 53, 807–813.

Woodward, S. A.   1984.   Texture and microstructure of heat-formed egg white, egg yolk and whole egg gels. Ph.D. Dissertation, University of Missouri, Columbia.

Woodward, S. A., and Cotterill, O. J.   1985.   Scanning electron microscopy of cooled egg white, egg yolk, and whole egg gels prepared by various methods. Food Sci. 50, 1624–1628.

Woodward, S. A., and Cotterill, O. J.   1986.   Texture and microstructure of heat-formed egg white gels. Food Sci. 51, 333–339.

Yeadon, D. A., Goldblatt, L. A., and Altschul, A. M.   1958.   Lecithin in oil-in-water emulsions. J. Am. Oil Chem. Soc. 35, 435–438.

17

Hard-Cooked Eggs

W.J. Stadelman

The hard cooking, peeling, and preservation of eggs at food processing plants is a rapidly expanding egg merchandising process. The eggs so prepared are distributed to the retail market, the food service industry, and food manufacturers. Although attempts were made shortly after World War II to open up this market based on a patent by Trelease *et al.* (1952), it was not until the last 15 years that a significant number of eggs have been sold as the hard-cooked, peeled product.

An egg hard cooked in the shell, according to Irmiter *et al.* (1970), should rate high in the following criteria: (1) shell does not break during cooking, (2) shell peels off easily and does not adhere to the coagulated albumen, and (3) yolk should be well centered and free of dark rings. Although research has been done on the factors affecting each of these criteria, there is still some question as to how to obtain the highest quality product.

MINIMIZING CRACKING OF SHELLS DURING COOKING

To minimize the loss of quality in hard-cooked eggs by cracking during cooking, it is essential that only sound-shelled eggs be used. Hale and Holleman (1978) investigated the effect of cooking method, egg temperature, shell piercing, age of eggs, and oiling on shell cracking during cooking. Pressure cooking (5 psi) increased the incidence of cracked eggs as compared with cooking at atmospheric pressure. Shell cracking was slightly increased by beginning the cooking process in

385

EGG SCIENCE & TECHNOLOGY,
3rd Edition

boiling water, especially with poor quality eggs. Shell cracking was highest in eggs with large air cells. Shell piercing was found to have little effect on shell cracking during cooking. Irmiter *et al.* (1970) reported a greater incidence of cracked eggs when cooking was started in cold rather than boiling water. They also tested the effects of drilling 2.5-mm holes in shells rather than piercing them. The eggs with drilled holes had a greater incidence of cracking during cooking. The temperature of the eggs at the start of cooking had little or no effect on shell cracking (Irmiter *et al.* 1970; Hale and Holleman 1978).

Carter (1973) found that the piercing of fresh eggshells had no effect on breakage during hard cooking. With 5-day-old eggs the incidence of cracking was reduced by piercing, and with 28-day-old eggs piercing the shell over the air cell completely eliminated cracking during hard cooking. He theorized that in fresh eggs the shell is porous so the hole had no effect. As eggs are held in storage, shell permeability decreases and the air-cell size increases. In 5-day-old eggs the air cells were not large enough to take the full expansion of the egg during cooking. In contrast the air cell at 28 days was large enough for expansion, allowing a rapid escape of heated gases from the air cell through the hole.

FACTORS AFFECTING EASE OF PEELING

Hard-cooked egg producers have developed a two-point grading system for their finished products—No. 1 and salad grade eggs. The No. 1 grade eggs are intact eggs with no torn albumen or exposed yolk. The salad grade includes all imperfectly peeled eggs. Swanson (1959) observed that freshly laid eggs were difficult to peel after hard cooking and that eggs should be held until the pH of the albumen reached 8.7 or more. Eggs held for 48 hr at a temperature of about 15°C gave the desired pH in the albumen. Exposing eggs to ammonium hydroxide fumes in a closed chamber for 10 min gave similar results. Meehan *et al.* (1961) reported difficulty in peeling until the pH reached 8.9. Reinke and Spencer (1964) found that an albumen pH of 8.7 to 8.9 or higher resulted in easy-to-peel hard-cooked eggs. They observed that the shell membranes of easy-to-peel eggs were more compact than the shell membranes of hard-to-peel eggs when studied microscopically.

Based on the above observations as to the pH of the albumen necessary to achieve easy peeling of hard-cooked eggs, it is obvious that treating the shell with oil will result in hard-to-peel eggs. All of the research workers reporting on the effects of pH on the peelability of hard-cooked eggs arrived at this same conclusion, beginning with the work of McIntosh *et al.* 1942.

In measuring ease of peeling on peelability two factors are involved:
(1) the time required to remove the shell and (2) the appearance of the
peeled hard-cooked egg. Hard *et al.* (1963) who studied the effects of
several preservation methods (see Chapter 4) on the subsequent peelability of hard-cooked eggs, they reported that the time required to peel
an egg after one of several treatments was as follows:

Treatment	Time (sec)
Untreated	11
CO_2 plus oil	19
Oil	21
Silicone grease	22
Thermostabilization	10

While these times are in no way indicative of commercial rates of
peeling they do show the negative and dramatic effect of shell treatment
with oil.

Fuller and Angus (1969) reported on the effects of cooking medium
and pretreatment of eggs on the peelability of hard-cooked eggs. They
cooked eggs in (1) tap water, (2) water with 1, 5, and 10% added sodium
chloride, and (3) the same levels of calcium chloride. Eggs were cooked
after 0-, 1-, or 2-days storage at 10° to 13°C. The addition of salts to the
cooking medium did not produce any change in the peelability of freshly
laid eggs, but sodium chloride improved the peelability scores of eggs
stored for 24 hours and possibly after 48 hours. They observed that good
peeling characteristics were obtained when the albumen pH was above
the range 8.6 to 8.9. Holding of eggs at 50°C for 24 hours or more resulted
in easy-to-peel eggs. According to Fuller and Angus (1969), holding eggs
at elevated temperatures for relatively short time periods greatly improved peelability (see Table 17.1).

Attempts have been made to develop procedures that would minimize
the need for aging of eggs or allow shell treatment with oil and still get
an easy-to-peel hard-cooked egg. The Poultry and Egg National Board
(Anon. 1966) published a procedure for hard cooking eggs that recommended puncturing the shell at the large end of the egg. It claimed that
this procedure "when used on eggs less than one hour from the nest,
produced hard-cooked eggs which could be easily and beautifully
peeled." The device used for puncturing the shell—a piercer—is described in an American Egg Board publication (Anon. 1981) as follows:
"Piercer. A sharp-pointed tool for making a small hole in an egg shell
before cooking. You can substitute a thumb tack or pin. Piercing the egg
lets air escape while the egg is cooking in its shell and may help make

TABLE 17.1. Relationship between Peelability Score and Storage
Temperature.[a]

Storage time (hr)	Average peelability score vs. storage temperature			
	10°–13° C	27° C	40° C	50° C
0	4.7	5.0	5.0	5.0
5.5	—	3.2	2.8	2.4
19	—	1.2	1.2	0
24	4.2	0.6	0.2	0

SOURCE: *Fuller and Angus (1969).*
[a] The lower the score, the better the peelability.

peeling easier." The effectiveness of piercing the shell in making for
easier peeling was investigated by Irmiter *et al.* (1970) and Hale and
Holleman (1978); both groups reported no beneficial effect on ease of
peeling.

Fry *et al.* (1966) exposed eggs to ionizing radiation prior to hard
cooking in an unsuccessful attempt to improve ease of peeling of fresh
eggs. Exposure of eggs to 100,000 rep of β-rays resulted in an immediate
loss of albumen structure according to Parsons and Stadelman (1957).
Fry *et al.* (1966) used 100 krad of γ-irradiation on eggs just prior to hard
cooking and 12 days prior to hard cooking; the irradiation treatment
had no effect on ease of peeling.

YOLK CENTERING AND SURFACE COLOR

A high-quality market egg must have a well-centered yolk. The effects
of storage position of the egg on yolk centering was studied by Campos
et al. (1975) and Cardetti *et al.* (1979), who found that yolk centering was
best in eggs that were stored on their side with the long axis of the egg
parallel to the earth's surface. The next preferred position was small end
up. Grunden *et al.* (1975), who investigated the effect of egg position
during cooking on yolk centering, reported that yolk position in hard-
cooked eggs is influenced by heredity (strain of chicken), cooking position, and age of egg. Brant and Sanborn (1963) found that age of hen,
egg position, agitation, and temperature during storage all affect yolk
position in raw eggs. Grunden *et al.* (1975) recommended cooking eggs
in an upright or standing position to achieve improved yolk centering.

Discoloration of the yolk surface of hard-cooked eggs is due to the
formation of ferrous sulfide produced at the interface of the yolk and the
albumen (Tinkler and Soar 1920). Iron from the yolk reacts with hydrogen sulfide from the albumen. Baker *et al.* (1967) confirmed this early
report and indicated that the severity of the discoloration could be
influenced by cooking temperature, cooking time, pH of the yolk, and
storage time before cooking. After cooking, the amount of discoloration

is influenced further by method of cooling. Discoloration of hard-cooked eggs is not a significant problem in commercial operations, where the eggs are cooked, peeled, and then generally packed in an organic acid solution. The pH change in the hard-cooked egg results in a disappearance of yolk discoloration. The pH of the hard-cooked white changes as much in 2 hr as the yolk pH in 2 days according to Acton and Johnson (1973). The hard-cooked egg and organic acid solutions reach equilibrium in about 6 days. Britton and Hale (1975) concluded that albumen pH was the key to yolk discoloration and ease of peeling; a low albumen pH caused less yolk discoloration but more difficulty in peeling.

The microbiological quality of hard-cooked eggs is determined by sanitation practices during cooling, peeling, and packaging. Acton and Johnson (1973) found staphylococcus and both aerobic and anaerobic spores in eggs pickled in a 3% acetic acid solution. Counts were low and no coagulase-positive organisms were found. They concluded that the pickled eggs were bacteriologically safe based on pH of the solution and low bacterial counts. Stadelman et al. (1982) found yeast and mold growth in packages of hard-cooked eggs packed in dilute citric acid solution after 1-week storage at either 4° or 22°C. No bacterial cells were detected by standard plate-count methods.

Oblinger and Angalet (1974) studied the microbiological growth on peeled and unpeeled hard-cooked eggs stored at 5° and 25°C. The 25°C storage temperature permitted rapid development of bacteria after 1 day of storage. Sliminess was visible on the peeled eggs after 4 days at 25°C. Both peeled and unpeeled hard-cooked eggs stored at 5°C failed to show any significant bacterial development during 24 days of storage. Based on these data, hard-cooked eggs used for Easter displays at room temperature should be eaten or converted to pickled eggs within one day or discarded.

The flavor of hard-cooked eggs was studied by Spencer and Tryhnew (1973). They held hard cooked eggs at 1.1°C for 3, 7, 10, 14, and 21 days before peeling and evaluating for sensory qualities. On a nine-point hedonic scale the flavor scores were as indicated in Table 17.2.

The appearance of hard-cooked, peeled eggs was studied by Nath et

TABLE 17.2. Flavor Scores of Hard-Cooked Eggs Stored at 1.1°C

Days of storage	Average score[a]	Flavor loss (%)
0	8.41	—
3	7.10	15.6
7	5.73	32.6
10	4.71	44.3
14	4.70	47.0
21	4.03	53.6

SOURCE: Spencer and Tryhnew (1973).
[a] A nine-point scale with high score preferred.

al. (1971), who found that hard-cooked eggs yield a greenish-yellow liquid exudate, derived entirely from the albumen. When eggs are stored in an organic acid solution, the exudate takes on a cloudy appearance (Hale *et al.* 1981). Hard-cooked eggs with exposed yolk material caused a further increase in turbidity of the acid solution as well as a decrease in microbial quality of the packed product.

HARD COOKING EGGS

The American Egg Board (Anon. 1981) recommends the following method for hard cooking eggs: put unbroken eggs in a single layer in a saucepan. Add enough tap water to come at least 1 in. above the eggs. Cover and bring rapidly to just boiling. Turn off heat. If necessary, remove the pan from the burner to prevent further boiling. Let stand, covered, for 15 to 17 minutes for large eggs. Adjust the time up or down by approximately 3 min for each size, larger or smaller. Cool immediately and thoroughly in cold water to prevent a dark surface on yolks.

Maurer (1975) compared the above method with several other procedures, including starting with warm (58°C) water, overcooking by boiling for 20 min, and starting with cool (5°C) or warm (21°C) eggs. He also studied eggs that had a small hole in the shell made by an egg piercer (Anon. 1981). In comparing results, Maurer (1975) found that all methods gave the same type of hard-cooked egg, except the 20-min boiling procedure which produced discolored egg yolks. The eggs with pierced shells exhibited excessive cracking during cooking.

In the commercial preparation of hard-cooked eggs the cooking is performed by immersion of the eggs on wet-heat resistant filler-flats in tap water, followed by quickly heating the flats by steam or gas to 90° to 95°C. Hale *et al.* (1981) used this commercial method with a 30-min hold time. Irmiter *et al.* (1970) recommended starting the cooking of eggs in hot water, but Hale and Holleman (1978) reported that starting with boiling water, as compared to tap water, produced a tougher albumen in the hard-cooked egg.

COOLING HARD-COOKED EGGS

Most researchers recommend that the eggs be cooled rapidly in clean tap water. Such cooling minimizes yolk discoloration as first reported in 1920 by Tinkler and Soar. If eggs are to be kept in the shell after hard cooking, the microbial shelf life will be enhanced by allowing the eggs to cool in air. Keeping the hard-cooked eggs in a dry condition will retard microbial spoilage.

PEELING OR SHELLING OF HARD-COOKED EGGS

The removal of the shell from hard-cooked eggs is a labor-intensive operation. Albumen pH is the major controlling factor in ease of peeling hard-cooked eggs as reported by Fuller and Angus (1969). These authors, as well as Reinke and Spencer (1964), reported that shell membranes of easy-to-peel eggs also appeared to be more compact than those of hard-to-peel eggs. It is possible that shell-membrane density is influenced by the albumen pH.

The instructions given by the American Egg Board (Anon. 1981) for peeling hard-cooked eggs are: "Cool the cooked egg promptly and thoroughly in running cold water (5 minutes isn't too long). Crackle the shell all over by tapping gently on a table or counter top. Roll the egg between the hands to loosen the shell, then peel it off, starting at the large end. Holding the egg under running water or dipping in water makes this operation easier." Equipment for peeling of eggs has been developed, but all machines thus far tear up too much albumen. For this reason, peeling of hard-cooked eggs is a hand operation. With the growing emphasis on the employment of handicapped persons in certain areas, it has been found that blind persons can do any outstanding job in peeling hard-cooked eggs.

Strict sanitation must be practiced in the peeling and further handling of hard-cooked eggs. Immediately after cooking, eggs are free of vegetative microbial cells. By application of strict sanitation this sterile state can be preserved and microbial contamination can be kept to a minimum (Figs. 17.1 and 17.2).

PACKAGING OF HARD-COOKED EGGS

Hard-cooked eggs prepared commercially are packed in containers with an organic acid and 0.1% sodium benzoate or potassium sorbate (mold inhibitors), as shown in Fig. 17.3. The most often used acid is citric acid. Hale et al. (1981) stored eggs in 0.2% and 0.85% citric acid with 0.1% sodium benzoate. Eggs packed in the more concentrated acid solution had firmer albumens and lower microbial counts after storage at 4°C. Stadelman et al. (1982) found that a solution of 2% malic acid with 0.5% sodium chloride gave good results when the packaged products were thermally processed using commercial techniques.

Various concentrations of one of several organic acids can be used. In addition to the citric and malic acids, already mentioned, fumaric, succinic, and lactic acids have been studied. It is generally agreed that if cost is included in the evaluation, citric acid is the preservative of choice.

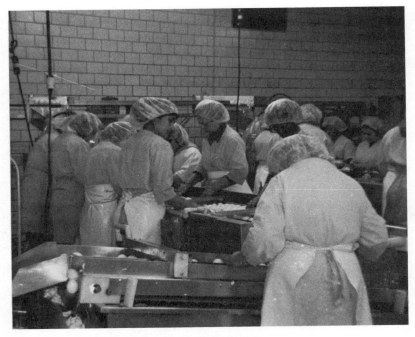

FIG. 17.1.   Women peeling hard-cooked eggs.
Courtesy of Country Queen Foods, Grand Rapids, Michigan.

The concentration of acid is usually about 0.1 N or about a 1.9% level. Hale *et al.* (1981) reported good results with 0.85% citric acid. The advantages of using a lower acid concentration are a less firm, more tender, albumen and a less acid taste. Fischer and Fletcher (1983) found that adding salt (NaCl) to the preserving solution resulted in improved flavor of the hard-cooked eggs after 30-days storage at 4°C.

When eggs are packed in a preservative, the weight of eggs is usually about twice the weight of acidic solution. The pH of the solution increases rapidly for the first few hours with equilibrium being attained in about 6 days (Acton and Johnson 1973). Stadelman *et al.* (1982) reported an increase in pH of the solution by about 0.3 pH units from 1 to 4 weeks, and a further slight increase of about 0.1 pH unit when eggs were kept for 5 months (Stadelman and Rhorer 1984).

When eggs are packed in a citric acid-sodium benzoate solution, it is essential that low refrigeration temperatures, down to −1°C, be maintained to achieve maximum shelf life. The production of shelf-stable hard-cooked eggs was studied by Trelease *et al.* (1952). Further details were given by Stadelman *et al.* (1982). For best results they recommend the packing of eggs in citric acid-sodium benzoate for 4 days and then repacking the eggs in a fresh solution of the acid-benzoate mixture in

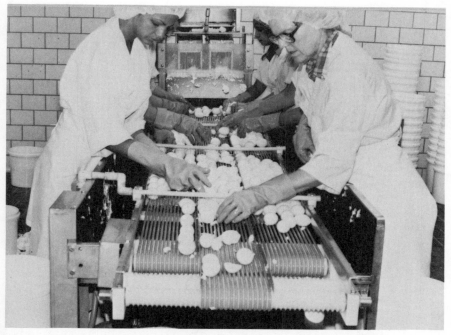

FIG. 17.2. Grading of hard-cooked peeled eggs into No. 1 and salad grades on the basis of smoothness of albumen surface.
Courtesy of Country Queen Foods, Grand Rapids, Michigan.

heat-retortable containers. The packages, cans, jars, or retortable pouches should then be subjected to either 99°C water (boiling) for 20 min or preferably to 121°C steam for 15 min. After cooling, the packages can be held at room temperature for more than 5 months. If temperatures are elevated to 38°C or more for 30 days, the egg albumen turns to a light tan color. The color becomes darker if temperatures are higher or if exposure is longer. After storage at 22°C for more than a year, only slight discoloration of albumen was found.

The thermally processed eggs do not change significantly during thermal processing or storage. When the package is opened, there is a sulfur odor which dissipates quickly. Otherwise, the eggs had no detectable off-odors or flavors (Stadelman *et al.* 1982).

USES OF HARD-COOKED EGGS

Hard-cooked eggs are extremely versatile. Slices and wedges can be used as garnishes on salads or cold-cut platters. Whole eggs make a delicious, nutritious snack. As a diced item, they are a frequent ingre-

FIG. 17.3.   Packaged hard-cooked eggs in 20-lb plastic buckets and pickled eggs in gallon jars.
Courtesy of Country Queen Foods, Grand Rapids, Michigan.

dient on salad bars. They are also the principal ingredient in egg salads. Furthermore, they may be pickled or spiced to yield a variety of additional products (Anon. 1981).

In commercial hard-cooked egg production the eggs are usually graded as No. 1 or as salad eggs. The No. 1 eggs are sold to the foodservice industry for use whole, wedged, or sliced. They are also used in the production of deviled eggs. Salad eggs are those with tears in the albumen or with missing chunks of albumen and exposed yolk material. Such eggs are used for dicing and subsequent freezing or for egg salad preparation.

Diced Eggs

Hard-cooked eggs, usually salad grade, are diced into cubes and then cryogenically frozen. Bengtsson (1967) found that egg-white blocks could be frozen to yield a satisfactory product when circulating nitrogen gas at −150°C or a liquid nitrogen spray was used. Slower methods of freezing resulted in an unsuitable product because of water loss on thawing, poor consistency, watery taste, and the unsatisfactory appearance of cut surfaces.

After freezing to a temperature of at least −10°C the diced eggs are packaged in moisture-oxygen resistant film and are stored at −20°C or lower temperature. The cubes should be free flowing at the time of removal from the freezer which indicates proper frozen storage. Storage of the diced eggs for up to 6 mon at $-20°C$ resulted in no deterioration in quality according to Bengtsson (1967).

Egg Salad

Hard-cooked eggs are chopped or ground, mixed with a salad dressing and seasonings to form a sandwich spread or the center of a stuffed tomato or a lettuce-leaf salad. Some delicatessens hesitate to sell egg salads because of reports that they are potential carriers of salmonella. Simmons *et al.* (1979) formulated an acidic egg salad that would not support salmonella growth when the acidic salad was stored at 22°C. Salad inoculated with over 6,000 colony-forming units of *Salmonella senftenberg* per milliliter were free of the organism after 60 hours at 22°C. The formula for this salad with pH of ingredients is:

Ingredient	Parts	pH
Chopped acidified egg	800	5.60
THS salad dressing	200	2.70
Sweet pickle relish	200	4.20
Mustard	50	4.00
Onions	75	6.20
Corn syrup	50	5.75

The pH of the final egg salad was 4.28. The onions were dehydrated with a 1:2 ratio of onion powder to water. With innovations such as acidic eggs and acidic salad dressings, egg salads are becoming more popular.

Deviled Eggs

According to the American Egg Board (Anon. 1981), deviled eggs are also known as stuffed eggs. Hard-cooked eggs are peeled and cut in half longitudinally. The yolks are removed, mixed with mayonnaise and seasonings, and then piled back into the depression in the whites from which the yolk had been removed. Newlin and Stadelman (1984) patented a set of molds to form cooked egg-white rings for deviled eggs.

Pickled Eggs

Pickled eggs are hard-cooked eggs marinated in brine. The brine is usually vinegar and pickling spices, but other marinades are also used.

TABLE 17.3. Recipes for Preparation of Pickled Eggs

	Type			
Ingredients[c]	Sweet and sour[a]	Spicy[a]	AEB[b]	Directions
Apple cider	1½ cups	1½ cups		1. Heat mixture to near
White wine vinegar	—	—	2 cups	boiling and allow to
Vinegar	½ cup	1 cup		simmer for 5–10 min.
Red cinnamon candy[d]	12 oz			2. Pour 1 cup of marin-
Pickling spice	1 tbsp	1 tsp	2 tsp	ade over 6 peeled,
Salt	2 tsp	2 tsp	1 tsp	hard-cooked eggs in
Garlic salt	1 tsp			quart jar
Garlic clove	—	1		3. Cover jar and place in
Onion, sliced, medium sized	—	½	1	refrigerator
Mustard seed	—	½ tsp		4. Marinate for a few hours or several weeks
Sugar	—	—	2 tbsp	

SOURCE: [a] *Maurer (1975)*; [b] *Anon. (1981)*.
[c] None of these recipes include beet juice, but this could be added, if desired.
[d] Added to impart color.

Frequently red-beet juice is added to the marinade to impart a red color to the egg. If marinated for only a few hours, the distinct red, white, and yellow colors can be seen when the egg is quartered or sliced. If the egg stays in the marinade for several days or more, the red color will migrate to the center of the egg. The simplest marinade would contain only vinegar and beet juice. Table 17.3 gives three recipes for pickled eggs.

McCready (1973) reported that for quick pickling of hard-cooked eggs the pickling solution should be at least 65°C when the eggs are added and should then be held at room temperature (about 24°C) for 24 hr. He also indicated that inclusion of sugar in the pickle formula resulted in a tougher pickled egg. Essary and Georgiades (1982) found that immersion of hard-cooked eggs in boiling pickling solutions resulted in a more tender pickled egg than immersion in pickle solutions at room temperature.

REFERENCES

Acton, J. C., and Johnson, M. G. 1973. Pickled eggs. 1. pH, rate of acid penetration into egg components and bacteriological analyses. Poult. Sci. *52*, 107–111.
Angalet, S. A. 1975. Physical and Microbiological Properties of Hard-cooked and Pickled Eggs (76-4214), J. L. Fry, and J. L. Oblinger, The University of Florida, Gainesville.
Anon. 1966. A World of Information about Eggs, Bull. E-23. Poultry and Egg National Board, Chicago, IL.
Anon. 1981. Eggcyclopedia. American Egg Board, Park Ridge, IL.
Baker, R. C., Darfler, J., and Lifshitz, A. 1967. Factors affecting the discoloration of hard-cooked egg yolks. Poult. Sci. *46*, 664–672.

Bengtsson, N. 1967. Ultrafast freezing of cooked egg white. Food Technol. *21*, 1259–1261.

Brant, A. W., and Sanborn, L. D. 1963. Some factors affecting broken-out yolk position. Poult. Sci. *42*, 227–231.

Britton, W. M., and Hale, K. K. 1975. Factors influencing peeling and yolk color of hard-cooked eggs. Poult. Sci. *54*, 1739 (abstr.).

Campos, E. J., Mellon, D. B., and Gardner, F. A. 1975. The effect of storage position on the interior quality of shell eggs. Poult. Sci. *54*, 1742 (abstr.).

Cardetti, M. M., Rhorer, A. R., and Stadelman, W. J. 1979. Effect of egg storage position on consumer quality attributes of shell eggs. Poult. Sci. *58*, 1403–1405.

Carter, T. C. 1973. The hen's eggs: Effect of a hole in the shell on incidence of splitting during cooking. Br. Poult. Sci. *14*, 485–491.

Essary, E. O., and Georgiades, M. O. 1982. Influence of pH, total acidity, pickling time, and solution temperature on shear values and sensory evaluation of pickled eggs. Poult. Sci. *61*, 2186–2193.

Fischer, J. R., and Fletcher, D. L. 1983. The effect of adding salt (NaCl) to the preservation solution on the acceptability of hard-cooked eggs. Poult. Sci. *62*, 1345 (abstr.).

Fry, J. L., Henrick, G. M., and Ahmed, E. M. 1966. Effect of irradiation on the peeling of newly laid eggs hard-cooked after irradiation. Food Technol. *20*, 1371.

Fuller, G. W., and Angus, P. A. 1969. Peelability of hard-cooked eggs. Poult. Sci. *48*, 1145–1151.

Grunden, L. P., Mulnix, E. J., Darfler, J. M., and Baker, R. C. 1975. Yolk position in hard-cooked eggs as related to heredity, age and cooking position. Poult. Sci. *54*, 546–552.

Hale, K. K., and Holleman, K. A. 1978. Factors affecting shell cracking, peeling ease and tenderness of hard-cooked eggs. Poult. Sci. *57*, 1187 (abstr.).

Hale, K. K., Potter, L. M., and Martin, R. B. 1981. Firmness and microbial quality of hard-cooked eggs stored in citric acid. Poult. Sci. *60*, 1664 (abstr.).

Hard, M. H., Spencer, J. V., Locke, R. S., and George, M. H. 1963. A comparison of different methods of preserving shell eggs. 2. Effect on functional properties. Poult. Sci. *42*, 1085–1095.

Irmiter, T. F., Dawson, L. E., and Reagan, J. G. 1970. Methods of preparing hard-cooked eggs. Poult. Sci. *49*, 1232–1236.

Maurer, A. J. 1975. Hard-cooking and pickling eggs as teaching aids. Poult. Sci. *54*, 1019–1024.

McCready, S. T. 1973. Temperature, percent sugar and pH effects on the flavor development and tenderness of pickled eggs. Poult. Sci. *52*, 1310–1317.

McIntosh, J. A., Tanner, R., Evans, R. J., and Carver, J. S. 1942. Cooking properties of eggs processed in mineral oil. U.S. Egg Poult. Mag. *48*, 345–347, 383.

Meehan, J. J., Sugihara, T. F., and Kline, L. 1961. Relation between internal egg quality stabilization and the peeling difficulty. Poult. Sci. *40*, 1430.

Nath, K. R., Vadehra, D. V., and Baker, R. C. 1971. The chemical nature of cooked egg exudate. Poult. Sci. *50*, 1611.

Newlin, J. L., and Stadelman, W. J. 1984. Deviled egg process. U.S. Pat. 4,426,400.

Oblinger, J. L., and Angalet, S. A. 1974. Storage stability of hard cooked eggs. Poult. Sci. *53*, 1415–1420.

Parsons, R. W., and Stadelman, W. J. 1957. Ionizing irradiation of fresh shell eggs. Poult. Sci. *36*, 319–322.

Reinke, W. C., and Spencer, J. V. 1964. Obeservation of some egg components in relation to peeling quality of hard-cooked eggs. Poult. Sci. *43*, 1355.

Simmons, S. E., deBartolucci, D. P., and Stadelman, W. J. 1979. Formulation and evaluation of a low pH egg salad. Food Sci. *44*, 1501–1504, 1509.

Spencer, J. V., and Tryhnew, L. J.   1973.   The effect of storage on peeling quality and flavor of hard-cooked shell eggs. Poult. Sci. *52*, 654–657.

Stadelman, W. J., and Rhorer, A. R.   1984.   Quality improvement of hard cooked eggs. Poult. Sci. *63*, 949–953.

Stadelman, W. J., Ikeme, A. I., Roop, R. A., and Simmons, S. E.   1982.   Thermally processed hard-cooked eggs. Poult. Sci. *61*, 388–391.

Swanson, M. H.   1959.   Some observations on the peeling problem of fresh and shell treated eggs when hard-cooked. Poult. Sci. *38*, 1253–1254 (abstr.).

Tinkler, C. K., and Soar, M. C.   1920.   The formation of ferrous sulfide in eggs during cooking. Biochem. J. *14*, 114–119.

Trelease, R. D., Sampson, G. O., and Strand, D. V. A.   1952.   Canning of hard-boiled eggs. U.S. Pat. 2,593,223.

Verstrate, J. A.   1980.   Chemical and Organoleptic Characterization of Flavor Changes During Storage of Hard-cooked Egg Albumen and Yolk (8025967). J. V. Spencer (Adviser). Washington State University, Pullman.

Nonfood Uses of Eggs

Ronald D. Galyean
Owen J. Cotterill

INTRODUCTION

The most significant use of eggs other than in human foods is, of course, for production of a new life. The egg is one of the miracles of nature in the way its design is perfectly suited for its primary purpose. In turn, man's efficient husbandry of poultry and egg production has allowed the development of a major food industry consisting of an annual production of more than 250 million pullet chicks placed in hatchery supply flocks each year, 200 million laying hens, and more than 400 million kilograms of egg products.

Historically, inedible eggs have played a part in the technological development of many industries. Many of the uses of inedible eggs were summarized by the U.S. Department of Agriculture (USDA) (1941). Past uses of egg products include albumen in photographic albumen paper, technical albumen as an adhesive in the bottling industry, and "tanner's egg yolk" as a leather dressing, particularly for suede and white leathers. Industrial egg albumen has been used for finishing certain types of glazed colored leathers, to season drum heads and banjo heads, and as an adhesive to attach gold lettering and gold and silver leaf to leather.

In each of these cases, eggs have filled an important part in the development of an industry. However, the unique functional properties of eggs or egg products have been replaced in many industries. For example, inexpensive plastics have replaced cork inserts in bottle caps, thus no longer requiring albumen as an adhesive; emulsion papers are now the common photographic media; strong synthetic adhesives have

399

replaced "technical" albumen as an adhesive. Thus, some of the non-food uses of eggs have been superseded by products of the same technology that were initially dependent upon the use of egg products.

Environmental concerns related to biological oxygen demand (BOD) and chemical oxygen demand (COD) have changed the egg industries' approach to by-product disposal. The combination of changes in marketing outlets for egg products and changes in environmental requirements on the egg industry have resulted in more emphasis to be placed on development of nonfood uses of eggs and egg-industry by-products.

Pet foods, shampoos, and vaccine production account for the majority of the current nonfood uses for eggs. Egg products also play an important role in bacterial culture media and the storage of semen for artificial insemination. This chapter is a discussion of some of these nonfood uses.

Eggs also have played an important part in the cultural and artistic life in the history of mankind, and continue to do so today. In that sense, the artistic use of eggs is also a nonfood use and will therefore be discussed in this chapter.

ANIMAL FEEDS

Pet Foods

Dog and cat foods represent a sizable outlet for eggs in today's supermarket and are becoming greater each year. A number of these foods include poultry by-products or parts that often include eggs. Several specialty items list whole eggs among other ingredients for the growing puppy and for other pets.

Feeds for Domestic Animals

Most uses of eggs by domestic animals are related to replacement of feedstuffs. Economic considerations are therefore paramount in the substitution of eggs for other feedstuff components. Except in isolated cases when eggs are in surplus, only eggshell waste or eggshell meal has been economically feasible as feedstuff replacement. These by-products are found in large quantities at broiler or egg-type chick hatcheries and at commercial egg-breaking operations. Eggshell waste is obtained either as a combination of the shell and liquid that normally clings to egg shells, or as the shell portion centrifugally separated from adhering liquid egg. Eggshell meal is generally a dehydrated form of eggshell waste.

Until the 1970s, only limited amounts of research were directed toward ascertaining the nutritive properties of eggshell waste or eggshell meal. Protein quality was the major concern with those egg by-products. Beginning in the 1970s, however, research was designed to determine the composition and quality of egg shell waste (Walton *et al.* 1973). Table 18.1 shows the proximate analysis of eggshell waste prior to centrifugation and after egg waste was centrifuged. The moisture level of product was moderately high both before (29.1%) and after (16.2%) centrifugation. The overall protein content on a dry weight basis is moderately low (7.56% and 5.31% dry weight basis, respectively). A study of protein composition was also conducted as a first step in determination of quality.

The composition of protein in egg white and in eggshell waste is shown in Table 18.2. That table is a comparison of the amino acid composition of protein from egg white and from centrifuged eggshell waste (the portion of eggshell after most adhering egg white has been removed). The most noticeable differences between egg-white protein and eggshell waste protein are in relative concentrations of the amino acids proline, glycine, isoleucine, leucine, tyrosine, and phenylalanine. The structural proteins found in egg-waste protein result in higher concentrations of proline. Additionally, some essential amino acids are found in lower concentrations than expected in egg-white protein. Basic amino acids in eggshell waste vary in their relationship to those found in egg white. Lysine concentrations are somewhat lower than in egg

TABLE 18.1. Gross and Elemental Composition of Eggshell Waste from Egg-Breaking Plants[a]

Components	With adhering albumen (%)	Centrifuged (%)	Washed (%)
Original moisture level (wet basis)	29.10	16.20	
Protein	7.56	5.31	5.15
Lipid	0.24	0.30	0.05
Ash	91.10	94.20	95.40
CaCO$_3$	90.90	91.80	93.10
CCE[b]	87.00	88.00	89.00
Calcium	36.4000	36.7000	37.3000
Iron	0.0020	0.0022	0.0023
Potassium	0.0970	0.0720	0.0600
Magnesium	0.3980	0.4000	0.4070
Sodium	0.1520	0.1260	0.1150
Sulfur	0.0910	0.0870	0.0430
Phosphorus	0.1160	0.1040	0.1170

SOURCE: *Walton et al. (1973).*
[a] Calculated on a dry basis.
[b] Neutralizing power expressed as CaCO$_3$.

TABLE 18.2. Amino Acid Composition of Eggshell Waste from Egg-Breaking Plants Compared to Egg White Expressed as a Percentage of Total Amino Acids Listed

		Eggshell waste		
	Egg white (Snider and Cotterill) (%)	With adhering albumen (6 samples) (%)	Centrifugal (2 samples) (%)	Washed (2 samples) (%)
Aspartic acid	10.11	9.84	9.59	9.11
Threonine	4.66	5.32	5.54	5.87
Serine	6.77	7.35	7.01	6.88
Glutamic acid	13.76	14.25	14.02	13.56
Proline	3.11	7.01	8.30	9.11
Glycine	3.19	5.76	7.01	7.09
Alanine	5.83	5.09	4.80	4.05
Cystine (includes cysteine)	2.49	4.64	3.69	7.09
Valine	7.93	6.11	5.90	5.87
Methionine	4.12	3.17	3.51	3.24
Isoleucine	5.29	3.85	3.51	3.03
Leucine	8.71	6.45	5.90	5.06
Tyrosine	3.81	2.83	2.77	2.43
Phenylalanine	6.92	4.30	3.32	2.02
Histidine	2.02	3.39	4.43	4.05
Lysine	5.91	4.19	3.69	4.05
Arginine	5.37	6.45	7.01	7.49
Total	100.00	100.00	100.00	100.00

SOURCE: *Walton et al. (1973).*

white, whereas histidine and arginine concentrations are somewhat higher.

An insight into the possibility of using eggshell waste as an animal dietary supplement was shown by Vandepopuliere *et al.* (1975). The nutritive value of eggshell meal from two methods of dehydration was tested with the result that eggshell meal fed to laying hens in diets of comparable protein, amino acid, and energy resulted in similar feed conversion and egg quality. The amino acid availability was comparable to that in diets prepared from a combination of wheat middlings and meat and bone meal or soybean meal. It also was found that the calcium from eggshell meal was as well utilized as that from ground limestone.

Thus, it appears that properly treated ground eggshell can be substituted in either laying or growing mash for oyster shell or high-grade limestone with satisfactory results for poultry. Both Arvat and Hinners (1973) and Vandepopuliere *et al.* (1973) reported that eggshells could be substituted as a calcium source in poultry rations. However, most states require that ground eggshell be processed to eliminate pathogenic organisms.

The microbial contamination of eggshell meal was summarized effectively by Vandepopuliere *et al.* (1975). They observed that the amino acid and mineral composition, as described by Walton *et al.* (1973), combined with a shell waste moisture content of 29% was an ideal growth medium for microorganisms. Thus, dehydration is used as one available method of controlling the growth of microorganisms and of utilizing available nutrients in eggshell waste. It has been shown in further studies by Vandepopuliere *et al.* (1978) that the numbers of surviving organisms in eggshell meal are dependent upon the temperature of the drier. A logarithmic decrease of surviving organisms was observed with increasing temperature.

As originally observed by Wilcke (1940), it is still true that the cost of eggshell processing as compared with other available sources of nutrients is the determining factor in its use as a calcium source or feedstuff substitute. The economic issue also has been addressed by Vandepopuliere *et al.* (1978). After developing a series of tables to describe cost based on a Missouri field installation, they concluded that the production cost of egg shell ranged from $13.58 to $39.82 per metric ton. The cost per ton decreased with increasing availability of eggshell waste.

Experimental Uses

Eggs are especially recognized as a rich source of high-quality protein. The white or albumen has a very high biological value in the amount and balance of amino acids. Egg white is used as a reference to compare proteins from other sources in the laboratory with animals such as the rat, mouse, chick, and others. The amino acid composition of whole hen egg has been used as the recommended profile for the Food and Agriculture Organization's (1965) chemical score for required amino acid in protein.

Individual egg-white proteins are now commonly used as molecular weight standards in sodium dodecyl sulfate polyacrylamide gel electrophoresis. Lysozyme, ovalbumin, and conalbumin are used to provide protein markers of molecular weights 14,600, 45,000, and 75,000 daltons, respectively.

FERTILIZER

Over the years many eggs and shells have been discarded by hatcheries as well as by processing plants. These have been dried and ground for fertilizer as well as feed. Within recent decades, however, environmental considerations such as air and water quality, including biological and chemical oxygen demands of waste products, have had to be taken into consideration.

Early studies by Lipman and Brown (1908) showed that albumen aids in the ammonification of soil, although it is too expensive for large-scale use. The rapid expansion of knowledge during the last two decades in commercial chemical fertilizers has been tremendous. In view of present-day costs, the use of egg waste for fertilizer (other than as a means for waste disposal) cannot be justified. The "organic gardener" might use these wastes to some extent. The use of eggshell as a soil "aerator" may be justified to some extent, although other factors such as soil pH also must be considered when shell is used in this manner.

BIOLOGICAL USES

Culture Media

Egg yolk is an excellent medium for the growth of microorganisms. Certain species of bacteria grow poorly on simple culture media, but thrive if small amounts of yolk are added. Egg yolk is used in a medium for the detection of *Clostridium perfringens* in foods (Hall *et al.* 1969).

Some egg products used as ingredients in commercially prepared media and media using egg products are as follows:

1. Egg albumen, soluble, dehydrated
2. Whole egg, soluble, dehydrated (available to bacteriologists who may desire to use for studies on bacterial metabolism)
3. Egg albumen, coagulated
4. Egg yolk, coagulated
5. Whole egg, coagulated (added to another medium for growth of anaerobic organisms)
6. Egg-meat medium (used to carry stock cultures of anaerobes and determination of proteolytic activity
7. Chick embryo tissue (a sterile, whole, minced chick embryo, freeze-dried for use in tissue-culture work)
8. Chick embryo extract (whole, undiluted extract of use in tissue culture procedures requiring embryo extract)
9. Petroff medium (glycerolated egg medium with beef infusion base to use with other media for isolation of tubercle bacilli)
10. Petrognani medium
11. Tellurite polymyxin egg yolk (TPEY) (developed for the isolation of *Staphylococcus*, helps in differentiating between coagulase positive and coagulase negative)

Medical and Pharmaceutical Uses

The fertile egg plays an important part in the manufacture of many vaccines. Dr. E. W. Goodpasture first realized the potential of the chick embryo for the propagation of various filtrable viruses. However, other scientists had experimented with the idea and prepared a few vaccines with varying degrees of success. In 1937, an epidemic of sleeping sickness infected 170,000 horses (40,000 died). Scientists used chick embryos to produce vaccines for this disease (Clarke 1940).

Cox (1952) listed the following viruses and rickettsia of animal infections propagated on embryonated eggs:

Routonneuse fever	Canine distemper
Eastern equine	Epidemic typhus
Fowl pox	Influenza
Laryngotracheitis	Louping ill
Mumps	Murine typhus
Newcastle disease	Pigeon pox
Psittacosis-LGV	Q Fever
Rabies	Rocky Mt. spotted fever
Vaccinia (cowpox)	Venezuelan equine
Western equine	Yellow fever

Other animal diseases treated with virus vaccines grown on chicken embryos include blue tongue disease in sheep, infectious bronchitis in poultry, and Rift Valley fever in sheep.

Purified Proteins

Several purified proteins derived from eggs are made available for research workers by biochemical supply houses. Those most recently listed include phosvitin, conalbumin, ovomucoid, albumen, crystalline (egg white), lysozyme chloride, and avidin.

Artificial Insemination

For a number of years a method for the preservation of bull semen has made use of egg yolk buffered at pH 6.75 as the pabulum for spermatozoa. By this means the fertilizing capacity of vigorous spermatozoa has been regularly maintained for periods in excess of 100 hr stored at 10°C (Phillips and Lardy 1940). Salsbury et al. (1941) used a dilute citrate solution, mixed with equal amounts of yolk, as a diluent to disperse the fat globules and other materials of the yolk. This technique improved the visibility of spermatozoa under the microscope.

MANUFACTURING

Leather

Eggs not suited for human consumption were packaged separately as "Tanner's Egg Yolk," although the product was usually a whole egg mixture. Formerly used as a dressing for leather and furs because of the excellent emulsifying capability of yolk components, Tanner's Egg Yolk is no longer used for this purpose. The trade name, however, has continued. Frozen and dried, Tanner's Egg Yolk was used to a limited extent in chinchilla food pellets (Koudele and Heinsohn 1964).

Cosmetics and Shampoos

Eggs have played an important part in the history of beauty culture, especially among women. According to specialists, the modern woman searching for ways to make the most of her natural endowment may well consider the egg as one of her allies. "Eggs," said the manager of one large beauty shop, "are wonderful for the hair. They not only cleanse the hair perfectly, but also leave it gleaming and soft with a beautiful natural sheen. Eggs often do wonders for the skin that needs tightening and cleaning up" (Snyder 1933). How to give an egg-mask facial, as well as an egg shampoo, is described in detail by Niles (1938).

Egg oil was used in Russia during the nineteenth century in the manufacturing of Kazan soap, noted as excellent for skin treatment. Egg yolk was also included in soaps in Germany (Romanoff and Romanoff 1949). Presently eggs are included in many commercially prepared shampoos sold in drug stores and supermarkets.

Adhesives

In the past an active demand developed for "technical albumen," a product recovered by centrifugally extracting the remaining albumen from crushed eggshells as they come from the breaking room. The albumen was then dried. This "technical albumen" was used to fasten cork inserts in metal caps on beverage bottles. In present years, however, with prices of whites relatively low, some dried albumen prepared for human consumption has been substituted for technical albumen (Koudele and Heinsohn 1964). About 1 lb of technical albumen is recovered from the shells for each 30-dozen case broken out.

ARTISTIC AND CULTURAL USES OF THE EGG

History

The historical and cultural uses of eggs and egg decoration has been documented in some detail in a text by Newall (1971). Egg decorating is an ancient art which originated before the time of Christ. Examples of decorated stones in the shape of an egg have been found on Easter Island. Although presently associated with the Easter season, many different religious and cultural associations with the decorated egg have existed. The egg has been related to many superstitions, myths, and beliefs over the history of mankind. Newall (1971) has documented examples of eggs related to myths of creation, witchcraft and magic, fertility, purity, and resurrection. As such, the egg has been used as a symbol for the cosmos, as the devil's food, as a charm to encourage reproduction, as a medium for the exorcism of devils, and as a symbol of rebirth. The egg has often been revered as a symbol of life, some early societies even prohibiting the use of eggs as food. Egyptian cultures associated the egg with reproduction, hanging eggs in temples to encourage fertility.

One of the earliest written records of decorated eggshells was during the thirteenth century reign of the English king, Edward I, who presented gold-leafed eggs to his court as gifts. In the sixteenth century Louis XV presented Madame Du Barry the gift of a gilded egg containing the "surprise" of a cupid. The decorating of eggs reached its zenith during the days of the Russian czars. Craftsmen were commissioned by wealthy Russians to prepare eggs ornamented with gold and precious jewels. Peter Carl Fabergé (1846-1920), a noted jewelry craftsman of the twentieth century, was the court jeweler for the czars. Fabergé's designs of decorated eggs have influenced the art for decades; many of his creations are still in existence as collectors' items.

Egg Decoration

Egg decoration, in a less costly way, is still practiced on a wide scale (Barnett 1966). According to L. M. Szilvasy (personal communication 1983), there are at least three different types of egg art being formally practiced by artisans and craftsmen in Europe and the United States. One of the most popular art types is the traditional ornamentation so valued in the past by European royalty. An example of modern egg ornamentation is shown in Fig. 18.1. In a comprehensive technical

FIG. 18.1.   An example of egg ornamentation.

manual, ornamental egg art is described in detail by Szilvasy (1976). Other methods of egg art include the "scratch work" technique that calls for dying eggshells with earth tones or with natural dyes, then scratching designs through the dye. Another method is the "wax resist" method in which a design successively is layered on the egg with beeswax, then the egg dyed with different colors. Each color is preserved by the wax during the dying process. The latter two methods are hundreds of years old, but still are used and provide beautiful examples of egg art.

Eggshell Mosaics

The use of eggshell fragments for the production of mosaics is a rarely used art form dating back to the old masters of fifteenth century Italy.

Tempera

The use of eggs in tempera painting is a practice that became common during the Renaissance in the hands of the old masters of Sienna and Florence. This technique is still in use by certain artists for special effects. The classic form of tempera painting includes the application of an egg yolk pigment to a specially prepared surface (Thompson 1936). As practiced today, tempera is an emulsion of egg yolk thinned with water which then is mixed with a powdered pigment to provide the desired color. The emulsion is characterized by its ability to dry quickly and is very stable. A detailed description of the techniques used in modern tempera art, including preparation of the tempera, is provided by Vickery and Cochrane (1973).

The Egg and the English Language

The English language reflects to a great extent the effect that eggs have had on people or the opinion that people have had of eggs over the centuries. The fragility of a relationship can be described as "like walking on eggs," while a studious individual can be described as an "egghead" (presumably a positive connotation). A likable person is "a good egg"; a child gone astray may be a "bad egg." If you really make a mistake you can "lay an egg" and you can make a foolish commitment by "putting all your eggs in one basket." How very seriously we take the egg in modern society may be summarized in the headline that is attributed to *Variety* newspaper's Sime Silverman concerning the stock market crash in 1929: "Wall Street Lays an Egg."

OTHER USES

The versatile egg has been utilized in varied processes, other than those described in this chapter, including the production of photographic plates, painters ink, spinning synthetic fibers, dye mordants in textiles, and in rubber manufacture. In many instances eggs have been employed for specialized purposes as listed above; however, they were later replaced by other materials (Romanoff and Romanoff 1949).

One of the more interesting nonfood uses of egg products is the animal repellent-attractant use described by Bullard et al. (1978A). They found that a fermented egg product (FEP) contained odiferous material which repelled deer and attracted coyotes. It was suggested that volatiles found in FEP were similar to those found in food accepted by coyotes (carrion) and refused by deer. They also noted that some of the same volatile fatty acids found in FEP were excreted in the anal glands of canids, suggesting that both chemical signaling and food aspects were involved in deer and coyote responses. Bullard *et al.* (1978B) later prepared a synthetic mixture that simulated the action of FEP in animal repellent-attractant properties. A practical use of the repellent properties of egg was reported by Manning (1982). Several Louisiana farmers sprayed mixtures of egg in water (60 dozen eggs in 200 gal. of water) in soybean fields and observed that deer did not cross the sprayed areas. Apparently the mixture achieved greater success in repelling deer than did commercially prepared deer repellents.

REFERENCES

Arvat, V., and Hinners, S. W. 1973. Evaluation of egg shells as a low cost calcium source for laying hens. Poult. Sci. *52*, 1996.

Barnett, B. D. 1966. The egg and art. Poult. Sci. *45*, 770–775.

Bullard, R. W., Leiker, T. J., Peterson, J. E., and Kilburn, S. R. 1978A. Volatile components of fermented egg, an animal attractant and repellent. J. Agric. Food Chem. *26*, 155–159.

Bullard, R. W., Shumake, S. A., Campbell, D. L., and Turkowski, F. J. 1978B. Preparation and evaluation of a synthetic fermented egg coyote attractant and deer repellent. J. Agric. Food Chem. *26*, 160–163.

Clarke, M. C. 1940. Found: A new use for eggs. U.S. Egg Poult. Mag. *46*, 644–646, 697–698.

Cox, H. R. 1952. Growth of viruses and rickettsia in the developing chicken embryo. Ann. N.Y. Acad. Sci. *55*, 236–247.

Food and Agriculture Organization 1965. Protein requirements. FAO Nutr. Meet. Rep. Ser. *37*.

Hall, W. M., Witzeman, J. S., and Jones, R. 1969. The detection and enumeration of *Clostridium perfringens* in foods. J. Food Sci. *34*, 212–214.

Koudele, J. W., and Heinsohn, E. C. 1964. The egg products industry of the United States. Kans. State Univ., Agric. Exp. Stn., Bull. *446*.

Lipman, J. G., and Brown, P. E. 1908. Ammonification in shale and clay soils. N.J., Agric. Exp. Stn. Rep. *29*, 129–136.

Manning, E. 1982. Egg spray keeps deer out of fields. Prog. Farmer, Midwest Ed. Nov.

Newall, V. 1971. An egg at Easter. Indiana Univ. Press, Bloomington.

Niles, K. B. 1938. Eggs for beauty as well as for food. U.S. Egg Poult. Mag. *44*, 676–677.

Phillips, P. H., and Lardy, H. A. 1940. A yolk-buffer pabulum for the preservation of bull semen. J. Dairy Sci. *23*, 399–404.

Romanoff, A. L., and Romanoff, A. J. 1949. The Avian Egg. John Wiley & Sons Co., New York.

Salsbury, G. W., Fuller, H. K., and Willett, E. L. 1941. Preservation of bovine spermatozoa in yolk citrate diluent and field results from its use. J. Dairy Sci. *24*, 905–910.

Snyder, C. G. 1933. Eggs and the quest for beauty. U.S. Egg Poult. Mag. *39* (9), 20–21.

Szilvasy, L. M. 1976. The Jeweled Egg. A Comprehensive Guide to the Design and Creation of Decorated Eggs. Association Press, New York.

Thompson, D. V. 1936. The Practice of Tempera Painting. Yale Univ. Press, New Haven, CT.

U.S. Department of Agriculture (USDA) 1941. Eggs and egg products. U.S., Dep. Agric., Circ. *583*, 81–85.

Vandepopuliere, J. M., McKinney, C. W., and Walton, H. V.   1973.   Value of egg shell meal as a poultry feedstuff. Poult. Sci. *52*, 2096.

Vandepopuliere, J. M., Walton, H. V., and Cotterill, O. J.   1975.   Nutritional evaluation of egg shell meal. Poult. Sci. *54*, 131–135.

Vandepopuliere, J. M., Walton, H. V., Jaynes, W., and Cotterill, O. J.   1978.   Elimination of Pollutants by Utilization of Egg Breaking Plant Shell-waste. Environ. Prot. Technol. Ser. EPA-600/2-28-044. U.S. Environmental Protection Agency. Cincinnati, OH 45268.

Vickery, R., and Cochrane, D.   1973.   New Techniques in Egg Tempera. Watson-Guptill Publications, New York.

Walton, H. V., Cotterill, O. J., and Vandepopuliere, J. M.   1973.   Composition of shell waste from egg breaking plants. Poult. Sci. *52*, 1836–1841.

Wilcke, H. L.   1940.   Egg shells are good poultry feed. U.S. Egg Poult. Mag. *46*, 617–618.

William, J. P., McCay, C. M., Salmon, O. H., and Krider, J.   1942.   A study of the value of incubated eggs and methods of feeding them to growing and fattening pigs. J. Anim. Sci. *1*, 38–40.

19

Egg Product, Process, and Equipment Patents (U.S.)

Owen J. Cotterill

All United States patents are listed in the *Official Gazette* of the Patent Office. It is published each Tuesday and contains one claim from each patent granted on that day. Copies may be found in most large public libraries. These listings are subject-matter indexed according to class and subclasses to assist locating specific subject areas. Also, some technical journals (for example, *Food Technology*) publish a listing of new food patents. Copies of specific patents can be obtained from the U.S. Patent Office, Washington, D.C. 20231.

Egg product and process patents are found in Classes 30, 99, 146, 209, 260, and 426. Although Subclasses 94, 113, 114, 182, and 210 are the most important, some patents are also found in Subclasses 1, 14, 17, 28, 56, 57, 65, 92, 123, 130, 139, 150, 170, 172, 177, 196, 199, 208, 327, 355, 427, and 440. Egg-breaking equipment is included in Class 146, Subclass 2. Class 426 includes mostly processes, products, and their compositions.

Much information exists in patents of interest to those involved in egg processing, research, product development, equipment manufacture, and teaching. Because of the limited listing or indexing of the patents on egg products, processes, or equipment, this information is not generally available. Approximately 4.5 million U.S. patents have been issued. Only about 0.01% of these pertain specifically to the egg-products industry.

The following list includes egg-product, process, and equipment patents; those based on shell eggs (preservation, hatching, cartons, etc.) are not included. It should not be assumed that this list is complete. The author would appreciate receiving additional numbers. Patents as far

413

EGG SCIENCE & TECHNOLOGY,
3rd Edition

Copyright © 1986 by AVI Publishing Co.
All rights of reproduction in any form reserved.
ISBN 0-87055-516-2

back as 1865 are listed in the first and second editions of this book. However, only those issued since 1960 are listed in this edition.

CHRONOLOGICAL LIST

The patent number, issue date, inventor(s), and title (a brief description may follow in parentheses) are given.

Number and date	Inventor(s)	Title and description
2,919,992 1/5/60	Gorman, J. M., & A. C. Keith	Egg white product and process of forming same. (Triacetin and Ca stearyl-2-lactylate)
2,920,966 1/12/60	Heinemann, B.	Egg-milk product. (Retards gelation)
2,929,150 3/22/60	Johnston, W. R.	Dehydration. (Sugar and corn oil as suspending liquid)
2,929,715 3/22/60	Sutton, W. J. L.	Protein composition. (A phosphate and ester whipping aid)
2,930,705 3/29/60	Janak, J., F. R. Humburg, & R. C. Hill	Processing egg yolks. (Canned baby food)
2,933,397 4/19/60	Maturi, V. F., L. Kogan, & N. G. Marrotta	Egg white composition. (Triacetin-foaming)
2,936,240 5/10/60	Kauffman, F. L., P. Park, & L. D. Mink	Treatment of liquid whole eggs. (Pasteurization)
2,950,204 8/23/60	Peebles, D. D.	Dried egg product and process of manufacture. (Dried egg white-lactose aggregate)
2,952,551 9/13/60	Long, F. L., & D. Long	Method of making a packaged frozen egg product. (Ham, onion, pickle, flour, egg loaf)
2,953,457 9/20/60	Sanna, F. L.	Method for drying food materials. (Spray— vacuum chamber, low temp.)
2,966,184 12/27/60	Willsey, C. H.	Egg-breaking machine. (Improved egg-cracking devices)
2,977,203 3/28/61	Sienkiewicz, B., R. B. Kohler, & M. Schulman	Agglomerating process. (Coffee eggs not mentioned)
2,978,335 4/4/61	Kidger, D. P., & R. J. Baeuerlen	Whipping agent and method of preparing the same. (Egg white plus glycols)
2,982,663 5/2/61	Bergquist, D. H.	Method of producing egg albumen solids with low bacteria count. (Liq. egg white heat treatment, high-temp. solids storage)

Number and date	Inventor(s)	Title and description
2,989,405 6/20/61	Stokes, J. L., M. N. Mickelson, & R. S. Flippin	Preparation of egg products. (*Salmonella* antagonism by *E. coli*)
2,999,024 9/5/61	Stimpson, R. H., & I. J. Hutchings	Method of preparing cooked egg yolk and cooked egg yolk product. (Yolk baby food)
3,009,818 11/21/61	Jokay, L., & R. I. Meyer	Process of preparing a dehydrated precooked egg product. (Cooked whole egg, freeze-dried)
3,014,805 12/26/61	Mitz, M. A.	Process for preparing an egg product. (Anaerobic yeast, extensive desugarization)
3,028,245 4/3/62	Mink, L. D., & J. H. Silliker	Treatment of egg material. (H_2O_2 treatment of egg white before drying)
3,039,107 6/12/62	Bradford, P.	Agglomeration of spray-dried materials. (Eggs not mentioned)
3,051,575 8/28/62	Chase, F. A.	Method of preparing stable egg-impregnated green coffee. (Improves flavor of brew)
3,060,038 10/23/62	Mancuso, J. J. & L. Z. Raymond	Method of preparing dehydrated eggs. (Diluted whole egg, cooking, adding raw egg, drum-drying)
3,062,665 11/6/62	Peebles, D. D., & P. D. Clary, Jr.	Method of producing a preserved egg white product. (Instant egg white—agglomeration)
3,073,704 1/15/63	Rivoche, E. J.	Egg food products and processes for preparing same. (Pre-cooked frozen egg)
3,076,715 2/5/63	Greenfield, C.	Dehydration of fluid fatty mixtures above normal coagulation temperature. (Concentration before drying to avoid coagulation)
3,077,411 2/12/63	Mitchell, W. A., & V. V. Studer	Method of dehydrating eggs. (Cooked whole egg-drum-dried flakes)
3,078,168 2/19/63	Bedenk, W. T.	Angel food cake. (One-step angel food cake mix)
3,082,096 3/19/63	Powers, F. C.	Method of packaging an angel food cake. (Angel cake baked in protective cover for delivery to consumer)
3,082,098 3/19/63	Bergquist, D. H.	Method of preparing powdered egg albumen. (Two-fluid spray nozzle)
3,093,487 6/11/63	Jones, E., O. H. Tracy, & C. B. Defenbaugh	Egg products and processes for preparing same. (Skim milk, oil, egg mixture, 77°C, pasteurized)

Number and date	Inventor(s)	Title and description
3,096,179 7/2/63	Finucane, T. P., W. A. Mitchell, & L. Z. Raymond	Two-package angel food cake mix
3,113,872 12/10/63	Jones, E., & A. R. Johnson	Method of treating shelled eggs. (Heat pasteurized-vacuum)
3,114,645 12/17/63	Blanken, R. M., R. T. Carey, & D. E. Mook	Method of making a readily dispersible egg yolk powder and product. (Heat treated)
3,115,413 12/24/63	Kline, L., J. J. Meehan, & T. F. Sugihara	Process of spray-drying eggs. (Instant egg, gas nozzle)
3,133,569 5/19/64	Shelton, L., K. G. Jones, & R. G. Bush	Egg breaking and separating head assembly. (Diagrams of breaking machine assembly)
3,136,641 6/9/64	Koesterer, M. G., & E. T. Fritzsching	Method of treating eggs. (Propiolactone pasteurization)
3,137,330 6/16/64	MacLagan, R. T.	Egg-handling machinery. (Egg-breaking machine)
3,143,427 8/4/64	Thies, K.	Method of processing ovalbumin. (Low pH, whey mixture)
3,146,110 8/25/64	Buddemeyer, B. D., & J. R. Moneymaker	Cake mix and acyl lactylic acid additives employed therein. (Eggs mentioned)
3,152,910 10/13/64	Sugihara, T. F., & L. Kline	Making cake doughnut mixes with yolk-containing egg products
3,156,570 11/10/64	Holme, J.	Heat treatment of albumen to reduce micro-biological population therein. (Soap add.)
3,160,507 12/8/64	Finucane, T. P.	Two-package angel food cake mix
3,161,527 12/15/64	Smith, C. F.	Treatment of egg albumen. (High-temp. past. of egg white solids)
3,162,537 12/22/64	Scott, D.	Process for removal of sugars by enzymatic action. (Enzyme fermentation)
3,162,540 12/22/64	Kline, L., & T. F. Sugihara	Drying of yolk-containing egg liquid. (Carbohydrates added)
3,166,424 1/19/65	Stewart, R. A., V. J. Kelly, & L. J. Duckworth	Egg yolk process. (Coag. yolk, infant food)
3,170,804 2/23/65	Kline, L., T. F. Sugihara, & J. J. Meehan	Preparation of dried egg white. (Egg white, carbohydrate, low pH)
3,199,988 8/10/65	Kozlik, R. F., & J. L. Swanson	Process for making a whippable composition. (Foam toppings formulation, egg can be one ingredient)

Number and date	Inventor(s)	Title and description
3,201,257 8/17/65	Hamon, R. C.	Egg albumen compositions. (Ethylene oxide pasteurization)
3,201,261 8/17/65	Carey, R. T., & R. M. Blanken	Process for preparing dried eggs. (Free-flowing mixture egg white, yolk, whole egg)
3,206,314 9/14/65	Shinohara, M., & Y. Mikata	Method for boiling eggs
3,207,609 9/21/65	Gorman, W. A., C. K. Stearns, & S. M. Weisberg	Egg product. (Low-calorie egg white)
3,210,198 10/5/65	Keller, H. M.	Process for preparing a whippable composition. (Dry mixtures including egg albumen for whipping)
3,212,906 10/19/65	Jones, E.	Process for preserving eggs. (Low pH past.)
3,216,828 11/9/65	Koonz, C. H., & E. J. Strandine	Removing shells from cooked eggs. (High pH)
3,219,454 11/23/65	Howard, H. W., J. B. Jacobsen, J. Hsueh, C. Selby, & T. W. Workman	Low-calorie diet. (Contains eggs and gums)
3,219,457 11/23/65	Ziegler, H. F., Jr., & H. J. Buehler	Method of improving the whipping characteristics of egg whites. (Okra gum)
3,222,194 12/7/65	Gorman, J. M. & V. H. Hannah	Method of preparing a dried egg product. (Cooked egg dishes, scrambled)
3,232,769 2/1/66	Miller, M. W.	Method of preparing eggs and food products having cooked eggs therein. (Low pH, cooked chopped egg)
3,251,697 5/17/66	Lineweaver, H., & F. E. Cunningham	Process for pasteurizing egg white. (pH 7, Al)
3,260,606 7/12/66	Azuma, S.	Enzymatic treatment of egg. (Low pH, hydrolyzed, non-coag.)
3,262,788 7/26/66	Swanson, A. M., & D. J. Fenske	Process for aggregating difficult-to-aggregate particles and the product thereof. (Instantized egg solids mixed with lactose powder)
3,268,346 8/23/66	Spicer, A., & W. H. Sly	Preparation of dried egg. (Emulsifying agents in dried whole egg to improve functional properties)
3,285,749 11/15/66	Shires, F.	Method for producing an egg product. (Egg roll)

Number and date	Inventor(s)	Title and description
3,287,139 11/22/66	Ganz, A. J.	Co-dried liquid egg white-carboxymethyl-cellulose composition, process for preparation thereof and food mix utilizing same. (Gum in egg white)
3,293,044 12/20/66	Torr, D.	Method of making an egg product. (Seasoned whole egg spread)
3,318,703 5/9/67	Geldof, J. J.	Method of producing substitute for truffles. (Contains egg yolk)
3,326,255 6/27/67	Klint, J. T.	Apparatus for mechanically cracking eggs
3,328,175 6/27/67	Cunningham, F. E., L. Kline, & H. Lineweaver	Egg white composition containing triethyl phosphate having enhanced whipping properties
3,340,067 9/5/67	Wallis, J. H.	Flour mix composition and batter prepared therefrom. (Contains dried egg yolk
3,343,963 9/26/67	Kjelson, N. A.	Impregnating edible protein fibers with three-component binder and product. (Albumen, gluten, defatted oilseed)
3,352,024 11/14/67	Mellor, J. D.	Freeze-drying process. (Whole egg pulp)
3,362,836 1/9/68	Scott, D.	Process for production of albumen. (High pH desugarization)
3,364,037 1/16/68	Mink, L. D., & E. J. Strandine	Hydrogen peroxide treatment of eggs. (At low and high temp. for pasteurization)
3,378,376 4/16/68	Sebring, M.	Egg white composition, method of preparing and process utilizing same. (Guar gum in egg white
3,383,221 5/14/68	Chin, R. G. L., & S. Redfern	Egg compositions containing soluble phosphorus compounds effective to impart fresh egg color. (Monosodium phosphate, pH 5.5 to 7.0 to retain color of scrambled eggs) eggs)
3,385,335 5/28/68	Halverson, J. E.	Egg separator. (Jogging trough for separation of thick and thin egg yolk
3,385,712 5/28/68	Dodge, J. W., J. M. Darfler, & R. C. Baker	Method of making an egg product. (Egg roll, yolk in center)
3,404,008 10/17/68	Ballas, M., & M. R. Painter	Apparatus and process for pasteurizing egg products. (Vacuum treatment before pasteurization)

Number and date	Inventor(s)	Title and description
3,408,207 10/29/68	Katz, M.	Process for making a freezable egg product. (Scrambled egg mix with pectin and soy protein)
3,409,446 11/5/68	Van Olphen, H. A.	Process for preparing an egg concentrate. (Concentrated egg and sugar mixture)
3,417,798 12/24/68	Shelton, L.	Egg-breaking and separating apparatus. (Egg holding and positioning)
3,448,782 6/10/69	Williams, H. V.	Egg-handling machine. (Suction cups for holding egg shell during breaking)
3,471,302 10/7/69	Rogers, A. B., M. Sebring, & R. W. Kline	Whole egg magma product. (Hydrogen peroxide pasteurization of whole egg)
3,475,180 10/28/69	Jones, R. E.	Low-calorie egg product. (Scrambled egg mix, etc.)
3,493,393 2/3/70	Shires, F.	Method of producing an egg product. (Egg roll)
3,503,501 3/31/70	Seaborn, P. E.	Method of and device for detecting cracks in eggs
3,510,315 5/5/70	Hawley, R. L.	Egg product. (Egg roll—proteolytic enzyme added to yolk and starch to white, reduces rubberiness and syneresis)
3,515,643 6/2/70	Ghielmetti, G., & C. Trinchera	Process for the production of lysozyme. (Ion-exchange)
3,520,700 7/14/70	Kohl, W. F., & J. C. Sourby	Process for pasteurization of egg whites. (Phosphates to reduce pasteurization requirements)
3,547,658 12/15/70	Melnick, D.	Conveniently packaged food. (Defatted dried egg with oil added, also various product formulations)
3,549,388 12/22/70	Kohl, W. F., & A. D. F. Toy	Process for pasteurization of egg products and products so prepared. (Addition of phosphate to past. temp. of egg white)
3,561,979 2/9/71	Pontius, W. I., J. G. Endres, & L. A. Van Akkeren	Preparation of shortening-dried egg products. (Shortening-dried egg-leavening mixture)
3,561,980 2/9/71	Sourby, J. C., W. F. Kohl, & R. H. Ellinger	Process for pasteurization of whole eggs. (Phosphate addition to lower pasteurization temp.)
3,563,765 2/16/71	Melnick, D.	Low cholesterol-dried egg yolk and process. (Solvent extraction)

Number and date	Inventor(s)	Title and description
3,565,638 2/23/71	Ziegler, H. F., Jr., R. D. Seeley, & R. L. Holland	Frozen egg mixture. (Frozen egg containing milk solids, starch and gum for scrambled egg, etc.)
3,573,935 4/6/71	Sourby, J. C., W. F. Kohl, & R. H. Ellinger	Process of pasteurization of whole eggs. (Pasteurization at pH 8.3 at 52°–63°C for 0.5–10 min)
3,594,183 7/20/71	Melnick, D., M. I. Wegner, & D. R. Davis	Egg food product and process for the preparation thereof. (Fat and cholesterol extracted from egg yolk solids)
3,598,612 8/10/71	Ng, W.	Method of improving color of cooked egg products. (Reducing green sulfide ring discoloration by pretreating yolk with edible oxidizing agent)
3,598,613 8/10/71	Hawley, R. L.	Process for manufacturing cooked egg yolk products. (Egg roll—blending raw yolk with ground cooked yolk)
3,607,304 9/21/71	Levin, E.	Stable dried defatted egg product. (Defatting egg with azeotropic mixture of solvent and water)
3,613,756 10/19/71	Snyder, R. C., Jr., & R. M. Di Tore	Egg-shelling apparatus and method. (Fluid under pressure pushes egg from shell)
3,615,705 10/26/71	Kohl, W. F., J. C. Sourby, & R. H. Ellinger	Process for the pasteurization of egg whites. (Addition of phosphate)
3,615,716 10/26/71	Poulos, J. F.	Apparatus for packaging eggs. (Tubular package)
3,619,206 11/9/71	Evans, M. T. A., & L. I. Irons	Modified protein. (Acylation of egg yolk protein with anhydride of dicarboxylic acid to make improved emulsion—heat without coagulation)
3,624,230 11/30/71	Robinson, A. H., Jr.	Process for continuously scrambling eggs. (Coagulation in jets of steam)
3,627,543 12/14/71	Epstein, J. J.	Pan dried glucose free egg white albumen containing means for inhibiting discoloration when heat treated and the process of preparing the same. (Addition of aluminum ion to retain color)
3,640,731 2/8/72	Kaplow, M., & R. E. Klose	Shelf stable egg products. (Carbohydrate and emulsions)
3,658,558 4/25/72	Rogers, A. B., M. Sebring, & R. W. Kline	Preparation of whole egg magma product. (H_2O pasteurization of whole egg)

Number and date	Inventor(s)	Title and description
3,682,660 8/8/72	Kohl, W. F., J. C. Sourby, & R. H. Ellinger	Process for the pasteurization of egg whites. (Phosphate addition)
3,684,531 8/15/72	Foster, R. D.	Method for processing eggs. (Removing hard-cooked eggs from shells)
3,706,575 12/19/72	Broadhead, S. A.	Lactalbumin phosphate as a replacement for egg white. (Replaced with lactalbumin phosphate)
3,711,299 1/16/73	Zeigler, H. F., Jr.	Egg white composition and process of making same. (Gums)
3,711,304 1/16/73	Hawley, R. L.	Roll centering method and apparatus. (Machine to center egg yolk core)
3,737,330 6/5/73	Kohl, W. F., J. C. Sourby, & R. H. Ellinger	Egg white pasteurization. (Polyphosphate in pH 8–10 egg white)
3,738,847 6/12/73	Bechtel, P. J.	Process for producing a canned egg in meat pet food. (Core of egg with ground meat outside)
3,753,737 8/21/73	Latham, S. D., R. D. Seeley, H. F. Reitz, & R. F. Reitz	Omelet preparing and process. (Cooking)
3,758,256 9/11/73	Terada, K.	Method for processing a vital, flavored egg. (Salt under pressure)
3,764,714 10/9/73	Driggs, L. W.	Method of peeling eggs. (Peeled, packaged eggs pasteurized at 180°F using radio frequency)
3,769,404 10/30/73	Latham, S. D., & R. D. Seeley	Egg composition. (Gums, monoglycerides, water)
3,778,425 12/11/73	Kandatsu, M., & M. Yamaguchi	Process for the manufacture of granular or powdery purified whole egg protein. (Defatted whole egg)
3,782,269 1/1/74	Latham, S. D., R. D. Seeley, H. F. Reitz, & R. F. Reitz	Omelet preparing machine and process. (Cooking method)
3,791,285 2/12/74	Mack, L.	Apparatus for molding and cooking egg products. (Egg around a center core of meat)
3,798,336 3/19/74	Hawley, R. L.	Method for centering solid bodies within a coaguble liquid. (Egg roll)
3,806,608 4/23/74	Perret, M. A.	Synthetic egg composition. (No yolk)

Number and date	Inventor(s)	Title and description
3,823,659 7/16/74	Hubka, V., H. Rehoff, & M. M. Andreasen	Apparatus for preparing a non-packaged egg product. (Egg roll)
3,830,149 8/20/74	Kaufman, V. F.	System for pasteurization. (Liquid, e.g., egg white, egg yolk, salt egg yolk, whole egg)
3,830,945 8/20/74	Scharfman, H.	Method and apparatus for processing eggs with microwave energy. (Microwave)
3,840,683 10/8/74	Strong, D. R., & S. Redfern	Egg product. (No yolk)
3,843,811 10/22/74	Seeley, R. D.	Low fat egg product. (Frozen product; 4–6% whole egg, no egg yolk)
3,843,813 10/22/74	Driggs, L. W.	Method and apparatus for treating eggs. (Hard-cooked eggs subjected to radio frequency energy to pasteurize and reduce greening)
3,843,825 10/22/74	Hawley, R. L.	Method of forming a packaged egg product. (Frozen—resembles half hard-boiled egg)
3,846,557 11/5/74	Mulla, M. S., & V. S. Hwang	Boil for synanthropic flies and method for making same. (Dry solid)
3,851,571 12/3/74	Nichols, J. F.	Apparatus and method for encapsulating eggs. (Egg-shaped container for freezing)
3,857,974 12/31/74	Aref, M. M., J. J. Stroz, & G. W. Johnson	Process for the production of frozen eggs. (Liquid nitrogen; popcorn-like particles)
3,857,980 12/31/74	Johnson, C. A.	Process for preparation and preservation of eggs. (Inedible eggs processed for animal feed)
3,859,907 1/14/75	Hatcher, G.	Boiled egg desheller. (Oscillating basket; H_2O sprayed from below)
3,863,018 1/28/75	Shires, F.	Method of producing an egg product. (Egg roll)
3,864,500 2/4/75	Lynn, C. C.	Process of preparing an egg yolk substitute and resulting products from its use. (High protein, low fat-no egg yolk, low cholesterol)
3,877,362 4/15/75	Epstein, J. J., & L. R. York	Method and machine for removing shells from hard cooked eggs. (Roller-type conveyor)
3,881,034 4/29/75	Levin, E.	Reconstituted egg product and method of preparing. (Removal of egg lipids by ethylene dichloride, adding vegetable oil, etc., and drying)

Number and date	Inventor(s)	Title and description
3,884,136 5/20/75	Towry, J. R.	Egg roll assembly machine. (Apparatus for assembling egg roll products)
3,911,144 10/7/75	Strong, D. R., & S. Redfern	Egg product. (Cholesterol-free liquid egg product with xanthan gum for stability)
3,912,433 10/14/75	Kwok, C. M.	Automatic egg roll making machine. (Egg-roll skin around food stuffing)
3,917,873 11/4/75	Kuroda, N., & S. Mogi	Egg white composition having improved whipping properties and process for producing the same. (Clyclodextrins)
3,928,632 12/23/75	Glaser, E., & P. F. Ingerson	Egg substitute product. (Egg white, polyunsaturated liquid vegetable fats, fatty acid, lactylate alkali metal salt)
3,930,054 12/30/75	Liot, R., & E. Liot	Method of making improved dried egg product. (Sucroglyceride added to whole egg, yolk, or egg white before drying)
3,941,892 3/2/76	Glasser, G. M., & H. Matos	Low-cholesterol egg product and process. (Frozen sunny-side up egg product)
3,950,557 4/13/76	Kheng, C. S.	Egg component separation. (Unshelled egg partially solidified before separation)
3,951,055 4/20/76	Woebbeking, R.	Egg blower. (Hard-cooked egg peeler)
3,958,034 5/18/76	Nath, K. R., & M. W. Newbold	Fractionated egg yolk. (Separated with specified proportions of fat, protein, and total cholesterol)
3,958,035 5/18/76	Stearns, C. K., & A. D. Singleton	Method of manufacturing omelet type egg product. (Addition of 1–5% flour, then a special cooking sequence)
3,965,270 6/22/76	Epstein, J. J.	Method of freezing cooked eggs. (Bath of Freon cooled with liquid N_2 for freezing cooked egg)
3,965,272 6/22/76	Epstein, J. J., & L. R. York	Method of freeze treating mayonnaise-containing products. (Freezing egg, ham, etc., salad with Freon)
3,966,991 6/29/76	Oborn, R. E.	Angel food cake mix. (Sodium bisulfite in egg white used in angel cakes)
3,970,763 7/20/76	Moran, D. P. J., & H. J. Pennings	Heat pasteurized cake batter. (Contains eggs)
3,973,482 8/10/76	Khee, S. P.	Appliance for cutting hard-boiled eggs. (Cut in half with a hand cutter)

Number and date	Inventor(s)	Title and description
3,974,294 8/10/76	Schwille, D., E. Sorg, & U. Sommer	Process for the production of protein-containing food additives. (Wettable water-soluble food proteins including eggs)
3,974,296 8/10/76	Burkwall, M. P., Jr.	Simulated egg and meat pet food. (Semimoist foods, some canned)
3,978,779 9/7/76	Petersen, G. C., O. Thisgaard, & S. Wiig	Apparatus for preparing edible products. (Egg roll cylinder)
3,982,040 9/21/76	Oborn, R. E.	Egg white composition. (Egg white plus sodium bisulfite for use in angel cakes)
3,987,212 10/19/76	Seeley, R. D., H. J. Hartmann, & D. R. Sidoti	Cholesterol free egg product. (Frozen egg product for scrambled eggs)
4,001,449 1/4/77	Reardanz, E. H., & J. N. Boudreau	Intermediate moisture animal food having identifiable meat and egg components. (Has character of eggs)
4,029,825 6/14/77	Chang, P. K.	Production of egg white substitute from whey. (Egg white replaced by adding Na lauryl sulfate to whey)
4,046,922 10/6/77	Burkwall, M. P., Jr.	Shelf stable, semi-moist simulated egg. (Involves starches but not freezing cooked-freeze-thaw-reheated products)
4,056,051 11/1/77	Brown, E. A.	Hard boiled egg extractor. (Shelling hard-cooked eggs, a small tool)
4,068,570 1/17/78	Lanoie, P. E.	Apparatus for producing pieces of eggs. (Cooking eggs, then cutting or dicing into pieces)
4,068,573 1/17/78	Romero, A. L.	Utensil for cracking eggs. (Tool for cracking shell eggs, individually)
4,072,764 2/7/78	Chess, W. B.	Egg yolk extender. (75% replacement per weight basis of yolk)
4,082,856 4/4/78	Zwiep, T. C., D. G. Newhouse, & J. D. Craner	Process and apparatus for shelling eggs. (Vibrator to remove shell and water separate shells)
4,095,518 6/20/78	Jones, F. W.	Sectioning apparatus. (For sectioning rounded food articles such as eggs)
4,103,038 7/25/78	Roberts, W. L.	Egg replacer composition and method of production. (Egg replacer)

Number and date	Inventor(s)	Title and description
4,103,040 7/25/78	Fioriti, J. A., H. D. Stahl, R. J. Sims, & C. H. Spotholz	Low cholesterol egg product and process. (Extracting cholesterol from yolk with oil)
4,107,335 8/15/78	Glickstein, M., J. M. Tuomy, & M. J. Shwert	Freeze-dried mix for spoonable salad dressing and method therefor. (Free flowing, contains egg yolk)
4,111,111 9/5/78	Willsey, C. H.	Egg breaking apparatus. (Separates white from yolk)
4,115,592 9/19/78	Bergquist, D. H., F. E. Cunningham, & R. M. Eggleston	Instant dissolving egg white. (Sucrose coated)
4,117,774 10/3/78	Wilburn, E. R., & J. A. Sorter	Egg sheller. (Curved wedge inserts between shell and hard-cooked egg)
4,137,837 2/6/79	Warren, W. H.	Apparatus for separating egg whites from egg yolks. (Spiral loop on yolk cup of cracker head)
4,137,838 2/6/79	Warren, W. H.	Means for breaking and separating eggs. (Similar to 4,137,837)
4,138,507 2/6/79	Iimura, O.	Coagulated egg-white foodstuffs. (Starch and carboxymethylcellulose to control cooked, freeze, thaw problem)
4,140,808 2/20/79	Jonson, N. B.	Low-calorie products of the mayonnaise and dressing type, and a method of producing them. (Buttermilk 45–55%, egg yolk 8–12%)
4,149,456 4/17/79	Gisonni, T.	Egg peeler. (Cylinder with hooks to remove shell from hard-cooked eggs)
4,157,060 6/5/79	Avery, R. W.	Viscosity controlled egg cooker. (Oscillating eggs during cooking)
4,157,404 6/5/79	Yano, N., I. Fukinbara, & M. Takano	Process for obtaining yolk lecithin from raw egg yolk. (Extraction with dimethyl ether)
4,161,548 7/17/79	Warren, W.	Making long yolk hard cooked eggs. (Stirring inside soft-cooked eggs with a bent needle to produce elongated yolks in hard-cooked eggs)
4,167,138 9/11/79	Warren, W. H.	Cracking head for an egg breaking machine. (Faster egg white drainage; adjustable knife)

Number and date	Inventor(s)	Title and description
4,176,593 12/4/79	Terzian, R. T.	Cooking device. (Omelet)
4,181,745 1/1/80	Growe, G. H., M. P. Patterson, & D. C. Egberg	Method for decorating the shells of eggs. (Selected portions of shell wet with vinegar, then sprinkled with granular dyeing medium)
4,182,234 1/8/80	Reed, W. H.	Non-destructing eggshell egg contents remover. (Egg blower to remove liquid from shell)
4,182,779 1/8/80	Chess, W. B.	Egg yolk extender. (Defatted soy flour, short patent wheat flour, soy oil with lecithin removed principal ingredients in extender—75% egg yolk replacement)
4,184,422 1/22/80	Grise, F. G., & N. Lovell	Mixing material in sealed containers. (Mixing by piercing egg with small needle)
4,191,102 3/4/80	Cope, C. J.	Egg-deshelling apparatus. (Deshelling hard-cooked eggs)
4,200,663 4/29/80	Seeley, R. D., & R. B. Seeley	Cholesterol-free egg product having improved cooking tolerance. (Scrambled eggs, omelets, egg patties, etc.)
4,209,536 6/24/80	Dogliotti, A.	Filled food product and method of making same. (Egg white as cellular structure building aid in pastry shell)
4,214,010 7/22/80	Corbett, C. R.	Replacement of whole egg in baked custard. (Whole egg partially replaced with whey protein, carboxymethylcellulose, and a lactylated shortening)
4,218,490 8/19/80	Phillips, D. J., D. T. Jones, & D. E. Palmer	Functional proteins. (Egg white, soy, milk protein interaction)
4,219,585 8/26/80	Herring, N. I.	Process for extracting oil from egg yolks. (Extracting egg oil from yolks with heat at 375°–400°F)
4,228,193 10/14/80	Schindler, J., & A. R. Nugarus	Method for making scrambled eggs. (Steam injection; *also see* 4,233,891)
4,233,891 11/18/80	Schindler, J., & A. R. Nugarus	Apparatus for making scrambled eggs. (Steam injection; *also see* 4,228,193)
4,234,619 11/18/80	Yano, N., I. Fukinbara, K. Yoshida, & Y. Wakiyama	Decholesterolized and defatted egg powder and method for producing same.

Number and date	Inventor(s)	Title and description
4,237,145 12/2/80	Risman, P. O. G., & N. E. Bengtsson	Method of preparing foodstuffs containing coagulating proteins and a device for performing the method. (Foodstuff paste enclosed in microwave transparent tube through an applicator)
4,238,519 12/9/80	Chang, P. K.	Egg albumen extender prepared from derived protein containing compositions and additives. (Whey protein)
4,244,976 1/13/81	Kahn, M. L., & K. E. Eapen	Intermediate-moisture frozen foods. (Addition dextrose plus fructose to egg yolk to reduce water activity)
4,244,982 1/13/81	Menzi, R., & G. Dove	Process for preparing a food mousse. (Egg white coagulates to an aerated low-density structure)
4,247,573 1/27/81	Murray, E. D., T. J. Maurice, & L. D. Barker	Protein binder in food compositions. (Heat-coagulating properties like egg white)
4,267,100 5/12/81	Chang, P. K., & M. C. Concilio-Nolan	Process for forming an egg white substitute. (Whey adjusted to pH 12, reduced to 5.0, meringues)
4,276,820 7/7/81	Joannou, C. J.	Automatic egg cooker. (Small home egg cooker with oscillatory system to measure doneness)
4,277,504 7/7/81	Radlove, S. B.	Premix product for a dietetic cake mix. (Contains egg white but no yolk)
4,296,134 10/20/81	Baldt, W.	Liquid egg blend. (Low cholesterol)
4,297,382 10/27/81	Hosaka, D. B.	Process for drying compositions containing derived protein-containing compositions and additives
4,308,290 12/29/81	Fiejii, N.	Method of peeling shell from boiled egg. (Flowing water, circular rotating cylinder)
4,311,089 1/19/82	Fiejii, N.	Method and apparatus for peeling of shells of boiled eggs.
4,321,864 3/30/82	Willsey, C.	Egg breaking and contents separating machine. (Cracker head)
4,331,070 5/25/82	Denk, V.	Apparatus for heating liquids, in particular foodstuffs and luxury foodstuffs for the purpose of pasteurization, sterilization and the like.

Number and date	Inventor(s)	Title and description
4,333,959 6/8/82	Bracco, U., & J. L. Viret	Decholesterization of egg-yolk. (Reduce pH, add oil and centrifuge)
4,341,808 7/27/82	Croyle, B. A.	Frozen raw custard. (Frozen quiche fillings)
4,344,359 11/17/82	Frechou, J., & G. Isambert	Automatic machine for peeling hard-boiled eggs
4,350,713 11/21/82	Dyson, D., D. Lees, M. Fenn, & K. S. Darley	Production of sponge cake. (Contains eggs)
4,351,852 11/28/82	Rule, C. E., C. Gilmore, & E. J. Stefanski	Low calorie cake batter or mix. (Contains eggs)
4,357,272 12/2/82	Polson, A.	Recovering purified antibodies from egg yolk
4,362,761 12/7/82	Chang, P. K., & G. E. Scibeli	Use of heat coagulated whey protein concentrate as a substitute for gelled egg white
4,364,966 12/21/82	Chang, P. K.	Blends of egg albumen and whey protein having improved gel strength. (Extruded heat treatment of desugared dried egg white improves gel strength)
4,369,196 1/18/83	Sukegawa, Y.	Fresh cheese like food products and a process for their preparation
4,371,555 2/1/83	Tully, P. R.	Method for dyeing eggs
4,376,134 3/8/83	Kumar, S.	Low cholesterol sausage analog and process therefor
4,379,174 4/5/83	Radlove, S. B.	Dietetic cake mix
4,382,973 5/10/83	de Figueiredo, M. P., York, L. R., & Long, J. L.	Process for the production of a frozen chopped egg product
4,383,365 5/17/83	Metzigian, M. J.	Egg slicer with interchangeable components
4,388,340 6/14/83	de Figueiredo, M. P., B. F. Guevara, J. L. Long, R. G. Morgan, & R. Y. Lawrence	Process for the production of a frozen chopped egg product
4,409,249 10/11/83	Forkner, J. H.	Egg product and process of manufacture

Number and date	Inventor(s)	Title and description
4,414,235 11/8/83	Takekoshi, S.	Process for preparing instant macaronis
4,414,240 11/8/83	Lee, C. R.	Process for lowering the thermogelation temperature of egg albumen
4,420,495 12/13/83	Hammer, J., L. Schiel, & H. Kratt	Method and apparatus for preparing foamy sauces or the like
4,421,770 12/20/83	Wiker, J. M., & F. E. Cunningham	Method of preparing high protein snack food from egg protein
4,423,084 12/27/83	Trainor, T. M., & D. R. Sullivan	Process for preparing salad dressings
4,425,367 1/10/84	Berkowitz, D., A. B. Bennett, J. L. Secrist, & D. A. Millett	Method of producing thermally processed egg products
4,426,400 1/17/84	Newlin, J. L., & W. J. Stadelman	Deviled egg process
4,428,971 1/31/84	Havette, B. E., & C. Hebert	Process for the preparation of a composition for frozen or deep frozen souffles
4,431,681 2/14/84	Hegedus, E., J. R. Frost, M. Glicksman, & J. E. Silverman	Process for preparing a high quality reduced-calorie cake. (Contains egg white)
4,433,001 2/21/84	Weimen, R. E., T. J. Kalowski, & R. A. Novy	Method for preparing scrambled eggs. (Cooked in rings)
4,438,564 3/27/84	Ashton, H. P.	Egg scoop or spoon
4,440,791 4/3/84	MacKenzie, K. A.	Cultured egg-milk product
4,442,132 4/10/84	Kim, J. C.	Light bakery products for diabetics and method for the preparation of these products. (Contains either egg white or whole egg)
4,451,490 5/29/84	Silverman, J. E., J. R. Frost, E. Hegedus, & M. Glicksman	High-quality, reduced-calorie cake containing cellulose and process thereof. (Contains egg white solids)
4,455,318 6/19/84	Maurice, T. J., E. D. Murray, & J. M. Agnes	Simulated adipose tissue. (Contains egg white, 1–2.5%)
4,469,708 9/4/84	Rapp, H., & W. R. Dockendorf	Egg product and process. (Cooked egg similar to scrambled; coated with crisp batter)

Number and date	Inventor(s)	Title and description
4,477,478 10/16/84	Tiberio, J., & M. Cirigliano	Acid preservation systems food products. (Salad dressing containing egg yolk)
4,488,479 12/18/84	Sloan, N. R., D. Remys, & R. Novy	Portable apparatus for cooking eggs on a heated cooking surface
4,511,589 4/16/85	Padly, Y., & R. Borgeaud	A process for the continuous pasteurization of eggs. (Pasteurization of eggs)
4,512,182 4/23/85	Rizvi, S. S., & D. W. Gossett	Apparatus and method for continuously measuring the coagulation, gelation, denaturation, etc., of food components, where the force generated by a single sample is sensed, measured, and recorded during the process.
4,514,432 4/30/85	Grizinia, M., & G. Milei	Hen-egg albumen substitute and method for its manufacture. (A hen-egg albumen substitute)
4,521,435 6/4/85	Peters, L.	Methods to reduce weight loss of a hamburger type meat patty by adding a juice-retaining coating that contains a dry powder of egg white.
4,522,117 6/11/85	Weimer, R. E., & R. A. Novy	Apparatus for preparing scrambled eggs. (A grill to scramble eggs having at least two rings)
4,524,082 6/18/85	Lioti, R.	Highly concentrated egg white or salted whole egg product and its method of preparation. (A process for preparing concentrated whole egg product which can be stored at ambient temperature)
4,524,083 6/18/85	Lioti, R.	Reserved liquid eggs and method of preparation. (A salt or sugar containing raw egg product which can be maintained at room temperature)
4,537,788 8/27/85	Valerie, P., & F. E. Cunningham	Egg jerky product and method of preparation. (A method producing a meatless jerky-type product from whole eggs)

Selected Bibliography of Doctoral Dissertations on Eggs and Egg Products

H. R. Ball, Jr.
Owen J. Cotterill

The following bibliography of doctoral dissertations is a selected listing. Citations were included when their subjects were concerned with the physical-chemical, microbiological, or quality attributes of eggs and egg products.

The reasons for compiling this bibliography are twofold: (1) to focus attention on these often overlooked sources of information, and (2) to provide sufficient detail for efficient retrieval of this information. Each citation includes the author's name, the date the degree was awarded, the title of the dissertation, the advisor(s), the institution awarding the degree, and when available, University Microfilms abstract order number. With the above information, copies of dissertations of interest may be obtained from the library of the degree-granting institution or from University Microfilms.

Aminlari, M.   1980.   Sulfhydryl proteins of avian egg-whites. University of California, Davis (81-5367).

Angalet, S. A.   1975.   Physical and microbiological properties of hard-cooked and pickled eggs. Advs. Fry, J. L. and Oblinger, J. L., University of Florida (76-4214).

Azari, P. R.   1961.   The resistances of conalbumin and its iron complex to physical and chemical treatments. Adv. Hill, R. M., University of Nebraska (61-5370).

EGG SCIENCE & TECHNOLOGY,
3rd Edition

Backus, D. A. 1963. The effect of certain variables on the interior quality of newly laid eggs during short term storage. Adv. Furry, R. B., Cornell University (64-3652).

Baker, E. L. 1956. The effect of sugars on some physical and chemical properties of egg albumen. Adv. Ball, C. D., Michigan State University (59-1318).

Baker, R. C. 1957. Chalazae of the domestic fowl-lysozyme content and factors affecting their prominence. Adv. Stadelman, W. J., Purdue University (58-13).

Ball, H. R., Jr. 1970. Catalase activity of egg white. Adv. Cotterill, O. J., University of Missouri (70-20,766).

Banwart, G. J. 1955. Microbiological and functional changes in dried egg albumen stored at elevated temperatures. Adv. Ayres, J. C., Iowa State University.

Bergquist, D. H. 1951. Functional properties of egg white as influenced by atomization and drying. Adv. Stewart, G. F., Iowa State University.

Bollenback, C. H. 1949. Removal of glucose from egg albumen by a controlled fermentation. Adv. Stewart, G. F., Iowa State University

Chandler, E. 1966. Radiation-induced off-flavor development in whole egg magma. Adv. Goldblith, S. A., Massachusetts Institute of Technology.

Chandler, H. K. 1964. An investigation of the use of vitamin K_5, a radio-sensitizer, to *Salmonella typhimurium* in liquid egg. Adv. Licciardello, J. J., Massachusetts Institute of Technology.

Chang, P. K. 1969. Behavior of proteins and lipoproteins in egg components upon heat treatment. Adv. Powrie, W. D., University of Wisconsin (69-9671).

Chapin, R. B. 1951. Some factors affecting the emulsifying properties of hen's egg. Adv. Stewart, G. F., Iowa State University.

Chicoye, E. 1968. Autoxidation products of cholesterol in aerated sols and irradiated spray-dried egg yolk. Adv. Powrie, W. D., University of Wisconsin (68-9067).

Christmann, J. L. 1976. Isolation and characterization of hen vitellogenin: Partial homology with yolk protein phosvitins. Johns Hopkins University (76-22,911).

Chung, R. A. 1963. The effect of different dietary fats and cholesterol on the chemical composition of the egg and body tissues of the hen. Advs. Stadelman, W. J., and J. C. Rogler, Purdue University (64-4571).

Clark, J. R. 1963. Characterization and comparison of several related iron-binding proteins: Avian egg white conalbumins and human transferrins. Adv. Feeney, R. E., University of California, Davis.

Cotterill, O. J. 1954. Influence of lysozyme on egg white quality. Adv. Winter, A. R., Ohio State University (59-2551).

Cunningham, F. E. 1963. Insolubilization of egg white proteins. Adv. Cotterill, O. J., University of Missouri (64-4795).

Dahlquist, F. W. 1969. The binding and catalytic properties of lysozyme. Adv. Raftery, M. A., California Institute of Technology (69-17,082).

Empie, M. W. 1982. Thermodynamics and kinetics of single residue replacements in 17 ovomucoid third domains: Effect on inhibitor interactions with 3 serine proteinases. Adv. Laskowski, M., Jr., Purdue University (DA82-25709).

Farnsworth, G. M. 1956. Estimates of genetic parameters influencing blood spots and other economic traits of the fowl. Adv. Nordskog, A. W., Iowa State University.

Fletcher, D. L. 1977. Factors affecting the measurement and utilization of xanthophylls in the egg yolk and broiler skin. Advs. Harms, R. H., and Janky, D. M., University of Florida (7806696).

Forsythe, R. H. 1949. Fractionation of the egg white proteins in media of low dielectric constant and low ionic strength. Adv. Foster, J. F., Iowa State University.

Frank, F. R. 1963. Some physical and chemical factors associated with shell strength. Adv. Burger, R. E., University of Minnesota (64-4094).

Fromm, D. 1954. The effect of washing and thermal variation of storage conditions on biophysical changes of market eggs. Adv. Margolf, P. H., Pennsylvania State University.

Froning, G. W. 1961. Some physical and chemical changes occurring in oiled and unoiled eggs during short periods. Adv. Swanson, M. H., University of Minnesota (61-3664).

Funk, E. M. 1951. Maintenance of quality in shell eggs by thermostabilization. Adv. Halpin, J. G., University of Wisconsin.

Galyean, R. D. 1975. Chromatographic and electrophoretic study of native and processed egg white. Adv. Cotterill, O. J., University of Missouri-Columbia (76-7489).

Gardner, F. A. 1960. The role of chemical additives in altering the functional properties of egg white. Adv. Cotterill, O. J., University of Missouri (60-4038).

Garland, T. 1973. Studies on egg yolk myelin figures and granule low-density lipoprotein. Adv. Powrie, W. D., University of British Columbia, Canada.

Gossett, P. W. 1983. Some physical and rheological properties of raw and coagulated pH-adjusted or succinylated egg albumen with respect to water retention properties. Adv. Baker, R. C., Cornell University (DA83-9450).

Gravani, R. B. 1975. The physico-chemical changes in egg yolk caused by the egg yolk separation factor of *Streptococcus faecalis* var. *liquefaciens*. Adv. Baker, R. C., Cornell University (76-8162).

Greco, P. A. 1943. The pasteurization of liquid whole egg. Adv. Stewart, G. F., Iowa State University.

Griffith, D. L. 1980. Non-destructive techniques to determine egg quality. Adv. Reo, V. N. M., University of Georgia (80-29121).

Grim, A. C. 1965. The effect of ionizing radiation on the flavor of whole egg magma. Adv. Goldblith, S. A., Massachusetts Institute of Technology.

Grunden, L. P. 1976. The effect of proteolytic enzymes on functional and physical properties of egg albumen. Adv. Baker, R. C., Cornell University (76-15,915).

Haglund, J. R. 1965. A fluorescent antibody method for detecting Salmonellae in egg and poultry products and in poultry processing plants. Adv. Ayres, J. C., Iowa State University (65-7615).

Hamann, F. H. 1974. Analysis of egg steroids. Michigan State University (75-14,747).

Hamid-Samimi, M.-H. 1984. Criteria development for extended shelf-life pasteurized liquid whole egg. Advs. Hamann, D. D. and Swartzel, K. R., North Carolina State University.

Hanning, F. M. 1954. Effect of sugar or salt upon denaturation produced by beating and upon the ease of formation and the stability of egg white foam. Adv. Lowe, B., Iowa State University.

Hanson, H. L. 1945. Effect of concentrating egg white on desirability of angel cake. Adv. Lowe, B., Iowa State University.

Hartdegen, F. J. 1967. Chemical modification of lysozyme. Adv. Rupley, J. A., University of Arizona (67-3963).

Hartung, T. E. 1962. Some factors affecting the penetration of the egg shell membranes by *Pseudomonas fluorescens*. Adv. Stadelman, W. J., Purdue University (62-3459).

Hasiak, R. J. 1972. Effect of several physical and chemical agents on the viscosity and structure of egg yolk. Advs. Baker, R. D., and Vadehra, D. V., Cornell University.

Helbacka, N. V. L. 1956. Studies on blood and meat spots in the hen's egg. Adv. Swanson, M. H., University of Minnesota (57-1135).

Hoover, S. R. 1940. A physical and chemical study of ovomucin. Georgetown University.

Hyder, R. N. 1970. Heat debilitation of *Salmonella typhimurium* in whole egg. Adv. Cotterill, O. J., University of Missouri.

Johnson, T. M. 1980. Egg albumen proteins interactions in selected food systems. Adv. Zabik, M. E., Michigan State University (81-12097).

Kim, K. O. 1980. Starch gelatinization, egg foaming and physical properties of sponge cakes as affected by stabilizing agents and surfactants. Kansas State University (81-11821).

King, A. J. 1983. Modification of egg white oleic acid and physiochemical properties of ovalbumin and lysozyme treated with oleic acid. Advs. Swaisgood, H. E., and Ball, H. R., Jr., North Carolina State University.

Kline, R. W. 1945. Dried egg albumen. I. Studies of the nonmicrobiological changes during storage. Adv. Stewart, G. F., Iowa State University.

Kocal, J. T. 1978. Characterization of the very low lipoproteins and apoproteins of egg yolk granules. Adv. Nakai, S., University of British Columbia, Canada.

Lee, H. W. 1977. Proximate composition and physico-chemical properties of turkey egg yolk. Adv. Cunningham, F. E., Kansas State University (77-18,611).

Lifshitz, A. 1963. The exterior structures of the chicken egg as barriers to bacterial penetration. Adv. Baker, R. C., Cornell University (64-1021).

Lo, Y. C. 1982. Moisture sorption properties, nonenzymatic browning of egg-humectant systems and storage stability of an intermediate moisture egg product. Adv. Froning, G. W., University of Nebraska-Lincoln (DA82-1542).

Marion, W. W. 1958. An investigation of the gelation of frozen egg yolk after thawing. Adv. Stadelmen, W. J., Purdue University (58-3174).

May, K. N. 1959. The effect of certain factors on moisture, protein, threonine and other components of hens' eggs. Adv. Stadelmen, W. J., Purdue University (59-4165).

McBee, L. E. 1974. Ion-exchange chromatography of egg yolk components. Adv. Cotterill, O. J., University of Missouri-Columbia (75-20138).

McKinney, C. W. 1977. Ion-exchange chromatography and electrophoresis of whole egg. Adv. Cotterill, O. J., University of Missouri-Columbia (79-3924).

Mellor, D. B. 1965. A study of *Salmonella derby* contamination of shell eggs. Adv. Banwart, G. J., Purdue University (65-8641).

Meslar, H. W. 1978. Egg yolk biotin-binding protein: Its purification, properties and comparison with egg white avidin. University of Delaware (78-16910).

Michaud, R. L. L. 1968. A physiochemical study of the iron-binding sites of siderophilin and conalbumin. Adv. Woodwork, R., University of Vermont (68-5119).

Miller, C. F. 1945. The comparative lifting power of magma from fresh and aged, pasteurized and dehydrated eggs when used in sponge cake. Adv. Lowe, B., Iowa State University.

Mohamed, S. Y. 1969. The effect of ionizing radiation on selected chemical, physical, and microbiological characteristics of egg white proteins. Adv. Gardner, F. A., Texas A & M University (69-14,156).

Mountney, G. J. 1957. Penetration and growth of *Pseudomonas fluorescens* in chicken eggs. Adv. VanderZant, W. C., Texas A & M University.

Najib, H. 1982. Influence of phosphorus, protein, sulfur amino acids, and linoleic acid levels on egg size and performance of hybrid egg-type hens. Adv. Sullivan, T. W., University of Nebraska-Lincoln (DA82-17551).

Nikoopour, H. 1967. Determination of the growth characteristics of *P. fluorescens* on selected egg white protein substrates. Adv. Gardner, F. A., Texas A & M University (67-9801).

Palladino, D. K. 1980. Chromatographic separation of acylated egg white proteins and biophysical characterization of succinylated lysozyme and ovalbumin. Advs. Ball, H. R., Jr., and Swaisgood, H. E., North Carolina State University at Raleigh (81-14603).

Pankey, R. D. 1967. The effects of dietary fats on some chemical and functional properties of eggs and on some chemical properties of the lipids of liver, blood and depot fat of the hen. Adv. Stadelman, W. J., Purdue University (67-16,696).

Payawal, S. 1944. Denaturation of egg proteins. 1. Effect of heat treatments on viscosity of liquid egg products. Adv. Stewart, G. F., Iowa State University.

Perry, F. D. 1934. Influence of rations and storage on the physical characteristics of eggs. Adv. Henderson, E. W., Iowa State University.

Pitsilis, J. G. 1978. Rheological properties of commercial egg products. Advs. Brooker, D. B., and Walton, H. V., University of Missouri-Columbia (79-06911).

Pope, C. W. 1959. The effect of nutrition upon the incidence of blood spots in chicken eggs. Adv. Schaible, P. J., Michigan State University.

Potts, P. L. 1981. Identification of an *n*-acetyl-beta-D-glucoseaminidase from egg cortical granules as a block to polyspermy in *Xenopus laevis*. Adv. Washburn, K. W., University of Georgia (DA82-06301).

Rasplicka, L. D. 1967. A rapid method of identification of Salmonellae in spray-dried whole egg solids. Adv. Spencer, J. V., Washington State University (67-9175).

Reagan, J. C. 1970. Improved methods for determination of certain organic acids in liquid whole egg and the influence of pasteurization on their concentration. Adv. Dawson, L. E., Michigan State University.

Reddy, C. V.   1964.   Influence of calcium in modern laying rations of shell quality and interior quality of eggs. Adv. Sanford, P. E., Kansas State University (64-7450).

Reinke, W. C.   1967.   Investigations on egg yolk gelation. Adv. Baker, R. C., Cornell University (67-9196).

Saari, A. L.   1963.   Lipoproteins in egg yolk. Adv. Powrie, W. D., University of Wisconsin (63-7668).

Sabet, T. Y.   1955.   Studies on egg washing and preservation. Adv. Mallmann, W. L., Michigan State University (55-1255).

Sauter, E. A., Jr.   1966.   Some electrophoretic components of egg white and their effects on egg quality. Adv. McGinnis, J., Washington State University (66-13581).

Schmidl, M. K.   1978.   Production, characterization, ultrastructure, and functional properties of egg albumin and corn zein plastein reaction products and their insoluble fractions. Cornell University (79-02293).

Schultz, J. R.   1966.   Dehydration-induced interactions of egg yolk lipoproteins and low molecular weight carbohydrates. Adv. Snyder, H. E., Iowa State University (67-5622).

Seideman, W. E.   1966.   Ion-exchange chromatography of egg yolk. Adv. Cotterill, O. J., University of Missouri (67-934).

Silverstrini, D. A.   1961.   Some physical and chemical aspects of mottled egg yolks. Adv. Dawson, L. E., Michigan State University (62-1676).

Simlot, M. M.   1963.   Fractionation of phosphoproteins and lipoproteins of egg yolk. Adv. Clegg, R. E., Kansas State University (64-2811).

Skala, J. H.   1961.   A study of certain physical and chemical properties of chicken eggs in relation to their initial albumen quality and to some characteristics of the hens which laid them. Adv. Swanson, M. H., University of Minnesota (61-2159).

Slosberg, H. M.   1946.   Some factors affecting the functional properties of liquid egg albumen. Adv. Stewart, G. F., Iowa State University.

Steinhauer, J. E.   1968.   Quantitative determination of certain short-chain acids in frozen whole eggs by gas-liquid chromatography. Adv. Dawson, L. E., Michigan State University (68-11,100).

Stevens, F. C.   1963.   Comparative biochemistry of avian ovomucoids. Adv. Feeney, R. E., University of California.

Stiles, P. G.   1958.   Studies of the cause and incidence of blood spots in chicken eggs. Adv. Dawson, L. E., Michigan State University.

Verstrate, J. A.   1980.   Chemical and organoleptic characterization of flavor changes during storage of hard-cooked egg albumen and yolk. Adv. Spencer, J. V., Washington State University (8025967).

Walsh, R. G. 1978. Chemical modification of disulfide bonds in avian ovomucoids; synthesis of active derivatives with new synthetic cross-links. University of California, Davis (79-05213).

Ward, J. B. 1962. The effect of certain dietary ingredients on the incidence of blood spots in chicken eggs. Adv. Schaible, P. J., Michigan State University (62-4471).

Wedral, E. M. 1971. Mechanism of bacterial penetration into the eggs of *Gallus gallus*. Advs. Vadehra, D. V., and Baker, R. C., Cornell University.

Weinstein, H. 1966. A study of functional properties of whole egg magma treated by thermal and ionizing energy simultaneously. Adv. Goldblith, S. A., Massachusetts Institute of Technology.

Wesley, R. L. 1966. Accelerated depletion of DDT residues from commercial laying hens. Adv. Stadelman, W. J., Purdue University (66-13,277).

Woodward, S. A. 1984. Texture and microstructure of heat-formed egg protein gels. Adv. Cotterill, O. J., University of Missouri-Columbia. (DA84-2566).

York, L. R. 1971. Microbial counts and organic acid quantitation as quality indices of egg products. Adv. Dawson, L. E., Michigan State University.

Index